Bird Feathers

Bird Feathers

A Guide to North American Species

S. David Scott
and Casey McFarland

STACKPOLE
BOOKS

Published by
STACKPOLE BOOKS
5067 Ritter Road
Mechanicsburg, PA 17055
www.stackpolebooks.com

Printed in the United States

First edition

Cover design by Caroline M. Stover
Range maps by Paul Lehman and Birdfellow.com
Drawings by Beatriz Mendoza
Photo on pages 1 and 23 by Jonah Evans

Library of Congress Cataloging-in-Publication Data

Scott, David, 1980–
Bird feathers : a guide to North American species / David Scott and Casey McFarland. — 1st ed.
p. cm.
Includes bibliographical references and index.
ISBN 978-0-8117-3618-3
1. Birds—North America. 2. Birds—North America—Identification. I. McFarland, Casey. II. Title.
QL681.S39 2010
598.097—dc22

2009049986

For my brother Michel, and the long days we spent
wandering Williamson Creek as children,
and my wonderful wife, Mikki, who has always
encouraged me to follow my dreams

S. D. S.

To my wonderful and unusual family,
the "McNeills": Beverly, Gary, Patrick, and Zora—
kind, loving, good humored, and intensely
curious about the workings of this world,
they all live lives that I admire

C. M.

CONTENTS

INTRODUCTION

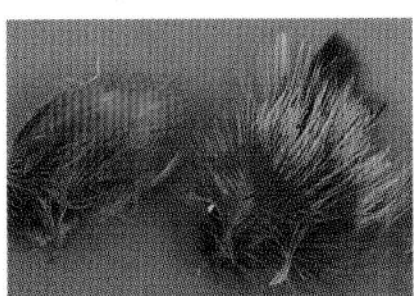

Nearly all of us, when wandering the park, the riverbank, the seashore, or our favorite path through the woods, have at some point looked down and discovered a feather. And nearly every one of us has for some reason picked that feather up and marveled at it in a quiet way. That reason, it seems, is both timeless and universal: since the age of our early hominid ancestors, we have been drawn to feathers as symbols of something powerful and mysterious. That draw still holds us: the child triumphantly catching a feather that's been tossed about on the wind, or the older couple who pick one up, dust it off, and lift it to the sky to examine its beauty, awed by its function.

People all over the world have always loved birds, and today birding skill and enthusiasm continue to grow. Over the years we have heard the same longing arise again and again: that there be a common, comprehensive resource not for identifying a bird itself, but for identifying a bird long gone from the feeder, the park, or the creek by the feather that it has left behind.

Like all animal sign, feathers are indicators of a bird's passage through a landscape, a bit of information that we can use to piece together a larger story. And like other animal sign, feathers can be very obscure clues, difficult to interpret. But the more we know to look for in a single feather, the more answers it will reveal.

This book serves as a resource to help identify feathers found in the field, but it also helps the reader better understand the astounding functions they perform. When you explore the design of feathers, what once was a simple flight feather is suddenly a wealth of information that tells a story about a particular bird's life. From the feather's shape we can see the specific job it performed, where on the wing it originated, and how it contributed to the shape of the wing itself. And when we can clearly picture the form of the wing, we can see in our imaginations how the bird flew: for long or short distances, fast or slow, bounding through the air or floating effortlessly on gusts of wind.

Within the pages of this book we have done our best to give a simple history and overview of birds, their structure, and their feathers, starting with an odd and intriguing creature called *Archaeopteryx* that plummeted into the waters of an ancient lagoon 150 million years ago. After an exploration of bird evolution, we examine the wing types of modern birds and the particular morphology of the feathers within them.

The making of this book has forever changed the way we look at a feather and the way we watch birds fly. Feathers have for us become immensely enjoyable doorways into the lives and workings of birds, expanding our admiration and appreciation of bird life further still. Our hope is that this guide has the same effect on you.

Conservation

By the end of the nineteenth century, many of North America's wild bird populations were either alarmingly depleted or at the brink of extinction: commercial hunting for food and plume had taken a staggering toll, and concern for birds was mounting. In 1900, in an attempt to slow this tragic loss of bird life, Congress passed the Lacey Act, which prohibited the trade of illegally taken game across state lines. The Lacey Act represented a powerful overall growth in awareness of bird conservation, though the fight was far from over. Other movements also emerged at the turn of the century: ornithologist Frank Chapman, of the recently formed National Audubon Society, suggested that Christmas "side hunts," the long-standing traditional competition to shoot as many birds as possible on Christmas day, become "Christmas counts." The idea caught on, and Christmas bird counts remain popular today. Though the Lacey Act slowed the mass exploitation of bird species, after nearly two decades it became evident that a greater legislative effort was needed to protect native bird populations. In response to this need, Congress passed the 1918 Migratory Bird Treaty Act, which decreed that "all migratory birds and their parts (including eggs, nests, and feathers) are fully protected." The Migratory Bird Treaty Act, along with the efforts of many conservationists, successfully saved many bird species from certain extinction in the United States, and it remains a powerful tool for conservation today.

Those who choose to explore the wonders of birds in the natural world should be aware that *it is illegal to possess the feathers of migratory bird species covered under the 1918 Migratory Bird Treaty Act without the proper federal and state permissions*. We believe that hands-on exploration of the natural world is essential to personal growth and worldwide conservation, and we have created this guide to encourage enthusiasm for birds and feather identification. But please explore this facet of the natural world responsibly: take photographs or make sketches, record measurements, and jot down notes in a journal to document your feather finds.

Part I

Feather Origins and Morphology

Origins

1

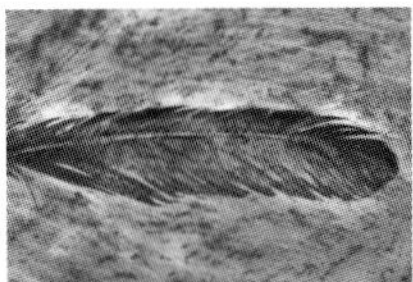

The incredible act of flight, developed first by insects some 350 million years ago, has been adapted and employed not only by birds but also by five groups of reptiles, a squid, two groups of frogs, three groups of fishes, and seven groups of mammals. Among all those that have taken to the air, it is birds that have advanced flight so exquisitely. Marching into new territory for the integumentary system, they sprouted from their skin a remarkably refined and complex apparatus—the feather—that, among other uses, enabled highly sophisticated flight. The perfection of feather design is extraordinary, and because of a frustrating lack of sequential evidence in the fossil records, we are still uncertain why or how it developed.

Until sixty-five million years ago, when Earth was warmer and largely blanketed in humid, low-growing forests, flying reptiles ruled the skies. With a thin membrane of skin pulled taut across long forelimbs, these creatures winged through open sky. This physiological adaptation effectively kept these reptiles airborne regardless of size; the largest species of pterosaur had a wingspan of forty feet. Though they were the dominant flying vertebrates at that time, they were not alone. Somewhat inconspicuously, the ancestors of birds were well on their way in exploring and advancing flight. It is estimated that ancient birds co-inhabited the world with their abundant reptilian counterparts for roughly 135 million years, since the early Jurassic Period.

In the late Cretaceous Period, a global catastrophe resulted in massive extinctions across the planet. Likely the consequence of a collision with a sizeable meteor, the ensuing devastation rid the world of *Dinosauria*, most of which had for some time already been losing ground on Earth. Many of the existing bird species died out as well, but a number survived—the relatives of the geese, ducks, loons, and other shorebirds that we know today. Vast open spaces across the globe were left relatively unoccupied, and the evolutionary process hastened to fill available niches. Bird life raced to the task, developing with impressive rapidity; within ten million years every order of birds that we know today, with the exception of small passerines, existed. Since the early Jurassic, when birds first made their appearance, some 150,000 species of birds have inhabited the Earth. Today, nearly two hundred million years later, the number of bird species inhabiting the Earth sits at around 9,672 (Sibley and Monroe 1990).

The History of a Feather

Solnhofen, Germany, 1860. Workers labored each day in a limestone quarry that some 150 million years ago was the muddy floor of a tropical lagoon. Once widely used by the Romans to construct buildings and roads, the quarry's stone was now excavated and split for use in lithographic printing. The smooth, thin slabs revealed the remnants of diverse life that once roamed those waters and their shores: knee-high dinosaurs, flying pterosaurs, beetles, moths, wasps, jellyfish, and shrimp—all fossilized exquisitely in the high-quality limestone. The quarry was known for delivering fossils both in great quantity and quality, and the workers there were well acquainted with the variety of creatures to be found. Local doctors often made a hobby of collecting fossils, and medical treatments were at times traded for fine specimens. So when one day a worker split a stone that revealed something new and astonishing—a feather—news spread quickly. There, embedded in rock well over one hundred million years old, was a fossilized feather not at all unlike one we would find today: a flight feather, possessing the same structural and aerodynamic characteristics of those of modern birds. The find was the first glimpse of the ancestors of birds, and one of seven important fossils to come from the limestone quarry in the following 130 years.

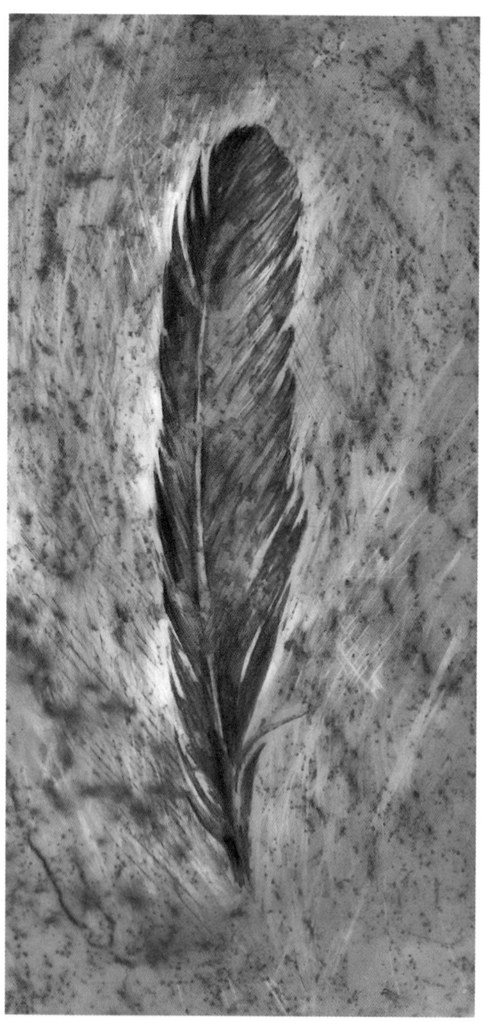

The first fossilized feather discovered in the Solnhofen quarries. Note the similarities between it and modern flight feathers.

Archaeopteryx.

> A few modern birds have claws similar to those of *Archaeopteryx* extending from their forelimbs when young and lose them as they mature. Nestling coots, moorhens, and hoatzins use these claws as an aid to clamber about in trees branches, suggesting that *Archaeopteryx* may have used them for similar purposes.

In 1861, just one year after the discovery of that single, fossilized feather, another remarkable find was made in Solnhofen: a skeleton of a chicken-size "bird," nearly complete, with its neatly fanned feathers forming a clear halo around its body. The first *Archaeopteryx*, or "ancient wing," had been unearthed.

Unlike modern birds, *Archaeopteryx* had a long, bony tail, a heavy skull laden with teeth, and wings equipped with three curved claws.

Though specimens from Solnhofen are the earliest fossilized birds yet unearthed, they are clearly not the first in their lineage. This leaves numerous questions as to the origin of birds and, equally important, feathers themselves. The feathers of *Archaeopteryx* are aerodynamic and pinnate, indicating that they had long been undergoing evolutionary influence. Who, then, came before *Archaeopteryx*, and why did feathers evolve in the first place? The most popular theory is that this ancient, prototypical bird evolved from theropod dinosaurs, but reasons as to why it developed feathers and eventually took to the air are still heavily debated.

The origin of the feather had long been considered synonymous with the origin of flight. In the last decade new discoveries have challenged and advanced theories about the ancestors of birds and the birth of feathers. These finds include a number of dinosaur fossils that are nonavian but nonetheless have simplified feathers sprouting from their dorsal axis or hanging from their forelimbs and tails, suggesting that the most primitive feathers came long before the origin of birds and flight (Prum 2002).

One longstanding theory was that the primitive feather was essentially an enlarged, complex scale protruding and hanging from the legs, tail, and body, possibly evolving in response to an increasing need to be airborne: the longer and more aerodynamic a "scale" grew, the better it allowed its owner to glide. More likely, though, is that feathers underwent a complex process filling other roles before they even came close to having lift-giving properties, and that they most likely evolved from a conical tubercle that branched with greater complexity over time rather than from a platelike structure. (Prum 1999, Brush 2000).

Flightless Birds

While it is important to consider the evolutionary advancement toward flight, it is equally noteworthy that many species appear to have digressed *away from* flight just a short time after feathers evolved to make flying possible. The benefit of flight, it would seem, would be the last thing that a feathered bird would want to lose. But flight is energetically costly, and if a bird's environment allows for it, it is apparently advantageous to give it up. For instance, the lack of terrestrial predators on particular islands created conditions that were ideal for the development of flightless species. Another advantage of losing flight is that size can go largely unchecked, which is advantageous for numerous species. Among other things, benefits of larger mass include an increased ability to maintain body temperature and the potential for greater land speed in some ratites (ostriches and others). The greatest threats to flightless birds have come at the hands of humans, who have introduced invasive predators and destroyed habitat.

While dinosaurs are for the most part considered to have been cold-blooded, there is evidence that some of the smaller species were actually homoiothermic, or *warm*-blooded, giving credence to a hypothesis that feathers first developed for insulation purposes. No other natural material matches the insulating properties of feathers; they are marvelously efficient. A small, highly active, warm-blooded reptile, expending energy by both hunting and maintaining its own body heat, certainly could have benefited from a downy covering.

Other factors may have influenced feather development, including water repellency, protection from solar radiation, defense, communication (primitive feathers could have been pigmented for a showy appearance), or camouflage. Most important to consider is that feathers may not have evolved for the sake of flying, but over time those early, simple versions were the stepping stones to flight.

From Feathers to Flight

The question still remains: *how* did the ancestors of birds go about putting the feather to use as a tool for flight? At some point, small dinosaurs adorned in feathers of some sort began "catching air" and refining the flight process. The first of two traditional theories is that *Archaeopteryx* was arboreal, slowly evolving its feathers and taking flight by leaping and gliding from trees. Others maintain that it was a ground dweller, or cursorial, dashing after prey on two legs, with its outstretched wings providing lift and maneuverability.

With just one exception, the *Archaeopteryx* fossils all lack a bony breastbone. This indicates that it was likely made of cartilage instead of bone and therefore was an unsubstantial aid to powered flight. Because a keel (projecting from a bony breastbone) is structurally necessary to anchor the massive breast muscles responsible for sustained flapping, the fact that it is lacking raises the question of how well *Archaeopteryx* could actually power itself through the air. Nonetheless, the feathers of *Archaeopteryx*—almost indistinguishable from those of modern birds—were aerodynamically designed to perform flight of some sort. Also possible is that *Archaeopteryx*, like lizards of the genus *Basiliscus*, was able to run on water. The development of flight feathers would have generated lift, increasing the distance that could be covered across a liquid surface, enabling escape from predators and travel between islands in the lagoon (Videler 2005).

For over a century, Solnhofen limestone quarries were the only locations that offered evidence of the earliest birds. But in 1995, in the Liaoning province in northeastern China, new bird fossils began to emerge from what was once a large lake some thirty million years younger than the Solnhofen lagoon. One of the fossils discovered was *Confuciusornis*, a specimen clearly farther along the evolutionary ladder. Though *Confuciusornis* still had grasping appendages extending from the wing (like those of *Archaeopteryx*), it had greatly reduced tail bones, long tail feathers in some specimens, and, most importantly, a beak. Even more advanced species have since surfaced in Spain, North America, Argentina, and Australia.

2

Avian Physiology: Built for Flight

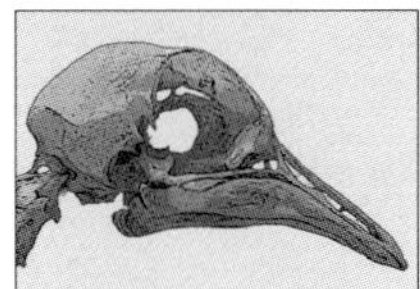

A number of flightless birds look quite a bit like their flying cousins: they have wings of typical size and feathers that are long and robust. And because it is easy to associate such feathers with an ability to fly, one might automatically assume they have the ability to simply take off if approached. But their feathers are useless for flight because of their physical build: their skeleton and musculature render them physically incapable of flying, an important reminder that it is much more than the feather that enables birds to roam the sky.

Before the advent of airplanes, people mimicked the design of birds in hopes of achieving the freedom of flight. A great many flapping and gliding contraptions were conceived and tested, endeavors that yielded only a sense of disappointment and a few good lessons about flight dynamics. Others involved in similar experiments were far less fortunate: condemned men in China sometimes met their end by being thrown from towers in such contraptions. These failures, of course, happened because the body of the bird is richly modified to enable flight; the skeleton, lungs, and muscles are radically adapted for a flying lifestyle. Humans, however, lack the immense muscles necessary for powered flight; they also have long, flexible spines and heavy bones.

Musculoskeletal System: The Skeleton and Musculature

Bird bones are strong in proportion to their weight, and many are hollow, reinforced with an internal crisscrossing strut system that provides stability. In addition to the obvious advantage of reducing weight, hollow bones are pneumatic, which means that the hollow air spaces within the bones are directly connected to the respiratory system, enhancing respiratory efficiency. Flight may have had some influence on the pneumatic skeleton, but it is not solely responsible for its design. A species of carnivorous dinosaur

(*Aerosteon riocoloradensis*) with telltale characteristics was recently discovered: it sports hollow pneumatic bones like those of birds, suggesting that the system may have evolved simply for more efficient breathing, to reduce weight, or to help keep the animal cool. While bird bones tend to be lightweight, individual bones are not always lighter than the bones of other vertebrates of similar proportions. Nonetheless, the skeletal structure as a whole is light and strong. The quintessential case is the frigatebird, whose entire skeleton is only *half* the weight of its feathers.

The tail, pelvis, and thoracic region of the spine are all fused to create sturdy platforms that reduce weight and provide the rigidity necessary to maintain posture of the body during flight. The ribs also have lateral extensions that overlap (called the uncinate process), bracing together to form a sturdy cage that protects the lungs and heart from the crushing force of beating wings. This interlocking rib feature, which until recently was thought to be a characteristic specific to birds, has also been discovered in nonavian dinosaurs (Zhou 2004).

The structure of a bird's neck makes it quite flexible. Because the forelimbs have little use beyond flight or swimming, the neck is crucial for the task of preening: the head must be able to reach most of its body to properly tend to feathers. While mammals have seven neck vertebrae, birds have eleven to twenty-five.

The furcula, the fused clavicles that we are familiar with as the "wishbone," is much like a flexible spring that aids flight: on the downbeat, muscles contract and squeeze the ends of the furcula toward each other. When the wings have reached the lowest point of the downbeat and begin to draw upward, the furcula "releases" and rebounds to its original position, much like a bow letting loose an arrow.

To provide sturdy support for the flight feathers, the bones of the "hand" are reduced and fused. The bone of the forearm, the ulna, has a series of bumps along its trailing edge to which the secondary flight feathers attach directly.

The keel, or breastbone, is extremely pronounced in most birds. Large and bladelike, it is a thin slab of bone along which the massive flight muscles are anchored. It also provides additional protection for the heart and lungs. Some flightless birds, specifically the ratites, have no use for a pronounced keel, and it is therefore absent. Others, such as penguins, use their "flight" muscles for swimming underwater, so they have retained their pronounced keel and powerful muscles.

The skull is extremely light as well. The heavy jaws and teeth of pro-avis were eventually replaced with a lightweight bill covered in a horny sheath. The teeth and jaws were replaced by the gizzard, a muscular organ that pulverizes food. Set low and far back in the body, it enables birds to essentially "chew" with a mechanism that does not hamper flight but makes it more efficient.

Like its skeleton, a bird's musculature is well adapted for flight. The muscles of most striking size are those of the breast, which power the wings. In

fused thoracic vertebrae

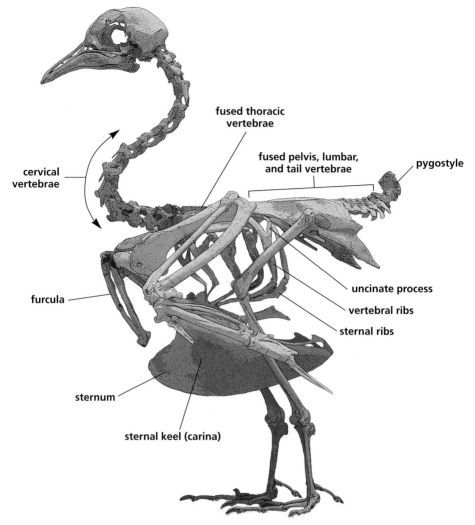

cervical vertebrae

fused pelvis, lumbar, and tail vertebrae

pygostyle

uncinate process

vertebral ribs

sternal ribs

furcula

sternum

sternal keel (carina)

The avian skeleton: Rock Pigeon.

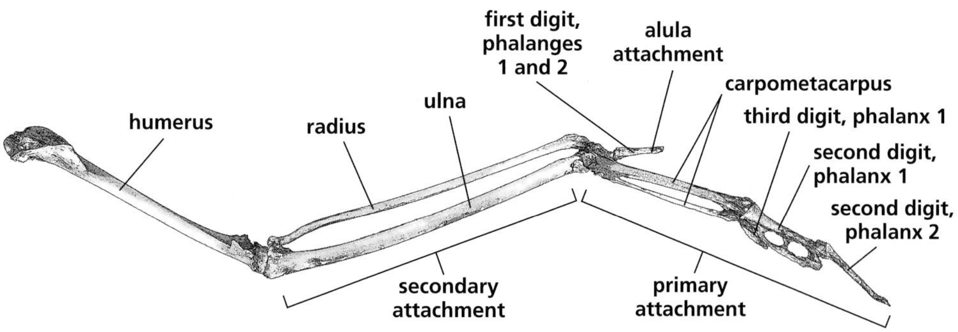

The skeleton of the right wing: Caspian Tern.

many birds, these bulky muscles make up a full third of its overall body weight. And in species such as grouse, which make powerful bursts through the brush, they may contribute nearly forty percent.

For the most part, musculature is centered on the underside of the bird, establishing a steady center of balance. In comparison, the back has little muscle mass but stabilizes the hard-working underside. The wings and neck are covered with a complex, lightweight network of muscles that provides optimum strength and control.

Flight has had its effect on the heart as well: this organ is the ultimate in endurance capability. Like mammals, birds have hearts with four chambers; however, they are often proportionately larger. A sparrow's heart is three times larger than that of a mammal of the same size. Bird hearts are also stronger and pump faster than those of mammals: the fastest rate belongs to the hummingbird, at one thousand beats per minute. Smaller birds have proportionately larger hearts than larger birds, and birds that migrate have proportionately larger hearts than nonmigratory species.

Respiratory System

The lungs of birds are surprisingly unlike those of most other land vertebrates. Due to their design and physiological complexity, they are among the most efficient on Earth. Bird lungs are small compared with those of mammals but connect to numerous inflatable air sacs that branch from the lungs into the muscles and even into the bones. And unlike mammals, which breathe in and out, their lungs inflating and deflating like a bellows, birds have a constant flow of oxygen in the lungs. They actually breathe in twice to complete one full respiration cycle: upon inhalation, air flows into the posterior air sacs, and then exhalation forces it from there to the lungs. On the second inhalation, air in the lungs, depleted of oxygen and full of carbon dioxide, flows from the lungs to the anterior air sacs, while fresh air flows down the bronchi to the posterior sacs once more. On the second and final exhalation of the cycle, the stale air is pushed up from the anterior air sacs and out of the bird. While mammals rely on a thick, muscular diaphragm to power their breath, the air sacs in birds are in part pumped by the action of hinged ribs and the furcula: with each beat of the wings, the lungs are compressed and then expanded, increasing ventilation during flight.

The respiratory system may owe some of its complexity and efficiency to high-altitude flight. Sparrows, when subjected to test chambers simulating twenty thousand feet, flitted casually about, demonstrating that they were largely unaffected and could ascend even higher. Mice, on the other hand, slumped helplessly in the same chamber, scarcely able to move. The highest altitude on record for a bird was documented when a Ruppell's Vulture met an ill fate in the engine of a passing jetliner at thirty-six thousand feet.

Feathers and Flight

3

Feathers far surpass fur or scales in terms of complexity and efficiency, making them the most advanced integumentary structures among vertebrates. Extremely light and flexible, yet strong enough to withstand the stress of flight, these unique and marvelously developed components of bird anatomy have enabled birds, above all other creatures on Earth, to conquer the open sky. Besides allowing for flight, feathers fill a wide array of roles: they provide insulation and waterproofing, protection from injury and ultraviolet light, camouflage, and coloration.

Anatomy of a Feather

Feathers, like scales, hair, nails, horns, and hooves, are made of keratin, a strong structural protein, and once fully grown become inert. Each feather on a bird's body is made up of numerous components. The bare, hollow stem at the lower end of the feather is called the *quill*, or *calamus*. At the base of the quill is a small opening called the *inferior umbilicus*; this provides an entry point through which blood vessel's nourish a developing feather. When a feather matures and no longer contains living cells, the inferior umbilicus closes and blood flow ceases. At the point where the feather vanes begin (see page 14), the quill becomes the central *shaft*, or *rachis*. Unlike the tubular quill, the rachis is square along its length. On most major flight feathers, a groove runs along the underside of the rachis, preventing kinking and therefore giving extra strength.

Extending from each side of the shaft is the feather *vane*, or *web*. The vane is made up of interconnecting barbs; along the length of each barb are tiny hairlike branches called *barbules*. Each of these barbules is covered in minute hooks called *barbicels* that interlock with each other upon contact. This system of connection allows the feather vane to flex while keeping the barbs connected, and is the reason that you can pick up a tattered feather and mend, or

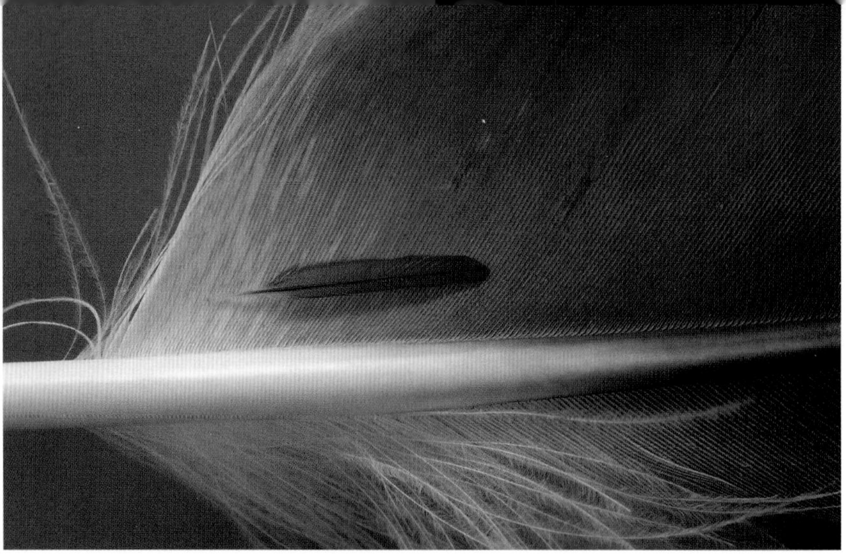

Though a hummingbird is 1100 times lighter than a Golden Eagle, the flight feathers of both bear the same design.

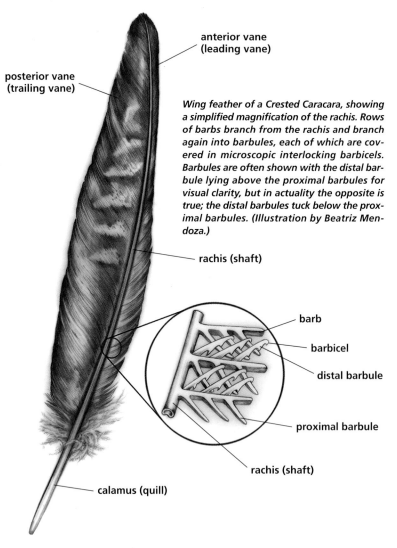

anterior vane
(leading vane)

posterior vane
(trailing vane)

Wing feather of a Crested Caracara, showing a simplified magnification of the rachis. Rows of barbs branch from the rachis and branch again into barbules, each of which are covered in microscopic interlocking barbicels. Barbules are often shown with the distal barbule lying above the proximal barbules for visual clarity, but in actuality the opposite is true; the distal barbules tuck below the proximal barbules. (Illustration by Beatriz Mendoza.)

rachis (shaft)

barb

barbicel

distal barbule

proximal barbule

rachis (shaft)

calamus (quill)

"zip together," the vane by simply pulling the split sections through a thumb and forefinger. Birds use their beaks to mend feathers in a method of grooming called preening. Because well-maintained feathers are virtually water and windproof, streamline the body, and keep flight most efficient, it is vitally important for birds to preen regularly.

Feather Types

Though most birds have many thousands of individual feathers covering their bodies, there are only six basic types of feathers. Each of these six types has characteristics and functions that make it unique and fairly distinguishable from the others.

Contour Feather

Contour feathers make up the majority of visible feathers covering a bird's body. Traditionally, flight feathers are included in this category; however, for the sake of morphology and identification, flight feathers have been given their own section in this guide. The contour feathers of the body have a relatively stiff shaft and vanes that are typically bilaterally symmetrical. These feathers form the smooth, compact outline of the bird that provides the streamlined, aerodynamic shape essential for flight. Contour feathers are organized on the bird's body like roof shingles, forming the protective waterproof barrier that makes up the bird's outermost layer of insulation. The base of the feather is normally downy, and in many species there is a clearly visible aftershaft, or *afterfeather,* which extends from the base of the shaft and can be as simple as a few fluffy barbs or as complex as a smaller, intact feather growing from the shaft. The afterfeather increases the insulating properties of the contour feather.

Semiplume

Semiplume feathers serve as an intermediate layer between outer contour feathers and the inner layer of down. Like contour feathers, they have a supportive shaft that runs their entire length, but they lack the barbicels, or hooks, to hold the barbs together. The result is a feather that has no vanes and presents a fluffy appearance. The primary role of semiplumes is insulation, but they also help define the bird's shape by creating a thick layer on which the contour feathers rest.

Down Feather

Down feathers are distinctive in that the feather shaft is either significantly reduced in length or missing altogether. As is the case with semiplumes, the barbs on down feathers lack hooks. Combine these characteristics and the result is a very fluffy feather with little structure. Down feathers insulate the

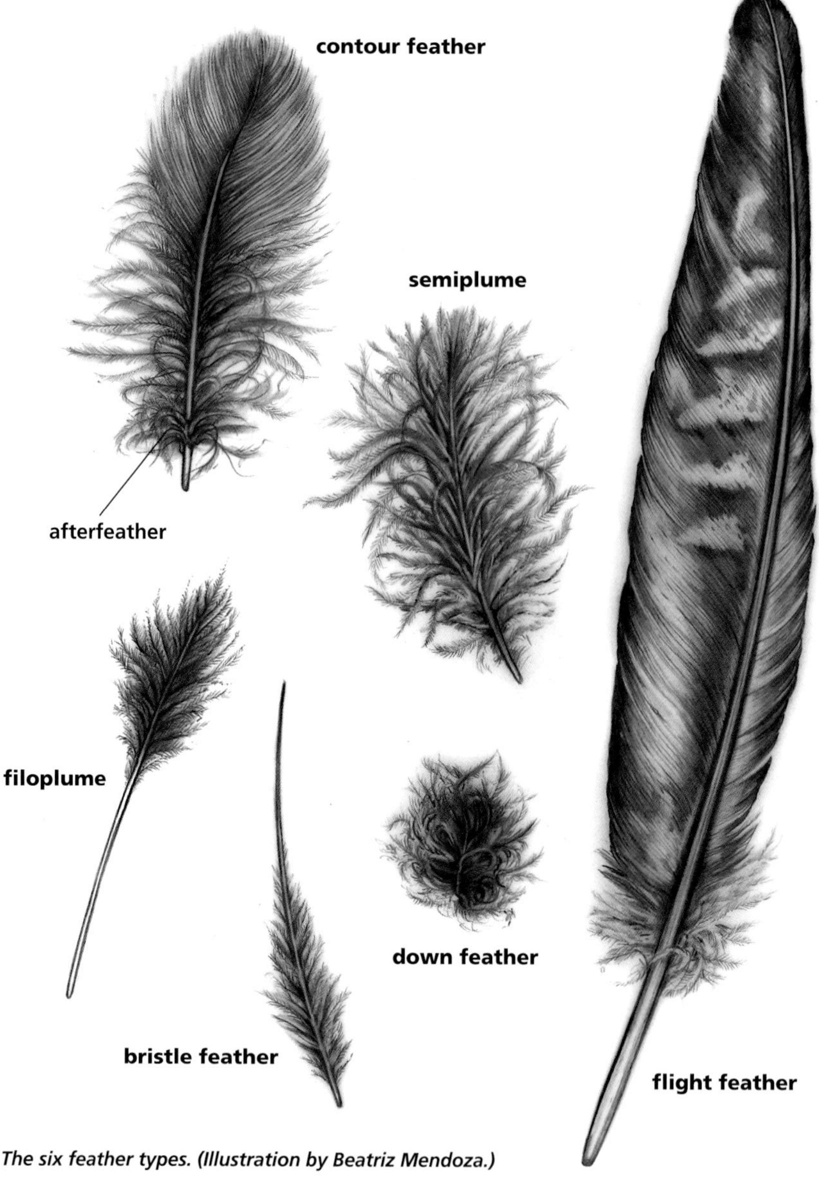

The six feather types. (Illustration by Beatriz Mendoza.)

bird; the down layer is unevenly distributed on a bird and varies in its bulk between groups of birds.

Filoplume

Filoplume feathers are structurally simple. They are made up of a stiff, hair-like shaft, with either no barbs or just a few at the tip. Ornithologists believe that filoplumes, found directly under or protruding just past the contour feathers, transmit the slightest movements and vibrations to nerve endings in the bird's skin, the same way whiskers provide sensory information for some mammals. These feathers grant the bird a sense of touch, allowing it to recognize when other feathers are out of place, and may also help a bird gauge its airspeed.

Bristle Feather

Bristle feathers are similar to filoplumes and are made up of a stiff, tapered shaft lacking barbs except at the base. Bristles are only found around the head and neck and serve a sensory function, much like the whiskers of a dog or cat. They also may serve to protect the eyes or help birds catch insects. These "whiskers" are especially prominent on species of nightjars as well as woodpeckers, which benefit from protection from debris while hammering holes in trees.

Flight Feathers

The simplest feathers to identify are those that make flight possible. These include the feathers that make up the wing, called *remiges* (singular: *remex*), and those of the tail, called *retrices*. Flight feathers, varied in shape and design, constitute wing shape, which largely dictates particular flight styles.

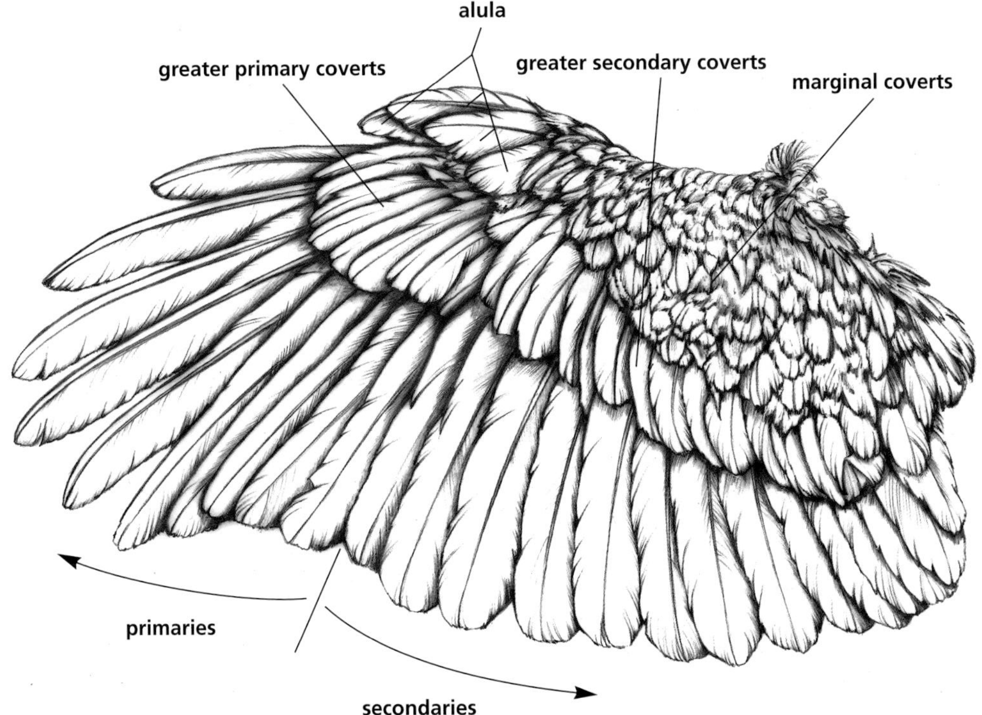

Basic topography of the upper (dorsal) side of the wing of a Harris's Hawk. (Illustration by Beatriz Mendoza.)

Note the attachment of primaries and secondaries to the skeletal structure of the bird's forelimb.

The two types of flight feathers of the wing are the *primaries,* attached to what is the bird's skeletal equivalent of our hand, and the *secondaries,* located along the forearm. Primary wing feathers generally have a long, relatively straight quill and one feather vane that is noticeably narrower than the other. The narrower vane indicates the leading edge of the wing. This feature can be used to easily determine which side of a bird's body the feather came from. The number of primaries varies by species, from ten to twelve; most birds typically have ten.

Various primary feathers. From left to right: Northern Flicker, Belted Kingfisher, Northern Pygmy-Owl, Yellow Warbler, and Merlin.

The short leading primary of a Pileated Woodpecker (top), shown in comparison with the adjacent primary.

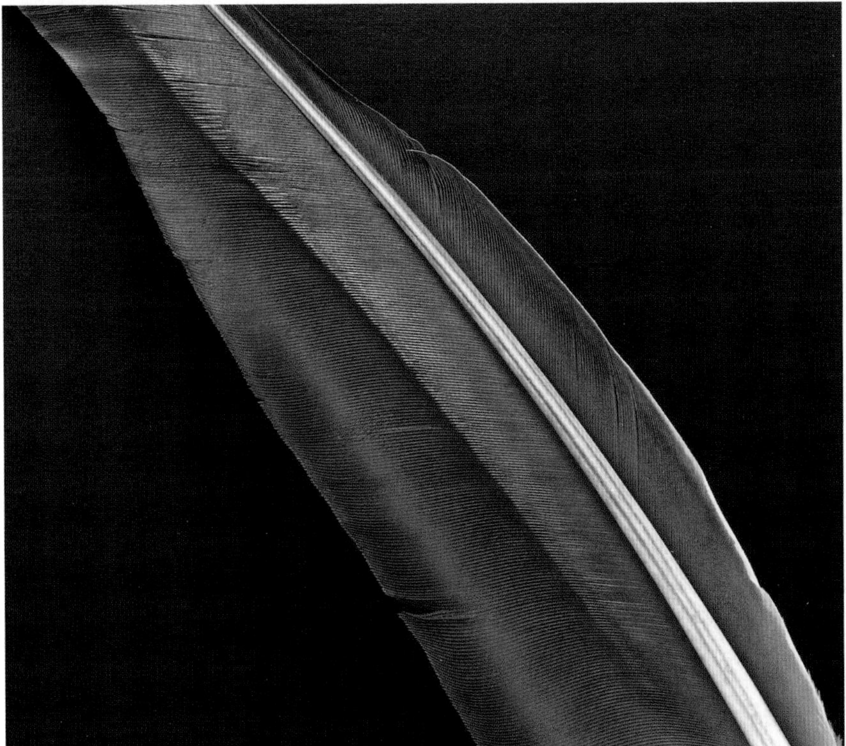

In ducks, geese, swans, and some other species, the barbs of the primaries are uniquely specialized: lengthened, flat, folded toward the quill, and tightly overlapped, they create a strong, glazed surface along the underside of the feather called tegmen. *These birds are heavy and fast fliers: the tegmen area presumably reinforces the structural integrity of their primaries, keeping the vane from splitting during flight. This shiny patch readily distinguishes these birds from other families.*

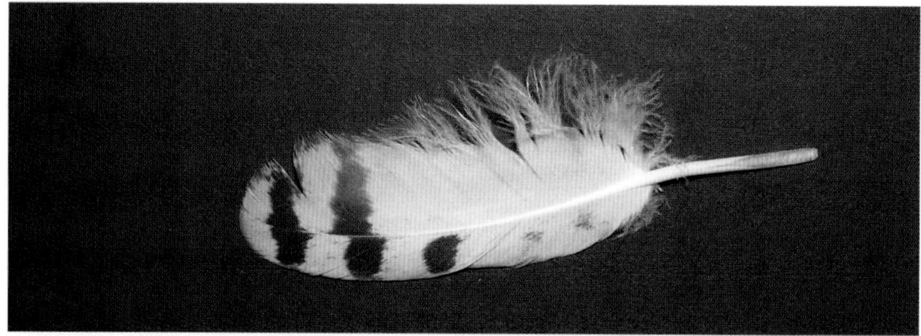

Top: *Greater primary wing covert from a Snowy Owl.* Bottom: *Greater primary wing covert alongside a primary feather from a Snowy Owl.*

Various secondary feathers. From left to right: Northern Flicker, Pileated Woodpecker, Belted Kingfisher, Yellow Warbler, and Merlin.

It's important to note that primary counts can be deceiving. Though this is not the case in all bird species, in many birds the leading primary is very reduced in length or vestigial and visible only with careful scrutiny. In some passerine species such as wrens, nuthatches, chickadees, and creepers, the leading primary is quite short. In the case of woodpeckers, the leading primary is even more reduced in size, so even though it is clearly visible, it can be missed in a primary count. In many bird species, this leading primary is so reduced that it is not readily apparent; it is a vestigial feather, or *remicle*, and concealed by the *coverts* (the feathers that partially cover the flight feathers). Some passerines are actually categorized as "nine primaried," but research examining many of these particular birds has revealed that they indeed have ten (Susanna, Hall 2005). This can also be true for birds with eleven or twelve primaries; be aware that a count of obvious, visible primaries across a wing may come up one feather short, as the last primary may be hidden from view.

Greater primary and greater secondary wing coverts are easily identifiable by the distinctive kink at the point where the shaft meets the quill. The kink in the feather allows the quill of the covert to attach directly to the side of the quill of the primary or secondary to which it is paired.

Secondaries can generally be distinguished from primaries by a greater curvature of the quill; a wider, more rounded tip; and vanes of more equal width. Body size and length of wing determine the number of secondaries, from as few as six on a hummingbird to as many as thirty-two along the long, narrow gliding wing of the albatross.

Tail feathers, which aid birds in flight control, are numbered in pairs from the center of the tail outward. The average number of tail feathers is twelve, but it varies by species from eight to thirty-two. As on the flight feathers of the wing, the narrower vane indicates the outside edge of the feather and faces away from the center of the tail. Tails of some birds are specialized beyond the function of flight. Woodpeckers use their stiff tails to prop themselves against tree trunks, giving them greater balance and leverage when hammering holes into the bark; creepers and some swifts use their tails as props as well. Other birds, such as the male peacock, have modified tails that fan out to create magnificent displays used for courting females. Because of specialization, these feathers often vary a great deal from species to species.

Pterylosis

Although birds appear to be evenly feathered, their feathers are distributed in distinct feather tracts, or *pterylae*. A bird with its all of its feathers sheared close to the base of the quill would show symmetrical patterns of tracts that either have or lack feathers. The bare regions in between are called *apteria* and are covered by adjacent feathers. It is likely that all primitive birds were evenly feathered. Some species, such as penguins and ostriches, still are, retaining the ancient characteristic.

Dynamics of Flight: Understanding How Birds Fly

A number of general comparisons can be made between birds and airplanes, but we must remember the bird is the quintessential opposite of a fixed-wing flyer, a wonderfully more complex and efficient flying machine. The wings of a bird can extend to support the body while airborne and then fold neatly against the body, hardly noticeable at all. Complex networks of long, thin muscles and tendons rotate, curl, shorten, or lengthen the wing as well as make minute manipulations to stabilize the bird, increasing and decreasing speed and height in an instant.

Bird flight is made possible first by the use of a basic shape called an *airfoil*. When air flows over a convex surface, like the wing of a bird or an airplane, the air moving across the top of the wing must move faster than the air mov-

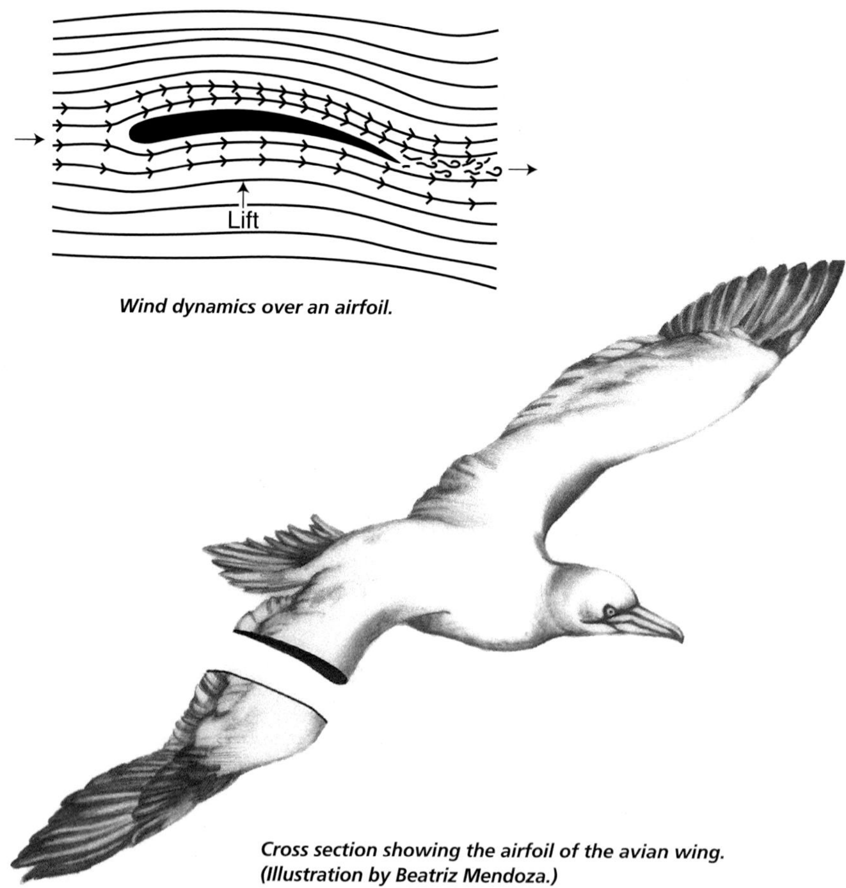

Wind dynamics over an airfoil.

Cross section showing the airfoil of the avian wing. (Illustration by Beatriz Mendoza.)

ing across the underside in order to reach the back side at the same time. The result is greater pressure below and lower pressure above, causing *lift*.

Unlike any other flying creature on earth, birds have made exquisite use of the airfoil. The overall structure of a bird's forelimb, or wing, is an airfoil: the leading edge is blunt and smooth, and the wing tapers and curves toward the trailing edge, creating convexity. Feathers themselves, especially wing feathers, are also airfoils. Their asymmetrical design and curvature create in cross section the perfect aerodynamic form.

The primary wing feathers act as propellers that pull the bird through the air while the secondaries provide most of the lift. Watching the flight of a crow, one will notice that the wing is not beating simply up and down, but reaching forward and "scooping" air—a motion similar to that of rowing a boat. Coverts, arranged on the surface of the wing, act to minimize wind resistance and aid in a smooth flow of air across the wing.

Wing Slotting

To continuously account for changes in air speed and direction, and to maintain the appropriate ratio of lift and drag, a bird makes slight adjustments to its wing and tail feathers while in flight. Many birds, especially those with slotted high-lift wings (see below), have specialized primary feathers that are *emarginated*: the feather vanes are uniquely shaped, varying from a slight taper along the feather's length to a dramatic portion neatly "cut," or notched, from the vane. This "slotting" shape is especially evident in the primaries of larger birds such as hawks or vultures—it gives the wing an overall "fingery" silhouette.

These slots become passages through which air flows at faster speeds, like water moving into a narrow channel. The result is minimized wind resistance and reduced pressure from above, which give the bird more lift. This "slotting" effect can also be created or accentuated by fully spreading the wing. Birds such as quail and pheasants make considerable use of the slotting effect; in order to make short, quick, and maneuverable bursts of flight, they take advantage of greatly reduced drag and increased lift across the wing.

For further manipulation of the wind across the wing, birds have an *alula*, attached to the bird's "thumb," stemming

Wing slotting on a Red-tailed Hawk in flight. Note the long fingery appearance created by emarginated feathers. (Photo by Jonah Evans.)

from the wrist of the wing. The feathers of the alula are called *alular quills*, and most birds have three or four. These feathers can be operated independently of the primaries and lifted to create a rush of air across the upper surface of the wing, which gives the bird precise control of lift and drag. Should the opportunity arise, examine the alula feathers of a bird in hand, noticing how they extend beyond the last joint of the wing. They are sturdy, prominent feathers, but often overlooked.

Various examples of emargination. Note the degree and location of "notching" on each feather. From left to right: Snow Goose, Black Vulture, Swallow-tailed Kite, and Red-shouldered Hawk.

In this Harris's Hawk wing, note the prominent alula and distinct slotting, both of which increase air speed across the wing and provide lift.

Flight and Wing Shape

Bird flight has evolved in response to a vast array of ecological and geographical challenges posed by various settings: the open seas, deserts, dense forests, shorelines, and grasslands. Each setting has affected the morphology of the species that live there, resulting in a wide variety of flight styles and wing designs. To better understand feather shape and function, it's necessary to consider wing shapes and the corresponding flight styles they are capable of performing.

The shape and relative size of a bird's wing can be described in terms of aspect ratio and wing loading. *Aspect ratio* is a term used to describe the ratio of wing length to width. The length divided by the width yields either a low number (low aspect ratio), as with the stubby wing of a quail, or a high number (high aspect ratio), as with the long gliding wing of a tern. *Wing loading* is the ratio of a bird's total body weight to the surface area of its wing. A heavy bird with relatively small wings, such as the cormorant, has high wing loading and must fly at faster speeds to remain airborne. The energy cost is thus much higher for species with high wing loading. The opposite is true in birds with low wing loading—many hawks, for instance, can maintain flight with relative ease. Because wing loading is simply a ratio of weight to wing surface area it does not always remain constant. For example, a migratory bird will have higher wing loading during some months of the year, when it has ample fat stores, and lower wing loading in leaner months.

Wing shapes can be placed roughly into four categories. The most common is the low-aspect-ratio *elliptical wing*, especially noticeable in game birds and passerines; it is found in birds that are typically very active and live in brushy, forested habitats. Often with a high degree of slotting, this wing type is perfectly designed for quick bursts of controlled flight. Elliptical wings are useful for maneuvering in densely vegetated habitats, but due to the low aspect ratio, the cost of flight is high. Wings falling into this category can vary greatly in shape and capability and for this reason they have been broken into two subcategories, *classic elliptical wings* and *elliptical game bird wings*. There can be a marked difference between the powerful, rounded wings of nonmigratory game birds and the sharper, more pointed wings of passerines, which are capable of long, sustained flight.

Birds that make long, gliding flights in open air have wings built in an opposite fashion: this *high-aspect-ratio wing*, like that of an albatross, is extremely long and slim and has no slotting. The narrowness of the wing and its remarkable length give it immense lift, but maneuverability and the ability to take off easily are poor at best.

The third type, the long, angular *high-speed wing*, is common to birds such as falcons, swifts, swallows, plovers, sandpipers, ducks, and terns. Often having a boomerang shape, high-speed wings are sleek, slim, and angled back at the wrist, with little to no emargination (with the exception of falcons). This design makes some birds capable of stunningly fast flight and awe-inspiring agility: peregrine falcons can approach 200 miles per hour in a dive, and the White-throated Needle-tailed Swift of India has been clocked at an impressive 219 miles per hour (Page and Morton 1995).

The fourth type is the *slotted high-lift wing*, also known as a soaring wing, of eagles, vultures, condors, swans, pelicans, hawks, and some owls. Broad, long, heavily slotted, and strongly built, with a moderate aspect ratio, the

---

Natural adaptation has shaped the wing morphologies of birds for different types of flight but has also resulted in more subtle differences between closely related species and even between different populations within species (Pérez-Tris and Tellería 2001).

One of the most impressive changes in some passerine wing shapes occurs after the first complete molt, when the wing becomes longer and more pointed (Pérez-Tris and Tellería 2001). Many juvenile passerines have shorter, rounder wings than those of their migratory parents, presumably due to the fact that such a wing design gives more maneuverability. Pointed wings such as those of adult passerines allow faster, more energy-efficient flight and so are thought to be an adaptation for migratory movements (Senar et al. 1994). Juveniles are more vulnerable to predators than adults, and avoiding predators apparently may outweigh migration performance during the first year of life. In old individuals, feather shape of the wing and tail may change as well (Moller and De Lope 1999).

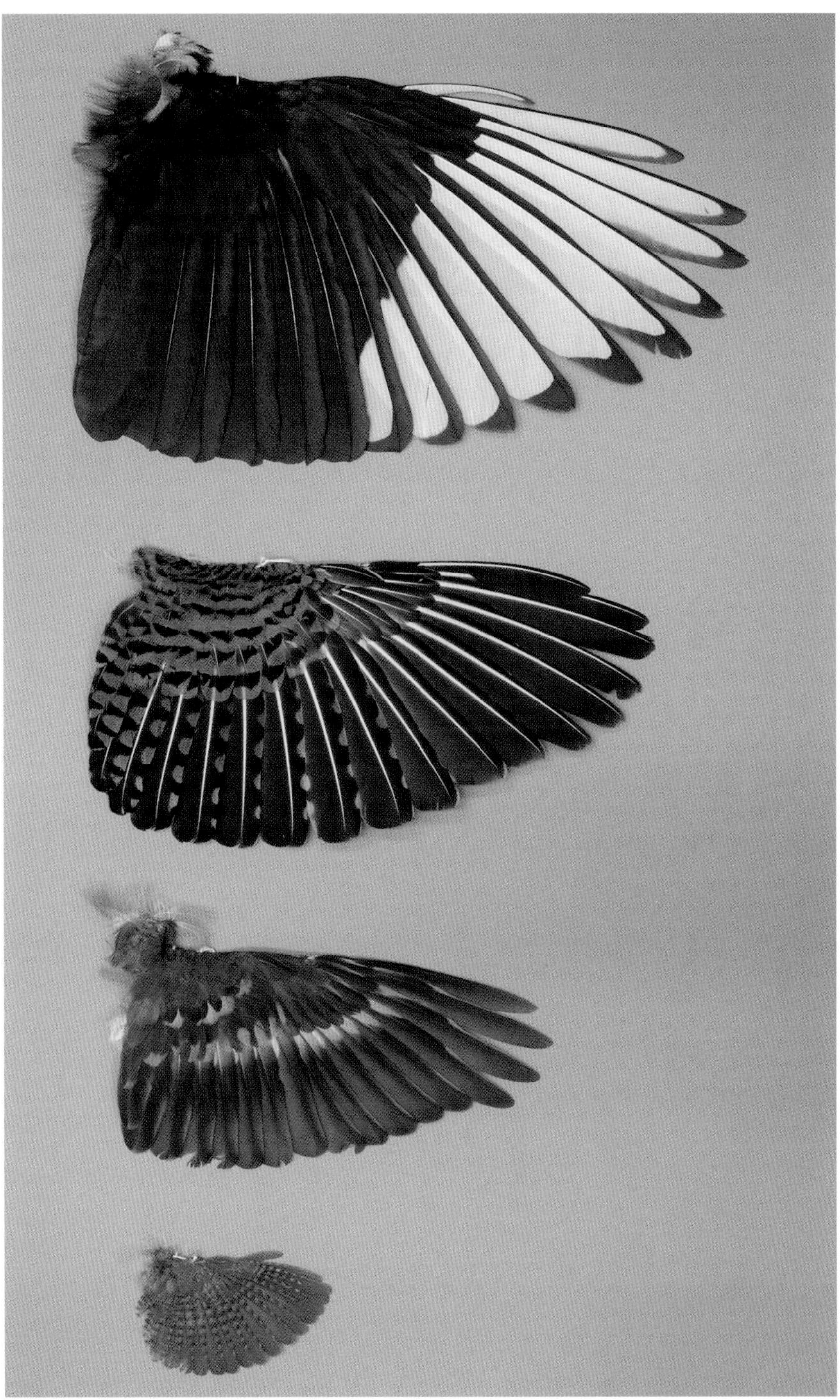

Elliptical passerine wings. From top to bottom: Black-billed Magpie, Northern Flicker, Varied Thrush, and House Wren.

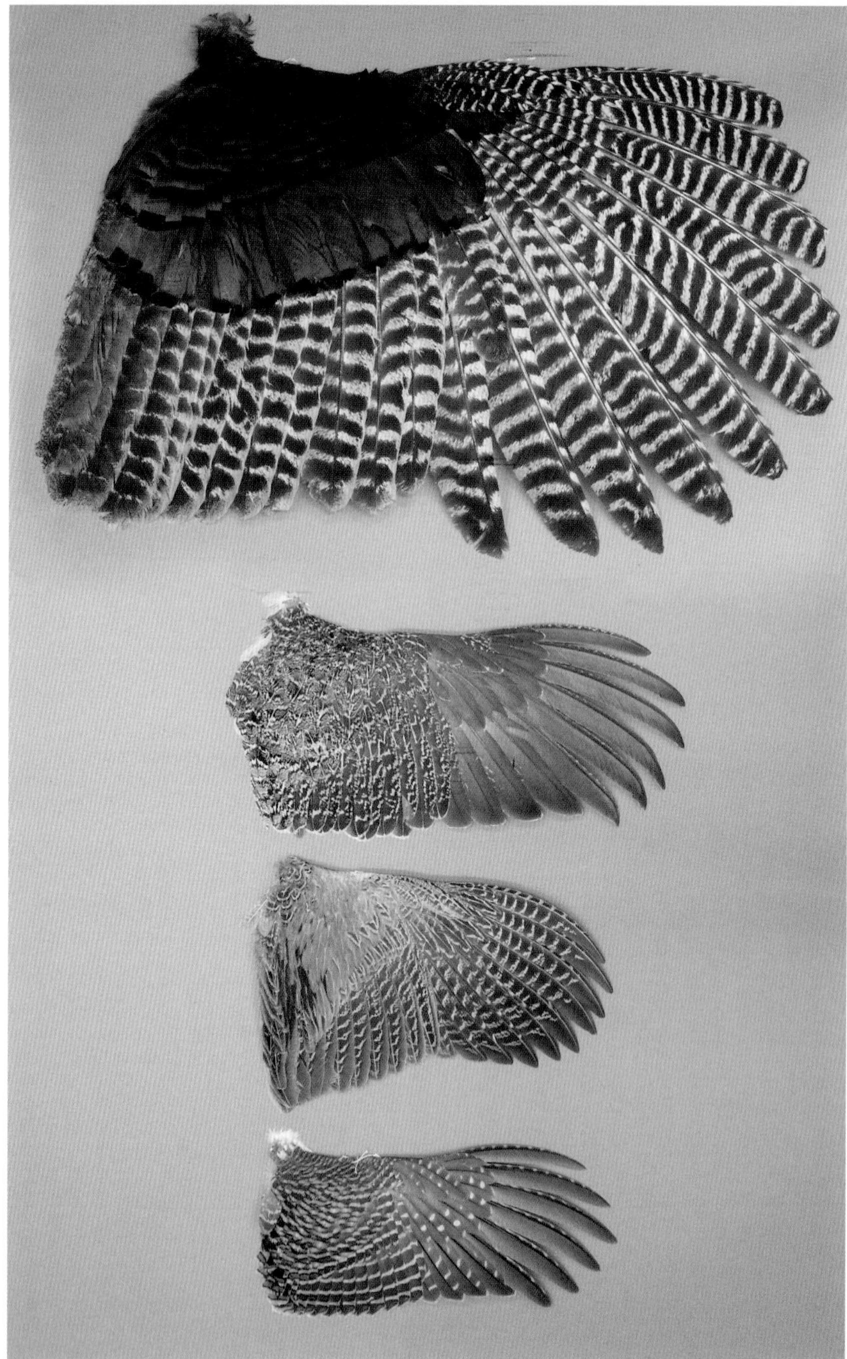

Elliptical game bird wings. From top to bottom: Wild Turkey, Greater Sage-Grouse, Ring-necked Pheasant, and Greater Prairie-Chicken.

High-aspect-ratio wings. From top to bottom: Black-footed Albatross, Greater Shearwater, and Bonaparte's Gull.

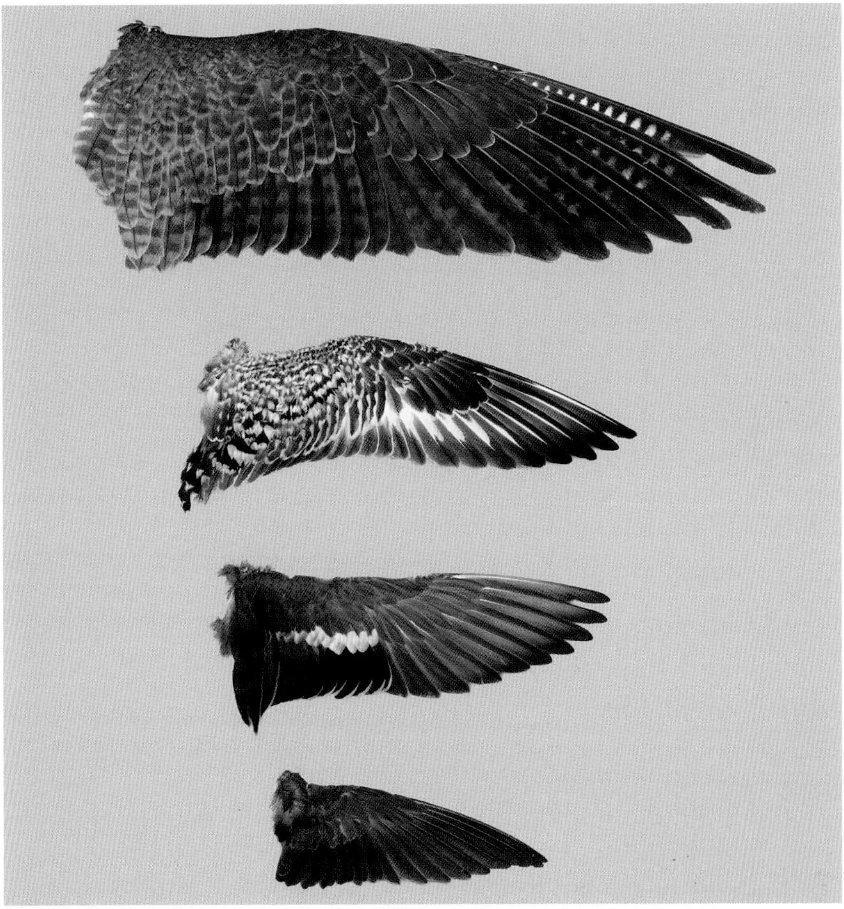

High-speed wings. From top to bottom: Peregrine Falcon, Black-bellied Plover, Green-winged Teal, and Purple Martin.

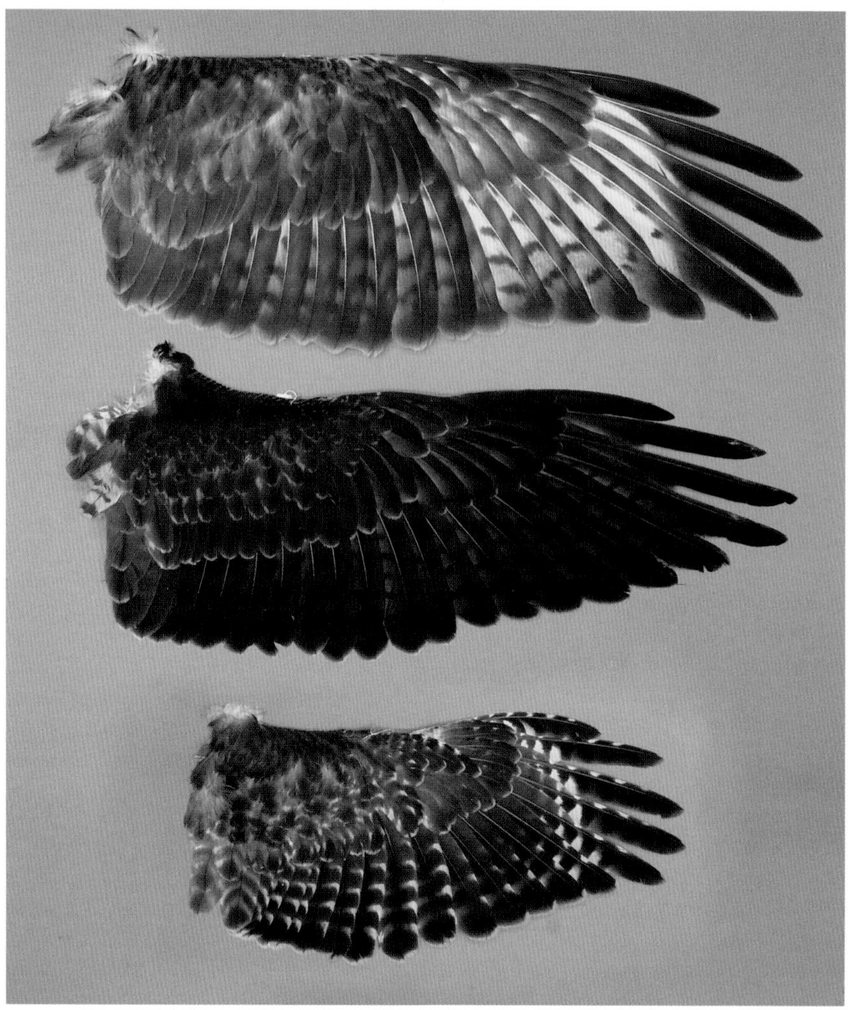

Slotted high-lift wings. From top to bottom: Swainson's Hawk, Ferruginous Hawk, and Red-shouldered Hawk.

design of this wing not only hoists these birds into the air and allows for easy, efficient flight on fixed wings but also provides the powerful lift necessary for carrying heavy catches.

Styles of Flight

When bird-watching, notice the array of flying styles you see employed: the sporadic, jumpy flight path of a woodpecker; the almost frantic, hurried flight of a mallard; the way a pelican easily coasts just above the crests of waves until they break, before peeling off at a right angle to ride the next wave.

Keep in mind that just because a bird can perform a particular style of flight, it does not mean that its wing is suitable only for that function. Many birds are capable of numerous flight styles. For instance, a pelican is a skilled surface glider but has a wing built like that of an eagle or vulture; its broad, heavily slotted wing is categorized as a slotted high-lift wing rather than a high-aspect-ratio gliding wing.

Flight styles can be broken into five categories, with each flight style including slight variations on the general theme. Many species can use more than one flight style, depending on their wing design and the ecological or geographical setting. The flight styles given here are based on a combination of those provided by Kerlinger (1995), Videler (2005), and Padulca et al. (2004).

Intermittent flight. Intermittent flight, which may be better defined as intermittent flapping, is employed by birds that have to flap their wings in order to remain airborne. This flight style consists of two basic strategies, *undulating flight* and *bounding flight*. Undulating flight, also called flapping and gliding, is common to a variety of birds, including starlings, gulls, swallows, crows, birds of prey, cranes, and herons. The bird flaps to a desired height and, on outstretched wings, descends slowly in an even coast until it needs to climb again. Very efficient for birds of prey, this type of flight requires little speed to stay aloft and therefore allows a slow, thorough search of hunting grounds below.

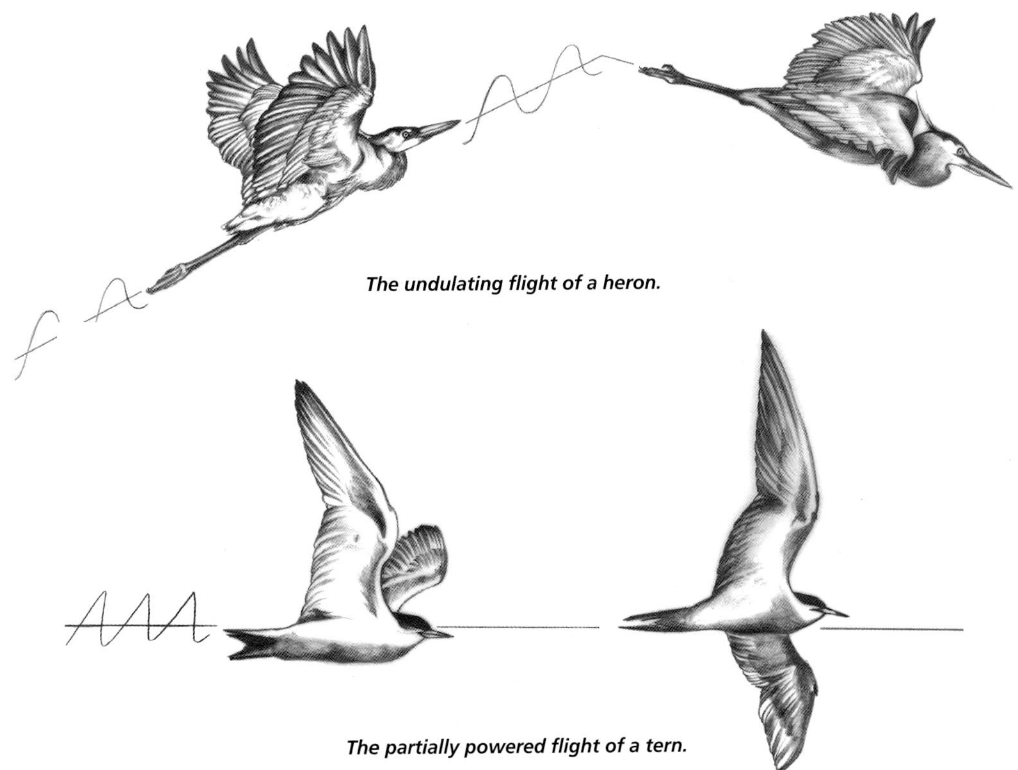

The undulating flight of a heron.

The partially powered flight of a tern.

Also falling into this category is *partially powered flight*, which differs from flapping and gliding in that there is little climb during flapping bouts, keeping the bird on a fairly straight trajectory.

Some of the smaller birds, like warblers, finches, and woodpeckers, have a wing shape that does not allow for significant glide and so use *bounding flight*: a few short, powerful bursts of flapping followed by the bird tucking its wings tight to the body and plummeting in a brief drop toward the ground. The momentum gathered at the bottom of the fall is then transferred into the next upward burst.

Powered flight. In powered flight, commonly seen in waterfowl, constant, rapid action of the wings is required to keep the bird aloft and moving forward. The moment the wing beats stop, the bird begins a quick descent; the heavier the bird, the faster the fall. This flight style is common among birds with high wing-loading or high-speed wings.

Nonpowered flight: gliding and soaring. Gliding and soaring flight encompass a number of strategies, all hinging on the same basic principles: the bird flies primarily on fixed wings, and it powers its travel by using the force of gravity and wind. In gliding flight, the bird loses altitude over the

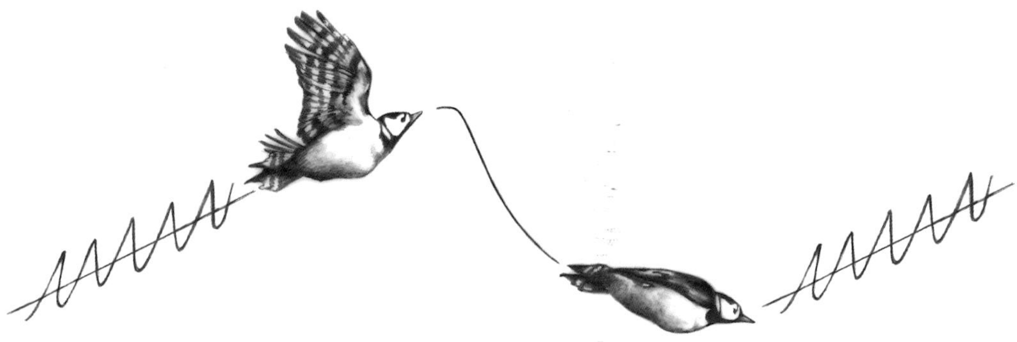

The bounding flight of a woodpecker.

The powered flight of a duck.

The gliding flight of a gannet.
(Flight style illustrations by Beatriz Mendoza.)

trajectory of its flight path, generally gaining speed as it descends. The speed is then used to make a quick climb to regain height and start the process over. In soaring flight, the bird gains altitude, or at least holds steady. Where wind meets obstacles like buildings, islands, waves, and hillsides, it funnels into eddies or ascending currents. Soaring birds extract "fuel" from these currents to power their flight, thus conserving significant amounts of energy that would otherwise be used for flapping.

When a particular portion of the Earth's surface heats faster than the area immediately surrounding it, a bubble of warm air, lighter than cooler air, rises as a thermal. Eagles, condors, vultures, and other large, heavy birds appear to be dependent on these thermals for much of their flight, using a gliding method called thermal soaring.

Gust soaring, or *sweeping flight*, uses gusts of wind pushed upward by waves. Birds such as pelicans and gulls use these gusts to power their flight by flying along the length of a cresting wave, gaining speed as it breaks. The birds then use this speed to pull up and glide over at a right angle to catch the next wave (Wilson 1975).

Hang gliding is also a familiar form of this flight style: a bird, likely a falcon or a hawk, "hangs" in midair with almost no flapping, supported by the updraft above the top of a cliff or other steep surface.

Molts and Plumage

When a new feather begins to grow, it first emerges as a "pimple feather," a small knot encased in a sheath that begins to stem outward from the *feather follicle*. Eventually, the sheath of the pimple feather splits, allowing the feather vanes to unfurl and take shape. When the growing feather has reached full size, blood flow to the feather stops, and the living cells within the feather die. Because a mature feather is not living, there is no sensory function in the feather itself. A feather lost by accident will immediately begin to grow a replacement, assuming there is no damage to the feather follicle, but a bird will have to wait until its next molt in order to replace feathers broken or worn by time, activity, and other stresses that the elements inflict on their plumage.

There are two widespread ways of describing plumages and molts. The more traditional method categorizes them as *winter* and *breeding* (or *nuptial*) plumages. In many cases, this system works well, but birds such as gulls sport subadult plumages that do not fit into either category until their third summer or later. Many birds go through a distinct series of pre-adult plumages, and some do not achieve full coloration until their fourth year. Still others breed in the southern hemisphere and return north in the boreal summer in their "winter" plumage. With so many exceptions to winter and breeding plumages, Humphrey and Parks (1959) created a new system for naming plumages using the terms *alternate* (breeding plumage) and *basic* (post-breeding plumage), which sufficiently avoids some of the mix-up.

The molt-cycle terminology created by Humphrey and Parks follows an average species from chick to its second winter.

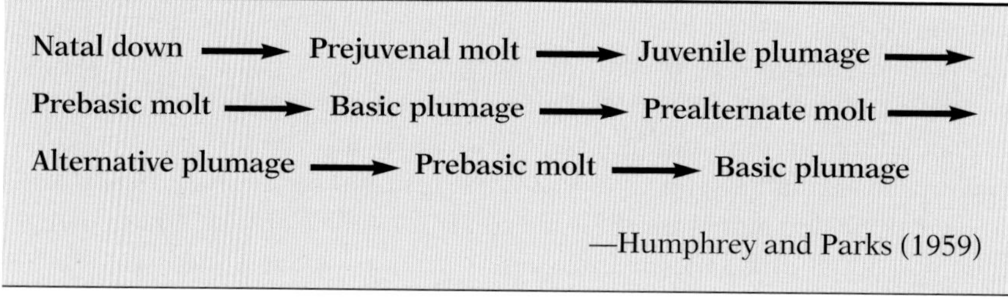
Natal down → Prejuvenal molt → Juvenile plumage →
Prebasic molt → Basic plumage → Prealternate molt →
Alternative plumage → Prebasic molt → Basic plumage

—Humphrey and Parks (1959)

Several molts occur in quick succession during the first few months of most birds' lives. The prebasic molt is significant in that the young bird will lose and replace *all* of its feathers at this stage. In a *complete molt* all of the feathers are replaced, whereas in a *partial molt* only some of the feathers are replaced. Birds molt in numerous fashions depending on a large array of variables, including breeding, ecological influence, and migration; birds largely time their molts so they do not coincide with heavily taxing events.

The molting process generally takes three to four months to complete, with a symmetrical loss of flight feathers across the body. Many birds, including woodpeckers, hawks, and swallows, molt once per year. Warblers and others molt completely once a year, then again, partially, before breeding season. A select few will molt completely twice every year.

A systematic loss of feathers usually still allows a bird to fly, but not all have that privilege: numerous species molt all wing and tail feathers at once, leaving some grounded for up to twenty-five days. Reasons for this are unclear, but larger species that are already heavy in relation to the size of their wings may find flight too taxing with the loss of even a few flight feathers. In studies of Harris's Hawks, flight performance was observed in relation to feather molts and showed that at the loss of a few primaries, flight speed decreased (Tucker 1991).

Plumage and Feather Wear

A common misconception is that all birds lose their winter (basic) wear and replace it with breeding (alternate) plumage. While this is true for many species, it is not true for all. Many passerines eventually gain their breeding plumage by simple wear and degradation of their winter, or basic, feathers. Over winter, the drab tips and edges of feathers that came in with the late summer molt wear away to reveal a starker, often vibrant coloration, resulting in the bold patterns of breeding plumage.

Feather Color

Unlike most terrestrial animals, whose drab, camouflaged fur, skin, or scales reflect a general necessity to stay hidden, birds are often boldly colored. Flight has helped birds avoid many of the dangers of being earthbound, allowing an opportunity to develop complex vocalizations and attractive coloration. The wide array of colors in birds can come from both pigmentation and the structure of the feather itself. In other words, some feathers, if ground to a powder, would retain their color, while others, their microstructure destroyed, would not.

There are three types of pigments found in feathers. The most common, *melanin*, is responsible for black, gray, and brown. Melanin is also associated

Worn gull feathers. Due to the lack of pigmentation, the white portions of the feathers wear considerably faster than the rest of the feather. Notice how the white tips and barring have worn away, while the dark portions remain fairly intact.

Fault Bars: Indicators of Stress

Similar to the rings of a tree, *growth bars* often show the rate of flight feather growth. These are markings across the feather vanes that are oblique to the shaft and alternate between light and dark. If a bird experiences a difficult or stressful period while growing feathers, deformities in the growth bars are evident. These deformities, termed *fault bars* or *hunger traces*, are identified by a lack of barbules on the barbs and appear as a thin, translucent line across the feather vane. Fault bars can be caused by any number of stressful experiences or nutritional duress and may damage the surface of the feather vane enough that the feather can tear, lose portions of the vane, or break away entirely (figure 4).

Looking for fault bars adds depth to feather finds: multiple flight feathers with identical growth and stress bar patterns are likely from a young bird in its first juvenile plumage. Feathers with fault bars from mature birds may be an indication of hardship or more stress-susceptible individuals. What might the fault bars be correlated with? An escape from a cat's paws? A severe cold snap?

Examine the various feathers here that demonstrate differences in fault bar appearance. They may be short "etchings" (figure 1) or run almost the entire width of the vane (figure 2); either can be deep enough to cause the loss of portions of the feather. In severe cases, the quill itself may show damage and deformity (figure 3).

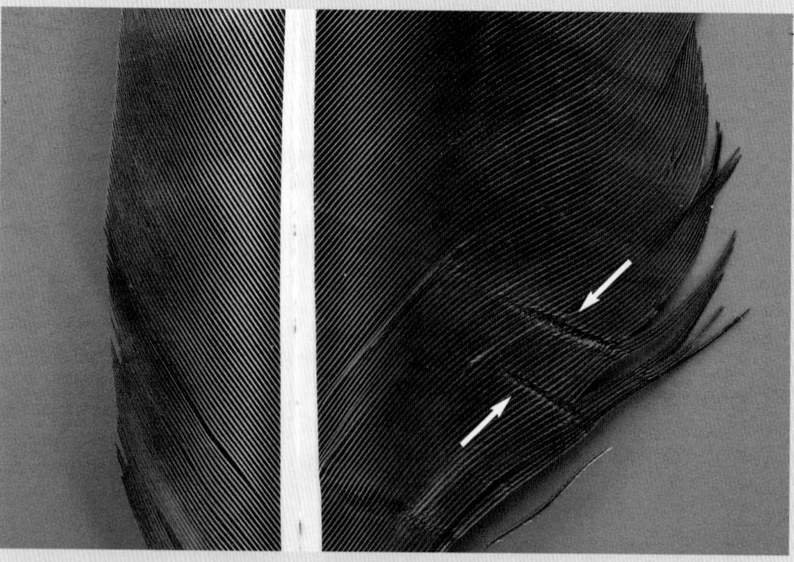

Figure 1: Fault bars can appear as short "etchings" . . .

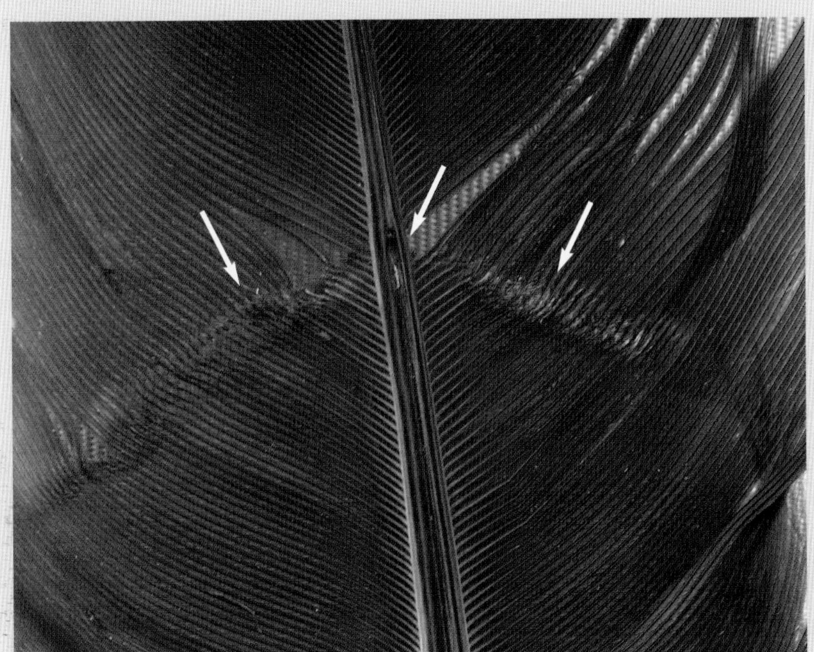

Figure 2: . . . or run almost the entire width of the feather.

Figure 3: Fault bars can cause feather deformities.

continued on page 38

Fault Bars: Indicators of Stress continued

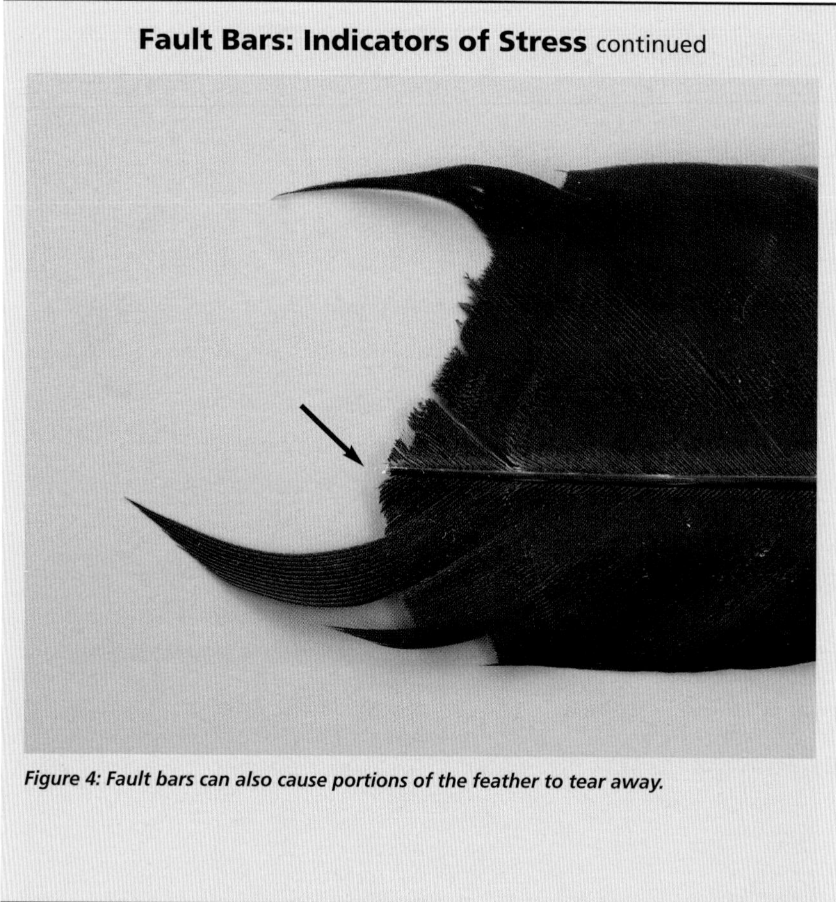

Figure 4: Fault bars can also cause portions of the feather to tear away.

with keratin, which provides structural integrity. Feathers without pigmentation are weaker and wear more quickly. The tips of some birds' flight feathers are black, though much of their overall plumage may be pale or white; the increase of pigmentation and keratin reinforces the tips of feathers, the part that is most taxed by flight.

Carotenoids produce brilliant reds and yellows and are usually found only in contour feathers. Carotenoid pigments do not occur naturally in birds, however; they are found exclusively in plant life. So a Northern Cardinal, for instance, eats bright red berries, and the pigments are processed in the liver, delivered to the bloodstream, and then transported to a developing feather. If lacking these carotenoids, a cardinal will lose its vibrant coloration.

The last type of pigment, and the least common, is *porphyrin*, which produces pinks, browns, reds, and greens. Unlike carotenoids, they are manufactured in the bird's body; the feather color is therefore unaffected by diet. (A similar pigment is psittacofulvin, found only in parrots.)

The beautiful blue hues of jay feathers, created not by pigment but by the feathers' structural composition.

Blue and Iridescent Feathers: Structural Color

Roll a Blue Jay feather between your fingers in sunlight and you will find that its coloration changes dramatically. Blue birds really aren't blue at all; blue, unlike other colors in feathers, is not produced by pigment. It was initially thought that the cells in the surface of the barbs and barbules of the feather created a complex pattern of refraction and reflection, bouncing only blue-light wavelengths back to the eye. Now, however, there are more refined explanations: in the spongy nanostucture of the feather, the matrix between keratin and microscopic air cavities scatters light waves in an orderly fashion, sending only the color blue to the observer (Prum 1998).

In iridescent birds, granules of melanin embedded in the barbules function as minute reflectors of numerous colors. In the microstructure of the feather, ordered layers of keratin or air cavities can each reflect different wavelengths of color as well. Like a prism, light is absorbed and scattered into component colors so that, depending on the viewing angle, you see either a rich array of color, or little of it, so that the feather appears black. These reflector structures can be spaced and shaped in a variety of ways on a single feather, giving a variety of colors, such as those seen on the tail feather of a peacock.

4 Flight Feather Identification

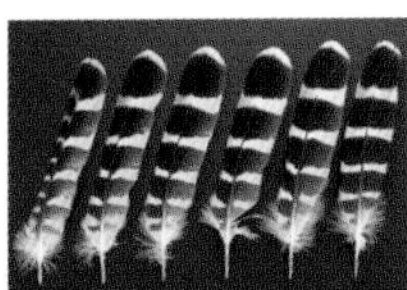

With birds and feathers, as with all things of the natural world, great diversity brings great challenge in identification. In the case of feathers, these challenges are magnified by the fact that variation occurs not only among families, but even within individual species. The feathers of a cormorant are strikingly different from those of its cousin the Anhinga; transient passerines may have differently shaped wings from nonmigratory passerines of the same species; and juvenile birds may have feather shapes that differ from those of their mature parents. Compounding this challenge is the fact that on the same bird, the size of the feathers, their shape, and their markings vary across the wing and tail.

Nonetheless, there are some strikingly similar characteristics that cross species and family lines. By delving into the morphology of a feather, one can glean an incredible amount of information about the feather and the way it is used, even if the species is not known. Flight feather morphology can tell us where the feather originated and what type of wing or tail it came from and can give insight into the bird's particular flight style.

When identifying the feathers of birds, it is useful to take into account things besides the color and shape of the feather itself. If you are not able to recognize the species immediately, it is best to start the task of identification within a larger context and work down from there. First look at the ecological context in which you have found this feather. Are you on a muddy riverbank, deep in the forest, along a creek, or in desert chaparral? Narrowing down the ecological context can take you a long way toward narrowing down the species possibilities.

Look around to see if there are any other feathers lying nearby, as finding multiple feathers from the same bird can aid in identification. It is common to find numerous feathers from birds that have fallen to predators, and though kill sites will sometimes present a tidy pile of feathers, more often than not the feathers have been strewn about by the wind or the predator, and one must search to find them.

After exploring your surroundings and gathering all clues available in the immediate area, take another look at the feather. What overall shape is it? Where do you think it came from on the bird's body? Using the information and photographs of wing types in the sections below, what can you tell about the feather from its shape? Is it a primary, secondary, or tail feather? Can you tell which wing type it might be from? Are there any colors on the feather that give you clues to the species or the bird's lifestyle?

Sometimes this search ends in the successful identification of a species, and sometimes it does not. Successful species identification can be challenging and in some cases, such as with many flycatchers, nearly impossible. However, if you use the resources in this guide to train yourself in the practical use of feather morphology, identification of individual species will become much easier, and your feather knowledge will be more complete.

Even when specific species identification is possible, one must often hold on to a feather mystery for a long time before finding any answers. It is for this reason we recommend that you keep a detailed journal of your learning and ongoing questions with regard to feather identification so that you can revisit them at a later date when you have more information and experience. Even after having the feathers of hundreds of birds in hand for the making of this guide, we have many feather questions that remain unanswered.

One thing is certain: it feels great to solve an old mystery, and, in fact, the older the question, the greater the reward.

Feather Morphology by Wing Type

Because feather color and pattern vary dramatically, it is very important to have a firm handle on basic feather morphology. The ability to distinguish a tail feather from a primary feather, for instance, is often the key to correctly identifying the species to which it belongs. In many instances, this distinction is more helpful in making accurate identifications than searching by color or marking.

The following section describes feather morphology for the four wing types explained earlier: the high-speed wing of birds such as swifts, sandpipers, falcons, and ducks; the high-aspect-ratio wing of albatross, gannets, gulls, and the like; the elliptical wing of game birds and passerines; and the slotted high-lift wing of birds such as eagles, vultures, hawks, pelicans, and swans.

For clarity, we have roughly categorized wing feathers into five different zones, starting at the tip of the wing and moving back toward the body of the bird: *leading primaries*, *midrange primaries*, *transitional primaries*, *secondaries*, and *trailing secondaries*.

The term *leading primary* in this text refers to the outermost (distal) three to six primaries of the wing, depending on the species. Following the leading primarics, the next three to five feathers are referred to as *midrange primaries*. The remaining primaries, usually one or two, are termed *transitional primaries*.

The majority of secondary feathers are simply called *secondaries*. The last several secondaries of the wing, closer to the body of the bird, are referred to as *trailing secondaries*. Traditionally, the last three or four feathers of some bird wings are referred to as "tertials"; however, in this text, tertials are included in the trailing secondary category.

Note that traditionally, feather counts begin at the wrist joint of the bird and number out in opposite directions; for example, primary number ten (on a bird with ten primaries) is the last, distal primary at the leading edge of the of the wing, whereas secondary number ten is close to the body of the bird. Assigning particular numbers to feathers is avoided in this text; instead feathers are categorized by morphology, starting at the leading edge of the wing and moving back toward the body.

Many bird species have a greatly shortened, or vestigial, primary feather on the leading edge of the wing. Because these feathers are vastly different than the rest of the primaries, they are not included in the wing feather descriptions given here.

High-Speed and High-Aspect-Ratio Wings

Feather morphology in high-speed and high-aspect-ratio wings are very similar; therefore, the descriptions of flight feathers from these two wing types are combined in one section.

Leading Primaries

The leading primaries of these two wing types appear slender and pointed and usually have a slight curve along the length of their shaft. The first or second leading primary is nearly always the longest, and because these two wing types use very little slotting, emargination on the leading primaries is either slight or absent, with the exception of *Falconidae* and many ducks. If feather vane emargination is present it is only present on the trailing feather vane and will occur near the tip of the feather. Once you are familiar with the various shapes of leading primaries in this category, they are fairly easy to distinguish from those of other wing types. The flight feathers of most high-speed and high-aspect-ratio wings are fairly flat, but the leading primaries of ducks have significant *camber* (cupped shape—see "game bird wings" on page 50).

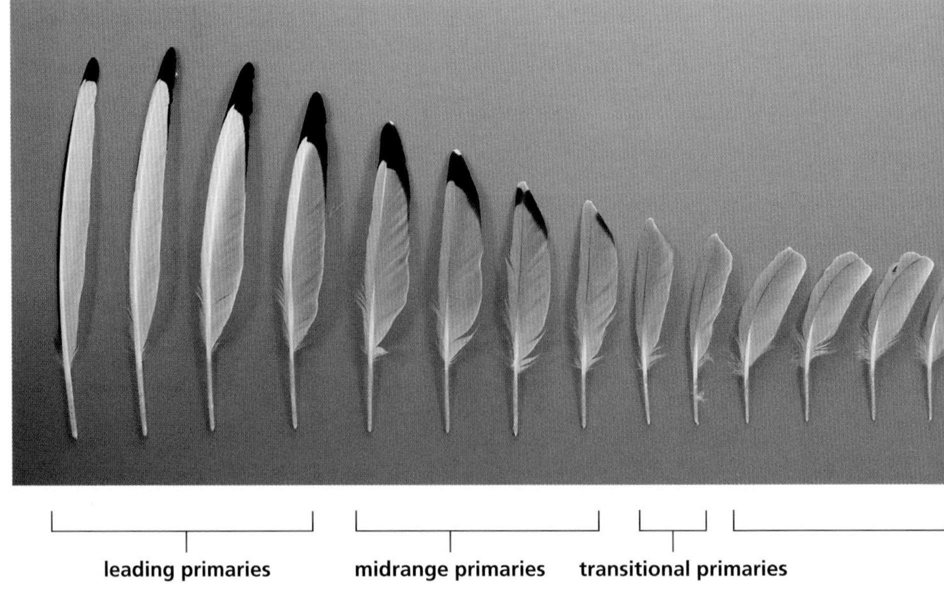

Exploded view of a Bonaparte's Gull wing showing feather morphology.

High-aspect-ratio wing of a Bonaparte's Gull.

Midrange Primaries

Midrange primaries usually begin to lose the subtle curve that is present in leading primaries, appearing straighter. Additionally, these primaries also become shorter, wider and blunter (around a forty-five-degree angle at the tip of the feather) as they near the transitional primaries. The change from leading primaries to midrange primaries in this wing type is much more subtle than the same change in other wing types.

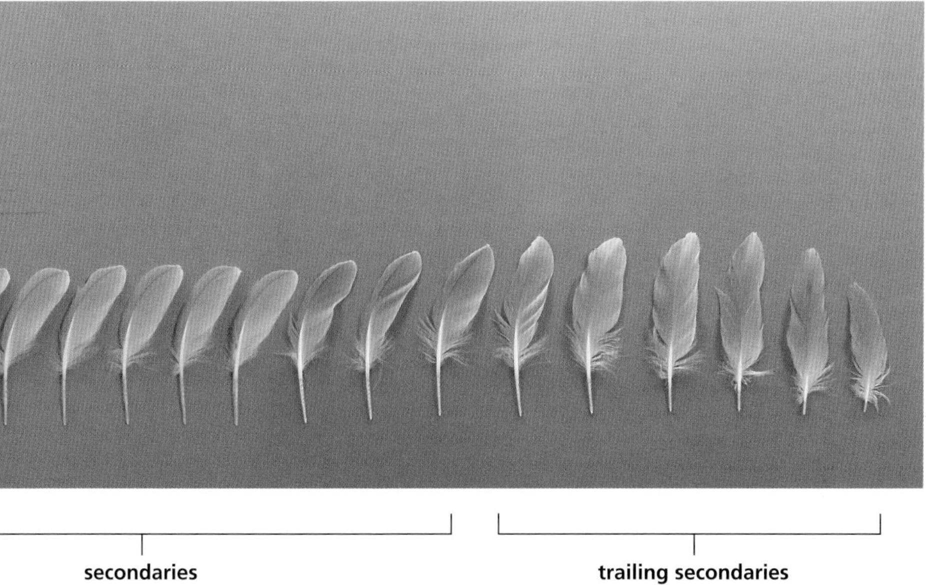

Transitional Primaries

The overall shape of a transitional primary can be either straight or slightly curved. On both high-speed and high-aspect-ratio wings, transitional primaries retain the sloping angular appearance of their trailing vane, unlike transitional primaries of passerine wings. The transitional primaries of many high-speed and high-aspect-ratio wings are mostly indistinct from the midrange primaries but are proportionately shorter.

Secondaries

High-speed and high-aspect-ratio wings show a clear transition from primary to secondary that can be easily identified on the wing and in the morphology of the feathers themselves. The secondaries on these two wing types are most easily distinguished by increased curvature of the feather shaft. The tip of the feather will usually appear rounded or notched, or the leading vane will have an angular appearance similar to the trailing vane of primaries. In most instances, once you learn the basic features of high-speed–high-aspect-ratio secondaries, they are easily distinguishable from the primaries.

Trailing Secondaries

The trailing secondaries on these two wing types are straight, flat, fairly evenly vaned, and less rigid than other flight feathers. They are generally longer than the rest of the secondaries, and though they are easily distinguishable from the other secondaries, they are often confused with scapular feathers.

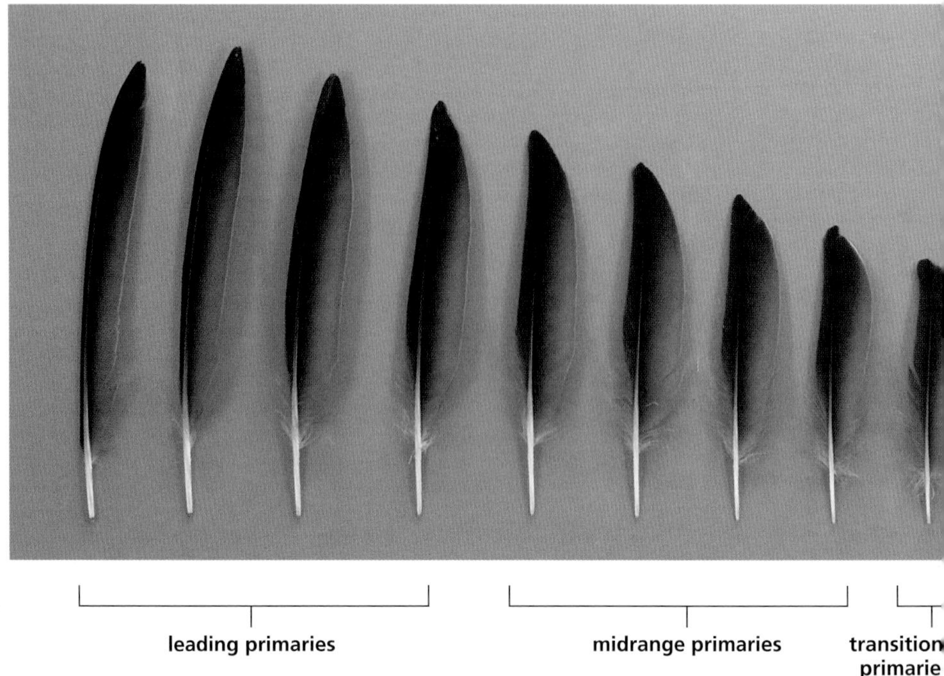

Exploded view of a Purple Martin wing showing feather morphology.

High-speed wing of a Purple Martin.

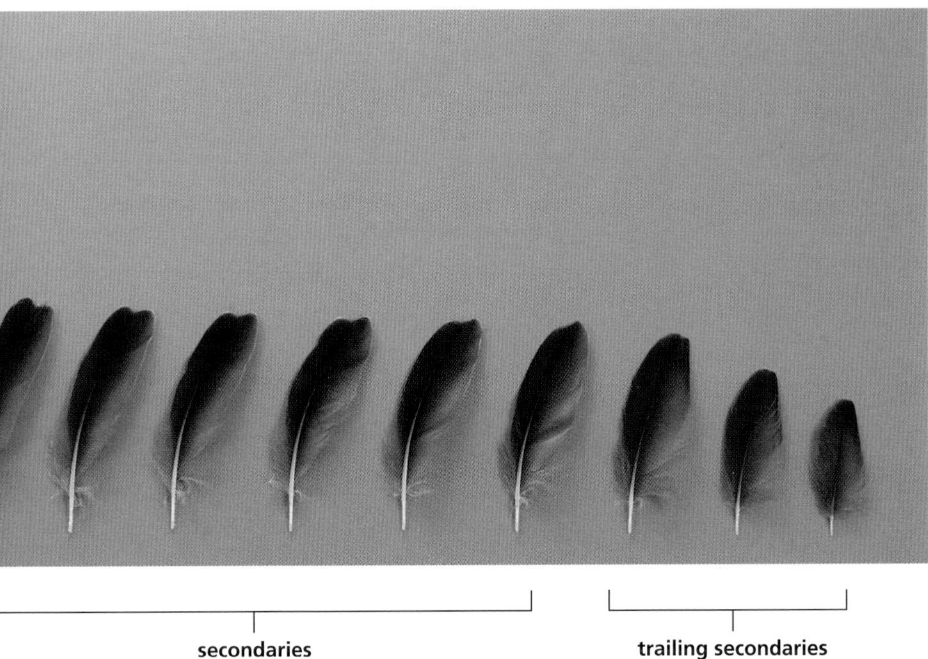

Classic Elliptical Wings

Leading Primaries

With the exception of those of some tyrant flycatchers, the leading primaries of most classic elliptical wings are slim and have tapered, rounder tips, lacking the sharply pointed appearance found in most high-speed, high-aspect-ratio leading primaries. The last leading primary is usually either the longest or equal in length to the longest primary. Though emargination in this wing type is never acute (with the exception of some owls), it is often present, and this can be used to determine how close to the leading edge of the wing a leading or midrange primary originated. If vane emargination is present it is generally found only on the leading feather vane; as is true with all leading vane emargination, the lower the "notching" occurs on the feather, the closer the feather was to the leading edge of the wing.

Elliptical wings use slotting to a much great greater extent than high-aspect-ratio and high-speed wings, but the slotting effect is due more to the overall wing and feather shape than to distinct vane emargination. Slotted high-lift wing types take advantage of both wing shape and extreme feather emargination for greater slotting.

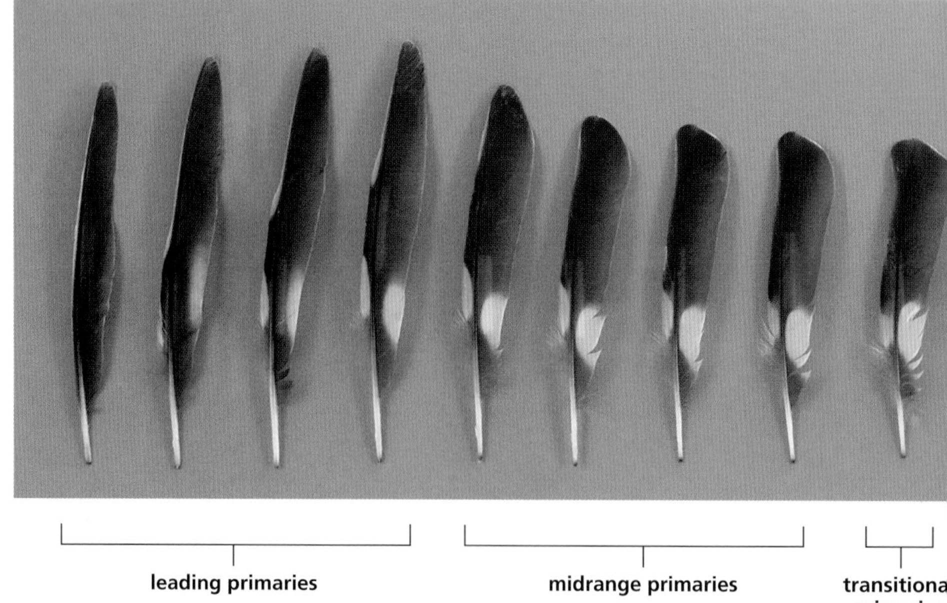

Exploded view of a Varied Thrush wing showing feather morphology.

Classic elliptical wing of a Varied Thrush.

Midrange Primaries

In midrange primaries, feather length begins to decline. The angle of the tip of the feather also gradually declines as the midrange primaries approach the transitional primaries.

Transitional Primaries

The transition from primary to secondary in the classic elliptical wing is much slighter than in high-aspect-ratio wings or the wings of upland game

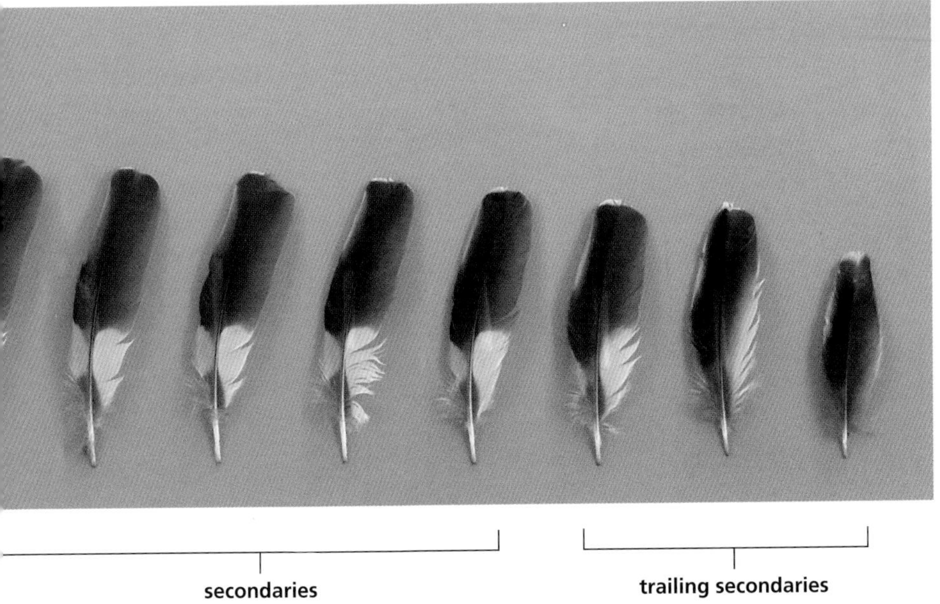

birds. The transitional primary is often nearly identical to the first secondary. The shaft of a transitional primary can be slightly more curved than the shafts of midrange primaries, and the angle of the tip of the feather is usually still present but subtle. It is sometimes possible to differentiate transitional primaries from leading secondaries, but we have not found a universal, reliable way of doing so.

Secondaries

Though secondaries of classic elliptical wings can vary greatly by family and species, they always have a slight curve to their quill. The tips of secondaries can take on a square, rounded, or notched shape, and an angle may be present on the trailing or leading vanes. The vanes of the secondaries will become more symmetrical as they progress back toward the body; however, passerine secondaries often do not display the vane symmetry present in the other wing types until they near the trailing secondaries.

Species such as warblers, small flycatchers, vireos, and others have asymmetrical secondary feather vanes and an overall curvature similar to the S shape in many tail feathers.

Trailing Secondaries

Trailing secondaries on the classic elliptical wing are identified by their flat, straighter quill, generally symmetrical vanes, and rounded or pointed tips. Unlike those of the wing types discussed earlier, the trailing secondaries of elliptical wings are often shorter than, and never longer than, other secondary feathers.

Game Bird Wings

Though upland game birds fall into the elliptical wing category, the morphology of their wing feathers is quite different than that of other elliptical wings. Because upland game birds make up a small portion of the elliptical wing category, they are referred to in this section as "game bird wings," distinct from all other elliptical wings.

The game bird wing of a Gray Partridge.

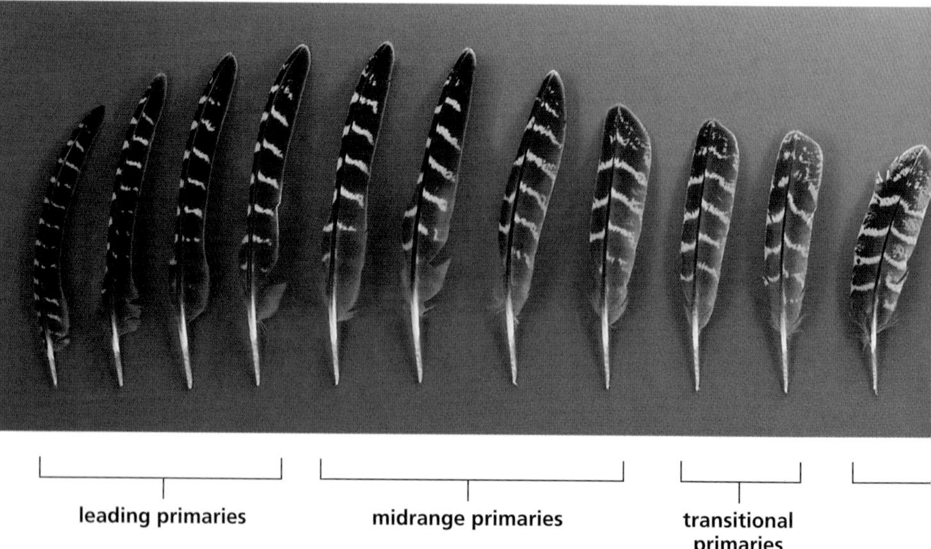

Exploded view of a Gray Partridge wing showing feather morphology.

Leading Primaries

The leading primaries of game bird wings are notably rigid and have a slim, sicklelike appearance. A unique feature of these leading primaries is that the feather vanes often *narrow* briefly about a third to half the way up the shaft before broadening again, giving the feather a slim "waist." Because this wing type relies heavily on wing slotting for bursts of explosive flight, there is often emargination beginning near the base of the leading vane. Even when emargination is not present, however, the overall long, slim design of these feathers creates large wing slots when the wing is spread. Another prominent feature of game bird wings is the distinct cupped shape, or *camber*, of the feathers, especially the leading primaries. Wings with a high degree of camber provide explosive lift and quick bursts of flight; anyone startled by the sudden takeoff of a nearby grouse can attest to the power of this design.

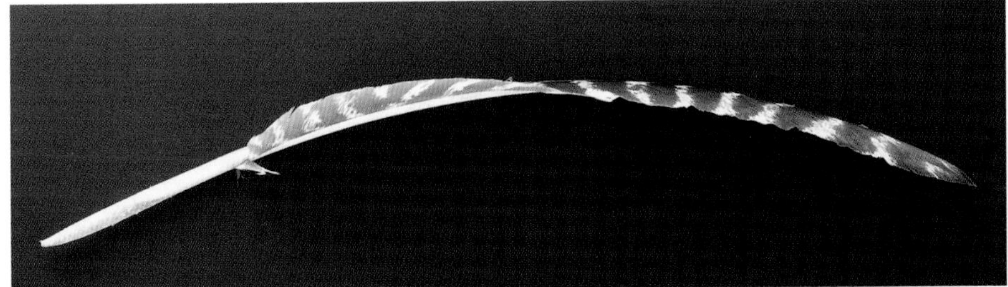

A high degree of camber can be seen in this Wild Turkey feather, which has been photographed in profile. In extreme cases such as this, the feather stands straight up on its leading edge when placed on a flat surface.

secondaries

trailing secondaries

Midrange Primaries

The longest feather on the game bird wing is usually found in the zone around the last leading primary and the first midrange primary. In contrast to the strong curvature of the leading primaries, the midrange primaries are usually slightly straighter. The feather vanes, though similar in appearance to those of the leading primaries, are wider and lack the pronounced "waist." When emargination is present, it occurs to a lesser extent; the emarginated, or notched, portion of the feather shortens as the primary feathers progress toward the body.

Transitional Primaries

The shaft of the transitional primaries remains fairly straight, and the leading vane is moderately to significantly wider than in the midrange primaries. The angle of the trailing vane grows less steep as the wing transitions to secondaries. Emargination is not present on transitional primaries.

With many bird species, trailing secondaries and scapular feathers can be easily confused with central tail feathers if the exact species is not known.

Secondaries

As in high-speed and high-aspect-ratio wings, the shift from primary to secondary in upland game bird wings is distinct. Secondaries in this wing type have significant curvature and fairly symmetrical vanes. Feather tips are rounded.

Trailing Secondaries

Trailing secondaries in game bird wings are even-vaned, flat, straight, and less sturdy than the majority of the other secondaries. Depending on the species they can be equal in length, shorter than, or longer than the leading to midrange secondaries.

Slotted High-lift Wings

Leading Primaries

Because this wing type relies heavily on slotting for additional lift, there is always emargination present on the leading and trailing vane of the leading three to six primaries. The amount of emargination can vary from moderate to extreme depending on the species and specific type of flight it performs. When the emargination of slotted high-lift wing feathers is acute, the leading and trailing vane of these primaries is much wider in the lower portion of the feather than it is in the upper portion. Camber is always present in wings and feathers of this type, though to varying degrees depending on the species.

Midrange Primaries

For the most part, emargination is absent on the midrange primaries, though they do still taper to varying degrees along their length. This overall tapered look usually lasts throughout the midrange primaries, but because the feathers shorten in length and become more blunted as they near the transitional primaries, the tapered effect becomes less distinct closer to the transitional primaries.

Transitional Primaries

Transitional primaries retain some of the angle on the trailing vane, and the leading vane continues to widen.

Secondaries

Secondaries of slotted high-lift wings are curved, in contrast with the generally straight primaries. The vane width of the secondaries is more symmetrical along the feather's length than it is in the transitional primaries. The tips of the secondaries are rounded whereas the primaries have a more tapered or angular appearance. The secondaries of this wing type are generally shorter than the primaries.

Trailing Secondaries

Trailing secondaries are usually equal to or shorter than other secondaries. Trailing secondaries on this wing type are straight and flat, with even vanes and rounded tips.

The slotted high-lift wing of a Swainson's Hawk.

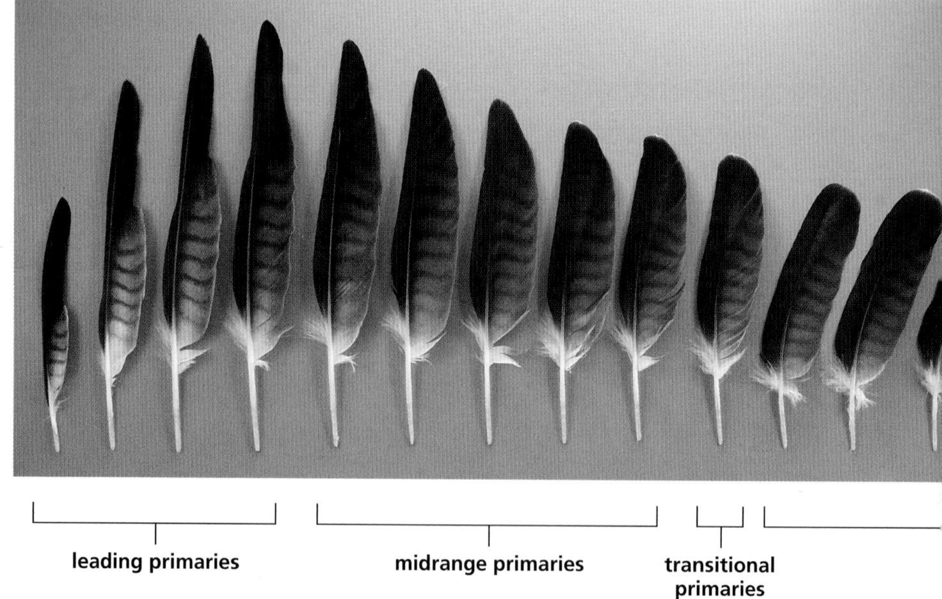

Exploded view of the slotted high-lift wing of a Swainson's Hawk showing feather morphology.

Tail Feather Identification

As noted previously, tail feathers can vary greatly from family to family and, in some cases, between species within the same family. This can make tail feather identification tricky at first. However, once you get a feel for the general characteristics of tail feathers they are often fairly easy to distinguish from primaries and secondaries. To help identify tail feathers, we have divided tails into two basic types: *curved tails* and *straight tails*. These two tail types identify common characteristics that hold true for most (but not all) species. In some instances, tail feather characteristics deviate from these types. For example, pheasants have characteristics of both: short, curved outer tail feathers and long, straight central tail feathers. Similarities in feather appearance of both tail types may also occur, specifically among feathers from the center of the tail. As a general rule, if the tail feathers are less than half the length of the longest primary, they are likely from a curved tail. If they are more than half the length, they are likely from a straight tail. To help differentiation in the field, compare feathers of each type shown in the images here to become familiar with their common characteristics.

Feathers of Curved Tails

Numerous birds, especially those with relatively short tails, such as ducks, geese, and shorebirds, have tail feathers that fall into this category. Most often, but not always, curved tail feathers correspond with high-speed and, in some cases, high-aspect-ratio wing types. These feathers are generally

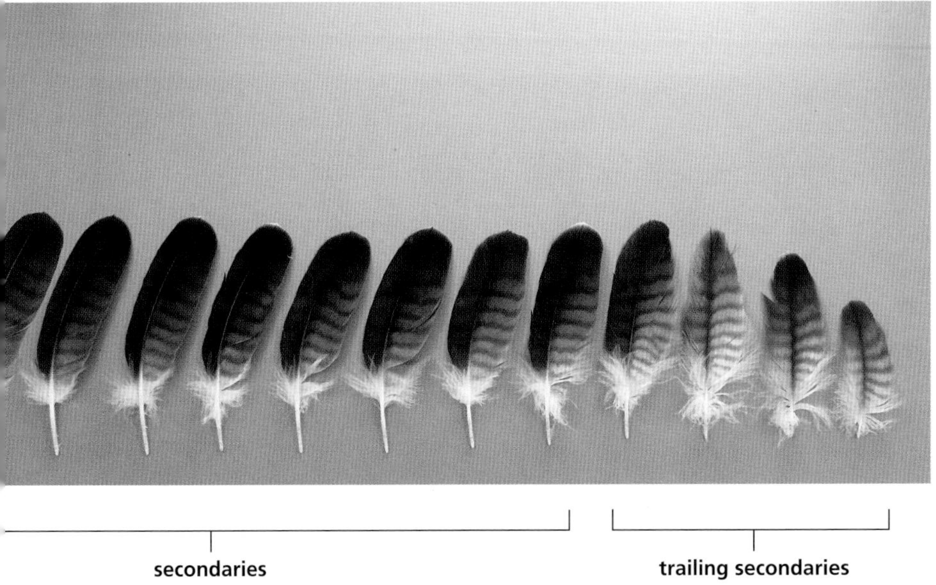

evenly curved down the length of the shaft, with the greatest curvature occurring in the outer tail feathers, which also have the most asymmetrical vane width. Central tail feathers are straighter and more symmetrical and range from slightly curved to straight. The tips of curved tail feathers are usually rounded or pointed, depending on the species. Even the straightest of the curved tail feathers can generally be distinguished from straight tail feathers by their overall shorter, wider shape.

Feathers of curved tails. From left to right: Western Sandpiper, Ruddy Turnstone, Brant, Marbled Godwit, and Burrowing Owl.

Typical morphology of the outer and central tail feathers of a curved tail. (Snow Goose)

Trailing secondaries and central tail feathers may also be difficult to differentiate from each other when the specific species and wing type/tail feather type is not known (see Savannah Sparrow on page 312).

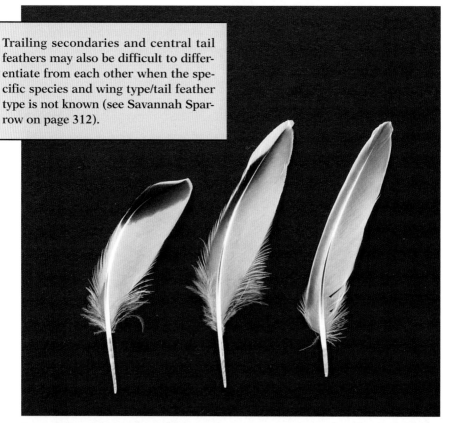

Some species with high-speed and high-aspect-ratio wings, particularly shorebirds, have tail feathers that share many of the same features as their secondaries and in some cases are difficult to distinguish from one another. Compare the secondary on the far left with the trailing secondary (center) and the outer tail feather (far right). (Western Sandpiper)

Curved tail feathers can be difficult to distinguish from the secondaries of high-speed and high-aspect-ratio wings, as they share many of the same characteristics. This is particularly true for shorebirds. For more on identifying characteristics of secondaries and tails, see the high-aspect-ratio and high-speed wing secondary description.

Feathers of Straight Tails

Tail feathers in this category are proportionately longer and straighter, like those of most passerines and raptors, and commonly correspond with the elliptical and slotted high-lift wing types. Though the shafts of the outer tail feathers do tend to curve, the curvature is generally isolated to the *lower region* of the shaft. Many tail feathers that fall into this category have what we refer to as an S-like curve along the feather length, especially noticeable in the outer tail feathers of

Feathers of straight tails. From left to right: Sharp-shinned Hawk, Pileated Woodpecker, Bullock's Oriole, Northern Cardinal, Elegant Tern, and Ruffed Grouse.

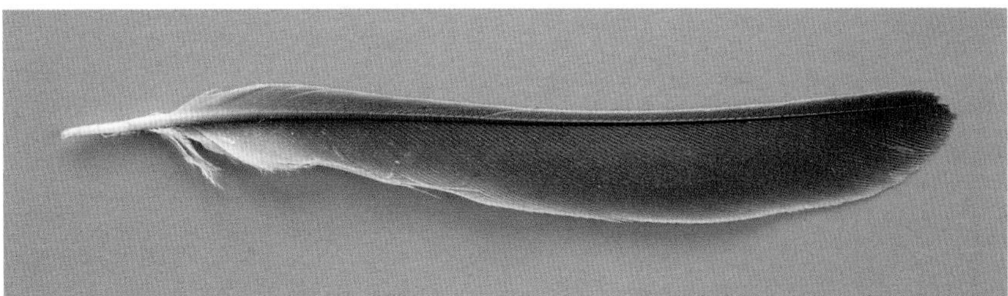

Note the S shape in this small passerine tail feather.

Typical morphology of the outer and central tail feathers of a straight tail. (Red-shouldered Hawk)

small passerines such as warblers. The S-curve of the shaft may be accentuated by the leading feather vane because it is often wider at the top and bottom than it is at the center. The outer tail feathers of larger birds such as hawks often exhibit an S-curve as well, though it is less pronounced.

One of the most reliable identifying characteristics for straight-tail feathers is a "step" in the quill that occurs just below the feather vane. This feature is only visible only when the feather is viewed in profile (see illustration). Sometimes this characteristic is evident in curved-tail feathers as well but is much more subtle.

The quills of a wing feather and tail feather, stripped of their vanes and viewed in profile. Note that the quill of the tail feather on the bottom has a distinct "step" near its base, a reliable identification characteristic for feathers of straight tails of any size.

Compare the two tail feathers in the photograph above (right). The feather on the left is from a straight tail and the feather on the right is from a curved tail. Note the curvature along the entire length of the feather at right. The curvature of the feather at left is isolated to the lower region of the shaft. The diagram (left) shows the quills of the same feathers.

Though difficult at times, tail feather identification will get easier with practice; the more feathers you find and identify, and the more you study the tail feathers in this book, the better you will become.

The charts given here are for quick reference, covering the basics of wing types and their corresponding feather morphology.

Type and Uses: Quick Reference Chart

Wing Type	Bird Examples	Quick Facts
Elliptical	Game Birds and Passerines, etc.	Quick bursts of speed; adept in densely vegetated habitats
High-Aspect-Ratio	Albatross, Shearwaters, Gulls, etc.	Slower or difficult takeoff; no wing slotting; less maneuverability
High-Speed Wing	Falcons, Swallows, Ducks, etc.	Sleek, slim, and angled back at the wrist; fast; little to no wing slotting
Slotted High-Lift (Soaring Wing)	Eagles, Vultures, Pelicans, etc.	Broad, long, heavily slotted; powerful lift for carrying heavy prey and efficient soaring flight

Type and Feather Morphology

Wing Type	Leading Primaries	Midrange Primaries	Transitional primaries	Secondaries	Trailing Secondaries
High-Speed *and* High-Aspect-Ratio	• Slender and pointed • Slight curve along length • Very rarely has emargination (exceptions: *Falconidae* and ducks) • Trailing vane tip is steep angle • High degree of camber in waterfowl	• Fairly straight • Shorter than leading primaries • Tip angle decreases to near forty-five degrees	• Straight or slightly curved • Sloping angular appearance of trailing vane remains	• Increased curvature of feather shaft • Feather tip appears rounded, notched, or leading vane will angle toward primaries	• Straight, flat, and more evenly vaned • Less sturdy than the majority of other secondaries • Often have floppy tip • Generally longer than the rest of the secondaries (may be confused with scapular feathers)
Elliptical Game Birds	• Notably rigid with slim, sickle-like appearance • Trailing vane narrows briefly partway up the shaft, then broadens again • Often clear emargination near the base of the leading vane • High degree of camber	• Straighter compared to the sicklelike primaries • Less emargination than primaries	• Leading vane is wider than in the earlier primaries • No emarginaton	• Distinctly different from primaries • Marked increase in curvature of feather shaft • Symmetrical vane • Tips rounded	• Straight, flat, and evenly vaned • Less sturdy than the majority of other secondaries • May be shorter or longer than, or the same length as, the other secondaries
Classic Elliptical	• Slim with more rounded, tapered tips (exception: tyrant flycatchers) • Last leading primary is usually longest primary • Emargination often present (though not acute)	• Feather length declines • Angle of trailing vane declines as feathers approach transitional primaries	• Slightly curved shaft • Angle of trailing edge still present but subtle • Hard to differentiate from secondaries	• Tip may be square, rounded, or notched • An angle may be present on trailing or leading vanes • Generally have asymmetrical vanes	• Flat, straight, or slightly curved shaft • Symmetrical vanes • Rounded tip • Never longer than secondaries
Slotted High-Lift	• Emargination always present on leading and trailing vanes of leading three to six primaries • Emargination varies from moderate to extreme • Camber present	• Emargination absent • Overall tapered shape	• Some angle retained on trailing vane • Greatly shortened • Blunted point	• Curved • More symmetrical vane width • Rounded tips	• Equal to or shorter in length than secondaries • Flat and straight • Even vanes and rounded tips

Hints and Tips

The following photos offer a quick reference for a few common feather characteristics. These features are not always present, but can be used as an aid to narrow your search.

Above: ***The barring patterns common to most hawks and owls. From left to right: Great Horned Owl, Great Gray Owl, Barn Owl, female Northern Harrier, Red-tailed Hawk, Red-shouldered Hawk.***

Right: ***Woodpecker primaries and secondaries are commonly black with white spots or white patches. Tails are notably rigid and often extremely pointed.***

Right: *Loons, pelicans, boobies, gannets, and waterfowl have remarkably long quills in comparison to those of other birds. Note the difference in the length of the midrange primary quill of a Snow Goose (on left) and a Red-tailed Hawk.*

Below: *The fuzzy or "furred" appearance of owl feathers. Also note the specialized leading edges of the leading primary feather. Both features help owls fly silently. Some hawks, kites, and nightjars have soft, furry feathers, but to a lesser extent.*

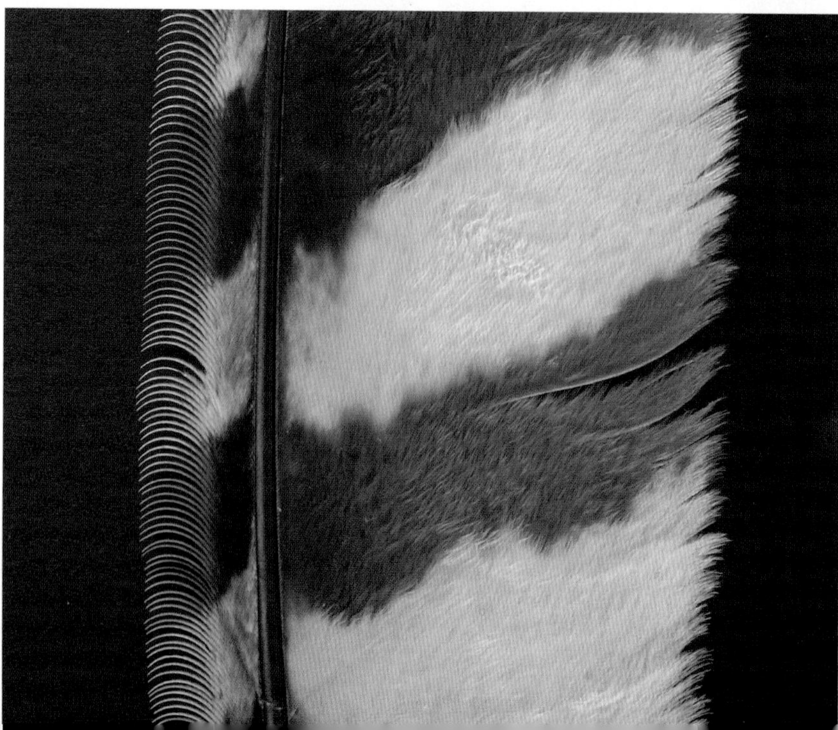

Mottled, camouflaged ground-bird feathers; the barring on flight feathers is sometimes similar to that of raptors. From left to right: Whip-poor-will, Ruffed Grouse, and Gray Partridge.

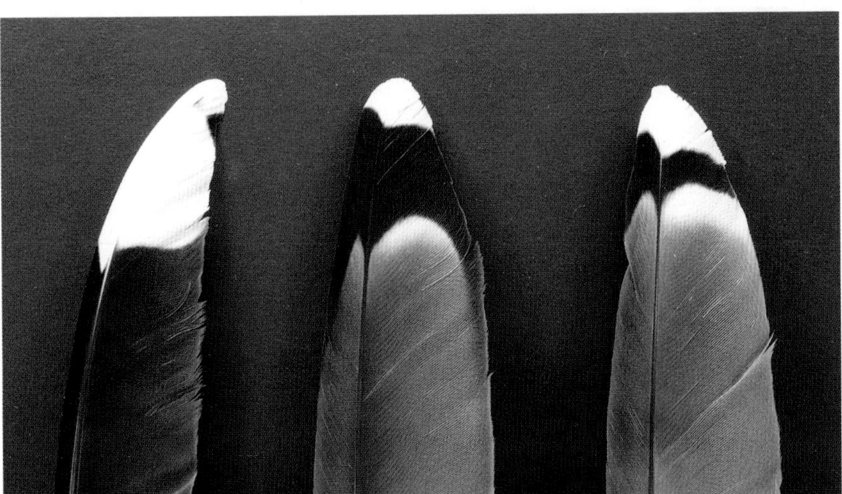

The bold primary feather tips common to gulls, which often display varying arrangements of black and white. (Western Gull)

Using this Guide

The second part of this guide presents photographs of flight and body feathers from 379 North American bird species grouped by order and family.

A number of common challenges encountered when identifying feathers are given here. Use feather morphology, ecological clues, and knowledge of the various markings and sizes of males, females, and juveniles to help determine a particular species. Sexual dimorphism (differences between males and females) can produce considerably different feather appearances in many species; male plumages can be extremely different from female plumages, and feather size can vary dramatically as well. In many cases, the flight feathers of males and females are very similar in color and marking, as most of the distinctions that set males apart are found on the body feathers and wing coverts. When you suspect that you have discovered the feather of a particular bird but it appears different from those displayed in this book, pay close attention to feather size and shape and carefully review both sexes in a reliable bird guide to further aid correct identification. Most of the feathers in this book are either from male specimens or either sex specimens where significant sexual dimorphism is not present. When females (or juveniles) with differing plumages or feather size are used it is noted.

Feathers can vary among individual species in color and shape as well (though shape to a lesser extent.) Each sampling in this book represents only one individual from one particular region of the country and might not match exactly with others of the same species, due to varying body size and color morphs.

Each species has been assigned a wing type, noted with a wing icon.

High-speed wing:

High-aspect-ratio wing:

Game bird wing:

Classic elliptical wing:

Slotted high-lift wing:

The wings of most birds fit neatly into a particular wing type, but a number of birds have wings and feathers whose morphology makes them challenging to place; in these instances, species have been labeled as "oddball."

Oddball wings are represented by a dotted line:

Note: Numerous birds have feathers that have significant twist along the shaft, especially common in the leading primaries. Most of the feathers of large birds with high degrees of twist or camber were fixed in place to give a full frontal view, but the twisted feathers of small passerines were too difficult to fix in place for photographs. The result is that the shape of some feathers in the guide may appear slightly different from that of a feather in hand, the vanes appearing narrower at the base of the shaft. Lay the feather on a flat surface to reproduce the shape shown in the guide.

Additionally, it is inherently difficult to capture the subtleties of a feather in a photograph, and no doubt some details are lost in the inability to hold each of these feathers into the sunlight to completely examine their features. Keep this in mind when comparing feathers in the field to the photographs.

Explanation of Feather Sampling

All of the wing feathers in the photographs are from the right side of the body and laid out in a consistent manner from left to right in the order they would appear on the wing. The tail feathers in the photographs, though not consistently from a particular side of the tail, are always displayed as they would be configured on the bird. The sampling of body feather clumps varies by species, depending on the variety of colors occurring on the bird and the condition of the samples that were available. Body feathers vary greatly in size on a single bird; measurements here are given only for general size reference.

For some species only a sparse sampling of feathers was obtainable, and in these instances effort was made to represent the species as completely as possible.

Feather descriptions are given for families that have useful, consistent identifying characteristics, and notes regarding similar species, unique features, or comparisons are included when applicable.

The following abbreviations are used in this guide to identify feathers represented in the photographs. All measurements are in decimal inches followed by centimeters in parentheses.

P:	Primary
S:	Secondary
TS:	Trailing Secondary
T:	Tail
WC:	Wing Covert
MC:	Marginal Wing Covert
UNWC:	Under-wing Covert
UPTC:	Upper-tail Covert
UNTC:	Under-tail Covert
SC:	Scapular
CR:	Crest/Crown
BR:	Breast
BE:	Belly
NA:	Nape
MA:	Mantle
TH:	Throat
NE:	Neck
FL:	Flank
RU:	Rump

Species descriptions also indicate if males are larger than females (M>F) or vice versa (F>M) and if the feathers shown came from a male, female, or juvenile, or if this is unknown.

Range maps show year-round, summer, winter, and migration ranges.

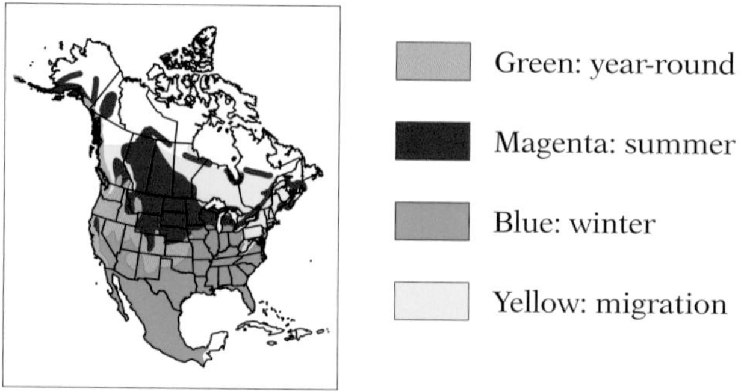

Part II

Feathers

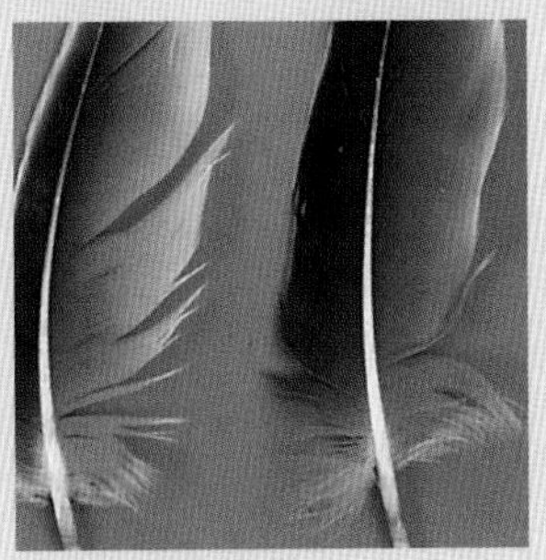

Geese, Swans, and Ducks *(Anatidae)*

Camber is always present to varying degrees. Waterfowl primaries have a distinct tegmen on the underside. Emargination is usually present in the leading three primaries of geese and swans. Emargination is sometimes present in the leading primaries of ducks, but to a much lesser degree than in geese or swans. With the exception of the leading primary, primaries, secondaries, and tail feathers have long quills. Duck primaries tend to be wider than other high-speed wings, and their secondaries are straighter.

Greater White-fronted Goose *(Anser albifrons)*

M>F

unknown
P: 11.7 (29.7)
P: 11.7 (29.6)
P: 9.1 (23.0)

S: 7.1 (18.0)
S: 6.7 (17.1)
T: 5.5 (13.9)
T: 5.5 (14.0)

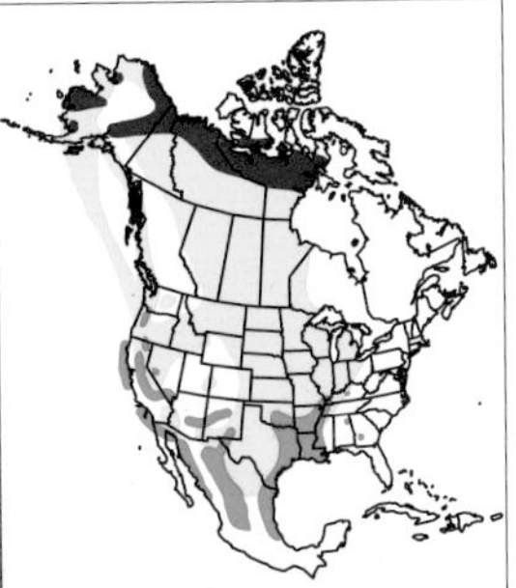

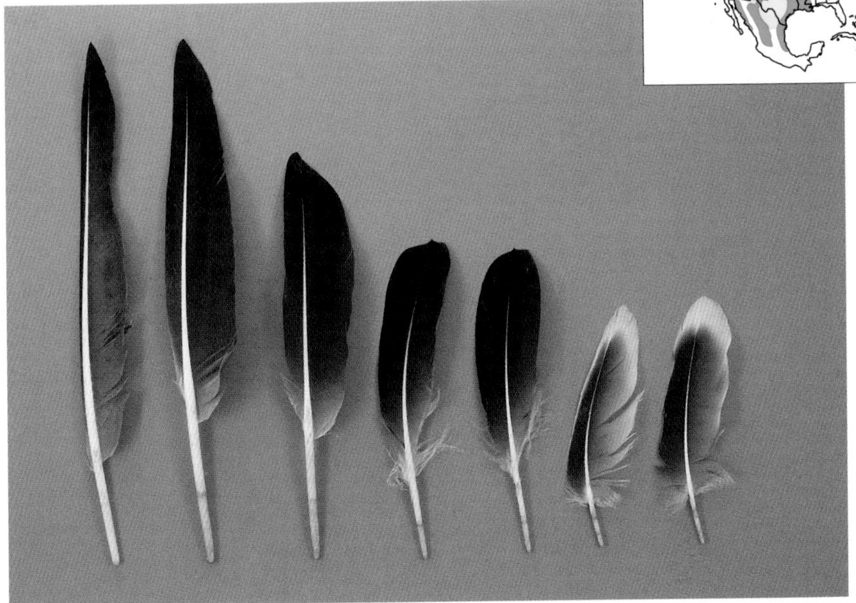

Greater Snow Goose *(Chen caerulescens)*

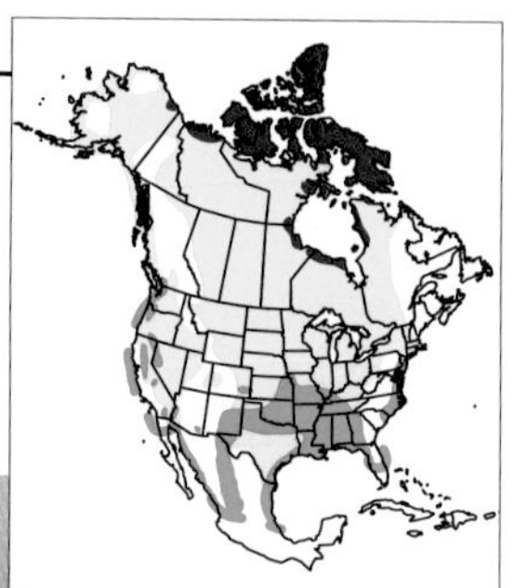

(white adult)
M>F

unknown
P: 11.7 (29.7)
P: 12.6 (32.0)
P: 10.8 (27.4)
P: 10.9 (27.6)

S: 7.3 (18.6)
S: 7.0 (17.7)
T: 5.8 (14.8)
T: 5.8 (14.8)
T: 6.0 (15.3)

MA: 2.4 (6.0)
BR: 2.0 (5.1)
UPTC: 3.3 (8.4)

Lesser Snow Goose *(Chen caerulescens)*

(dark adult)
M>F

unknown
P: 11.4 (28.9)
P: 12.3 (31.3)
P: 9.9 (25.2)
P: 7.5 (19.1)
S: 6.9 (17.5)
TS: 7.7 (19.5)
TS: 5.8 (14.8)

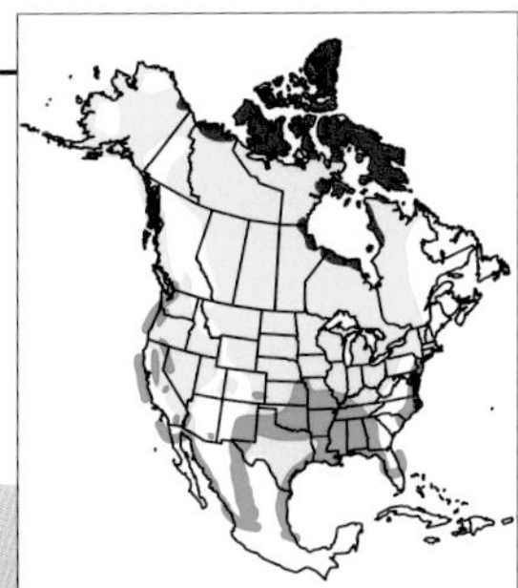

FL: 2.9 (7.4)
MA: 2.2 (5.6)
NA: 1.1 (2.7)
T: 5.9 (14.9)
T: 5.6 (14.1)

Ross's Goose *(Chen rossii)*

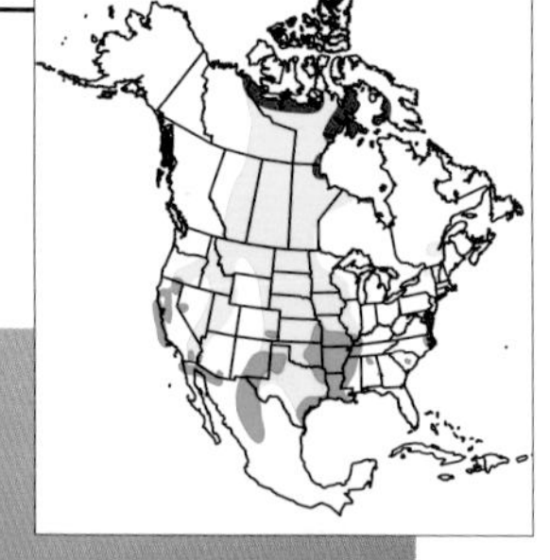

unknown	**P:** 8.5 (21.5)
P: 10.9 (27.6)	**S:** 6.7 (17.1)
P: 11.5 (29.1)	**S:** 6.3 (16.0)
P: 10.9 (27.8)	**TS:** 7.0 (17.8)

T: 5.4 (13.6)
T: 5.6 (14.1)
BR: 2.5 (6.3)
BE: 2.3 (5.9)
RU: 1.1 (2.9)

Brant *(Branta bernicla)*

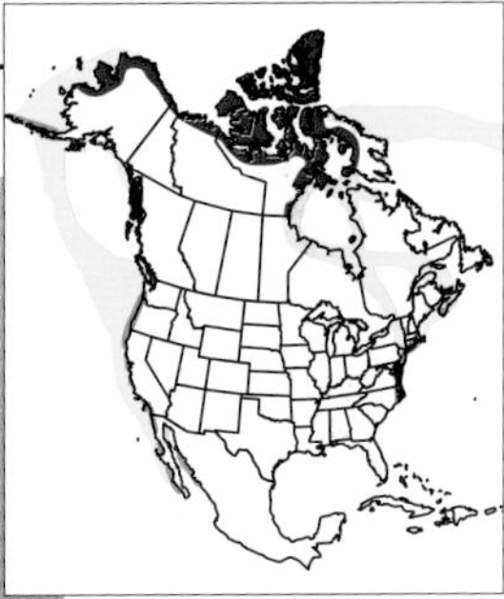

M>F

unknown
P: 9.3 (23.7)
P: 8.2 (20.9)
P: 5.9 (14.9)
S: 5.0 (12.8)
S: 4.8 (12.2)
T: 4.2 (10.6)
T: 4.1 (10.5)
NA: 1.4 (3.6)
BR: 2.0 (5.0)
RU: 2.2 (5.6)
FL: 3.3 (8.3)

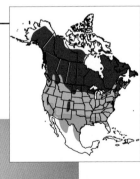

Canada Goose *(Branta canadensis)*

M>F

male
P: 13.3 (33.7)
P: 14.7 (37.4)
P: 12.0 (30.5)
P: 10.2 (25.8)
S: 8.9 (22.7)
S: 9.3 (23.6)
T: 6.7 (16.9)
T: 6.6 (16.7)
T: 6.3 (15.9)

Mute Swan *(Cygnus olor)*

M>F

unknown
P: 15.5 (39.4)
P: 16.3 (41.5)
P: 17.3 (44.0)
P: 14.4 (36.7)
P: 13.4 (34.0)

S: 12.2 (31.0)
S: 11.9 (30.1)
TS: 11.8 (29.9)
TS: 10.9 (27.8)

T: 8.0 (20.3)
T: 8.4 (21.4)
BR: 3.0 (7.5)
MA: 4.3 (10.8)

Trumpeter Swan *(Cygnus buccinator)*

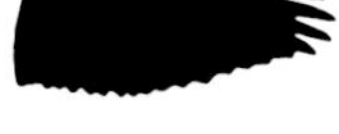

M>F

unknown
P: 17.0 (43.1)
P: 19.8 (50.3)
P: 17.3 (43.9)
P: 15.5 (39.4)

continued on page 76

Trumpeter Swan continued

S: 9.0 (22.8)
TS: 14.4 (36.7)
TS: 13.9 (35.2)
T: 8.8 (22.4)
T: 9.0 (22.9)
MA: 5.0 (12.8)
BR: 3.8 (9.6)

Tundra Swan *(Cygnus columbianus)*

M>F

unknown
P: 12.1 (30.8)
P: 13.3 (33.7)
P: 11.7 (29.9)
P: 9.7 (24.6)
S: 8.0 (20.2)
S: 8.6 (21.8)

T: 6.1 (15.4)
T: 5.8 (14.7)
T: 6.1 (15.5)
BR: 3.1 (7.9)
BE: 2.1 (5.3)

Wood Duck *(Aix sponsa)*

M>F

male
P: 6.1 (15.5)
P: 6.0 (15.2)
P: 5.5 (14.0)
P: 4.8 (12.1)
S: 3.9 (9.8)
S: 3.8 (9.7)
TS: 3.8 (9.7)
T: 4.1 (10.3)
T: 4.4 (11.2)
FL: 3.1 (7.8)
CR: 1.9 (4.8)
NA: 1.3 (3.3)
NA: 0.8 (2.0)
SC: 3.8 (9.6)
TC: 2.3 (5.9)
MA: 1.3 (3.3)
FL: 1.5 (3.7)

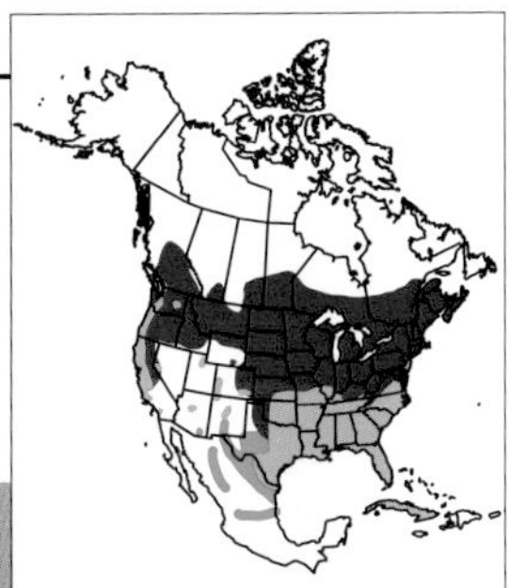

Gadwall *(Anas strepera)*

M>F

male	**S:** 4.3 (10.9)	**BR:** 2.0 (5.2)
P: 7.6 (19.2)	**TS:** 4.6 (11.6)	**BE:** 1.6 (4.1)
P: 7.8 (19.9)	**T:** 3.7 (9.5)	**MA:** 1.5 (3.7)
P: 7.2 (18.2)	**T:** 3.9 (9.9)	**RU:** 3.5 (8.9)
P: 5.4 (13.8)	**T:** 4.0 (10.2)	**WC:** 1.9 (4.9)

American Wigeon *(Anas americana)*

M>F

male, non-breeding	**S:** 4.4 (11.2)	**NA:** 1.8 (4.5)
	S: 4.4 (11.3)	**BR:** 1.7 (4.4)
P: 7.7 (19.5)	**TS:** 5.8 (14.8)	**FL:** 2.5 (6.4)
P: 6.5 (16.6)	**T:** 4.3 (11.0)	**RU:** 2.1 (5.3)
P: 5.2 (13.3)	**T:** 5.2 (13.1)	**SC:** 3.4 (8.7)

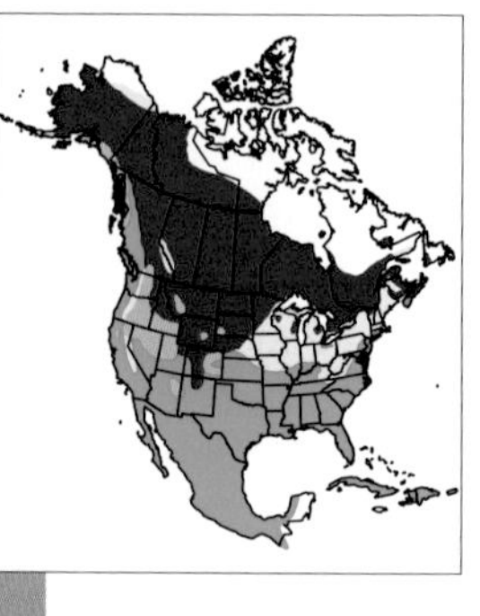

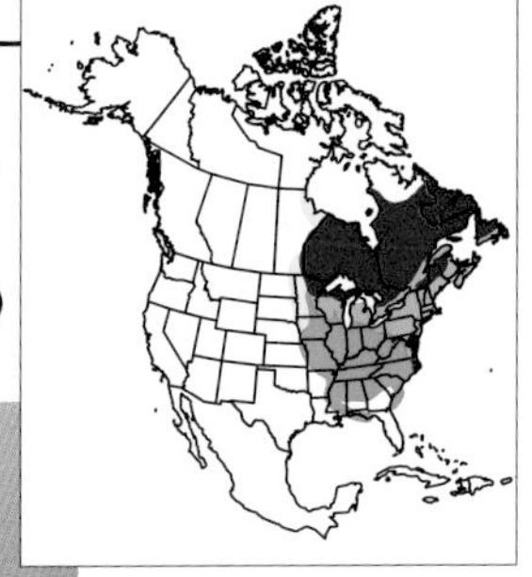

American Black Duck *(Anas rubripes)*

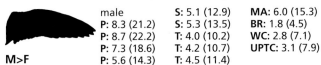

M>F

male
P: 8.3 (21.2)
P: 8.7 (22.2)
P: 7.3 (18.6)
P: 5.6 (14.3)
S: 5.1 (12.9)
S: 5.3 (13.5)
T: 4.0 (10.2)
T: 4.2 (10.7)
T: 4.5 (11.4)
MA: 6.0 (15.3)
BR: 1.8 (4.5)
WC: 2.8 (7.1)
UPTC: 3.1 (7.9)

Mallard *(Anas platyrhynchos)*

M>F

male, curly center tail feathers are iridescent green, secondaries iridescent purple
P: 8.5 (21.5)
P: 7.8 (19.7)
P: 6.7 (17.0)
S: 5.2 (13.1)
S: 5.0 (12.7)
TS: 5.8 (14.8)
T: 4.3 (11.0)
T: 3.8 (9.6)
T: 4.1 (10.5)
T: 4.4 (11.2)

continued on page 80

Mallard continued

NA: 1.1 (2.8)
BR: 1.5 (3.7)
BE: 1.5 (3.7)
RU: 2.4 (6.2)
MA: 1.9 (4.9)
WC: 2.7 (6.8)

Blue-winged Teal *(Anas discors)*

M>F

male
P: 5.6 (14.1)
P: 5.7 (14.6)
P: 4.1 (10.4)

S: 3.1 (7.9)
S: 3.3 (8.3)
T: 3.0 (7.7)
T: 3.3 (8.3)

BR: 1.4 (3.5)
MA: 1.6 (4.0)
RU: 1.3 (3.4)
WC: 1.3 (3.3)
TS: 4.5 (11.4)

Cinnamon Teal *(Anas cyanoptera)*

M>F

male
P: 5.6 (14.1)
P: 5.9 (15.0)
P: 5.3 (13.4)
P: 4.7 (11.9)

S: 3.2 (8.2)
TS: 4.2 (10.6)
T: 3.1 (7.9)
T: 3.4 (8.7)

BR: 1.6 (4.0)
MA: 1.6 (4.1)
WC: 1.7 (4.3)
FL: 2.8 (7.2)
SC: 5.0 (12.8)

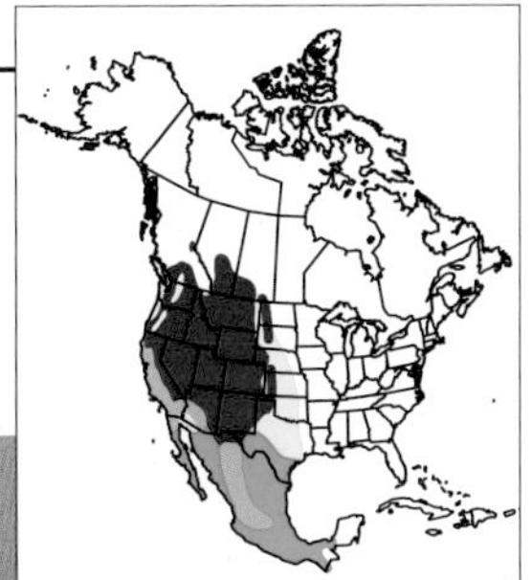

Northern Shoveler *(Anas clypeata)*

M>F

male
P: 7.3 (18.6)
P: 6.2 (15.7)
P: 4.8 (12.2)
S: 3.8 (9.7)

TS: 4.4 (11.1)
T: 4.0 (10.2)
T: 3.9 (9.9)
SC: 5.6 (14.3)
NA: 1.8 (4.5)

BR: 1.0 (2.6)
BE: 1.7 (4.4)
WC: 1.7 (4.3)

Green-winged Teal *(Anas crecca)*

M>F

male
P: 5.2 (13.3)
P: 5.4 (13.8)
P: 3.9 (9.8)

S: 3.2 (8.1)
S: 3.3 (8.3)
T: 3.3 (8.3)
T: 3.0 (7.7)

BR: 1.0 (2.6)
NA: 1.0 (2.6)
FL: 2.1 (5.3)
UNTC: 2.0 (5.1)

Canvasback *(Aythya valisineria)*

M>F

male
P: 6.4 (16.3)
P: 6.8 (17.2)
P: 5.8 (14.7)
P: 5.1 (12.9)

S: 4.0 (10.1)
S: 4.1 (10.4)
T: 3.1 (8.0)
T: 3.1 (8.0)

NA: 0.7 (1.8)
BR: 1.3 (3.2)
MA: 1.7 (4.2)
NE: 1.1 (2.8)
WC: 3.4 (8.6)

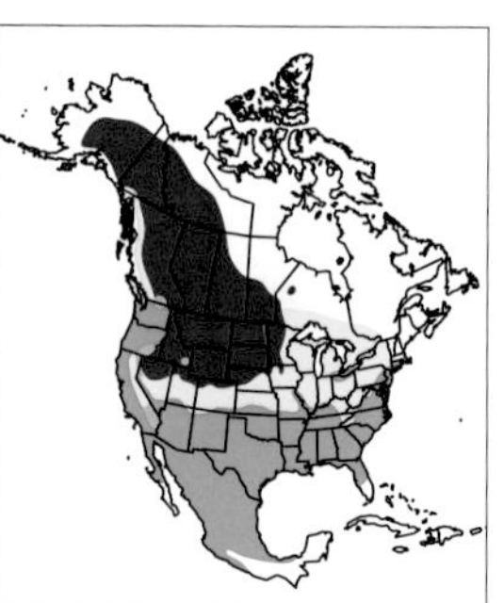

Redhead *(Aythya americana)*

M>F

male
P: 6.3 (16.1)
P: 6.6 (16.7)
P: 5.9 (15.0)
P: 4.8 (12.1)
S: 3.7 (9.5)
S: 3.7 (9.4)
T: 2.8 (7.1)
T: 2.8 (7.0)
BR: 1.3 (3.4)
BE: 1.3 (3.4)
CR: 0.6 (1.5)

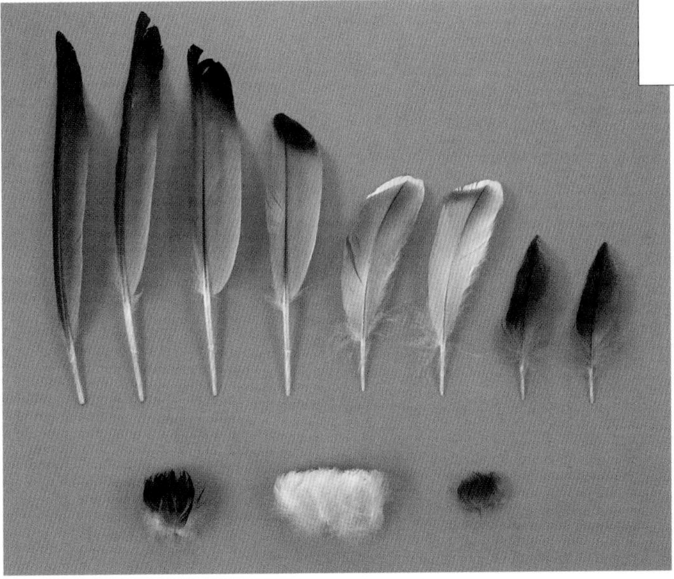

Ring-necked Duck *(Aythya collaris)*

M>F

male
P: 5.7 (14.5)
P: 5.6 (14.3)
P: 5.1 (12.9)
P: 4.0 (10.1)
S: 3.5 (8.9)
TS: 3.7 (9.5)
TS: 4.3 (11.0)
T: 2.9 (7.3)
T: 2.8 (7.0)
MA: 1.5 (3.8)
BR: 1.5 (3.8)
BE: 1.1 (2.8)
WC: 2.1 (5.4)
UNWC: 2.6 (6.7)

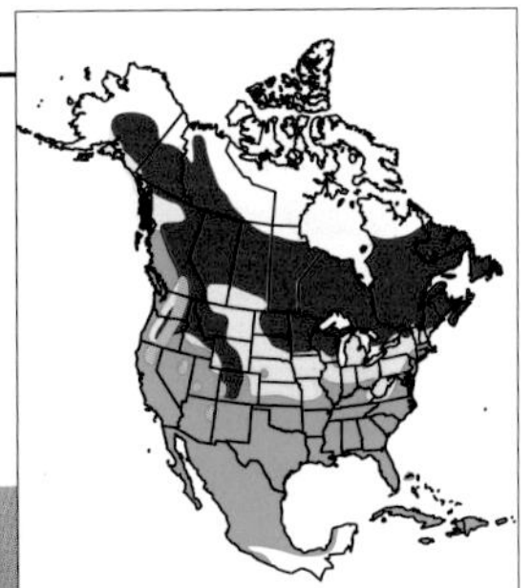

Greater Scaup *(Aythya marila)*

M>F

male
P: 6.4 (16.2)
P: 6.1 (15.4)
P: 4.3 (11.0)
S: 3.5 (9.0)
S: 3.6 (9.1)
S: 3.6 (9.2)
TS: 4.6 (11.8)
T: 2.8 (7.0)
T: 2.7 (6.9)
NA: 1.6 (4.0)
BR: 1.7 (4.3)
BE: 1.3 (3.2)
MA: 1.4 (3.6)
RU: 1.8 (4.6)

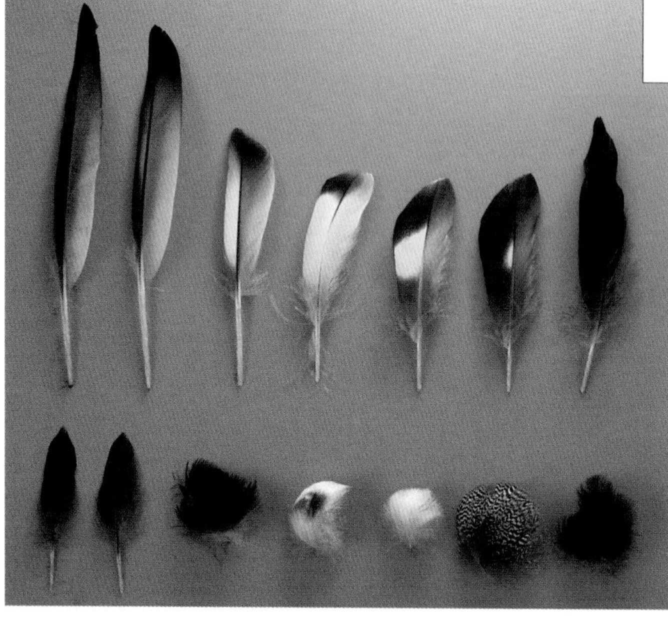

Lesser Scaup *(Aythya affinis)*

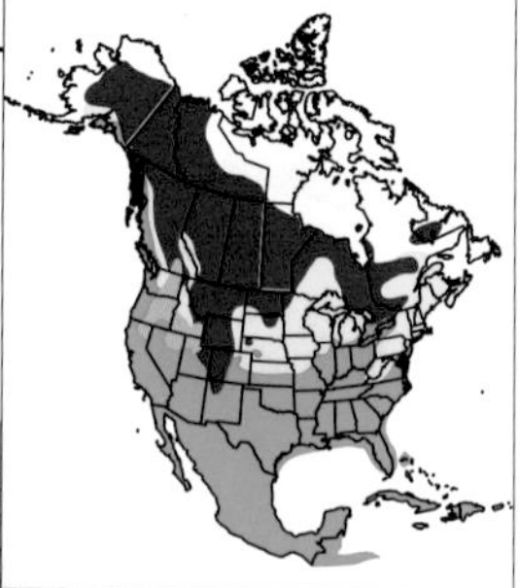

M>F

male
P: 5.6 (14.2)
P: 5.8 (14.7)
P: 5.2 (13.3)
P: 4.3 (10.9)
S: 3.5 (9.0)
S: 3.3 (8.3)
TS: 4.4 (11.1)
T: 2.5 (6.3)
T: 2.4 (6.2)
BR: 1.4 (3.5)
BE: 1.3 (3.3)
SC: 2.8 (7.0)

Common Eider *(Somateria mollissima)*

M>F

male
P: 8.0 (20.2)
P: 8.4 (21.3)
P: 6.9 (17.5)
S: 5.7 (14.4)
TS: 5.0 (12.8)
TS: 5.2 (13.1)
T: 3.8 (9.7)
T: 3.9 (10.0)
BE: 1.8 (4.5)
MA: 2.5 (6.3)
RU: 1.8 (4.5)

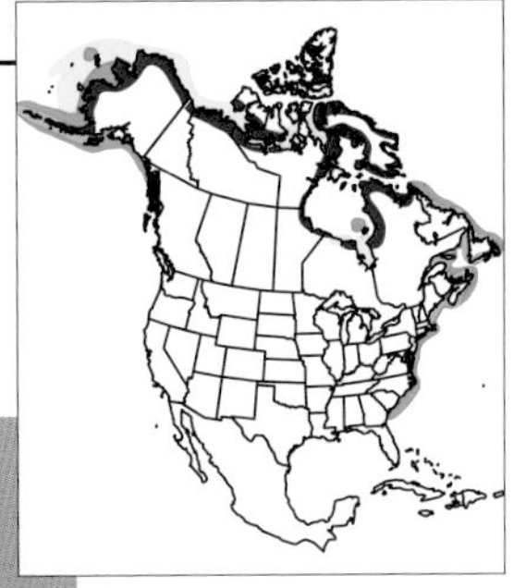

Surf Scoter *(Melanitta perspicillata)*

M>F

male
P: 7.0 (17.8)
P: 7.0 (17.8)
P: 5.6 (14.1)
S: 4.3 (10.8)
TS: 4.0 (10.2)
T: 3.2 (8.1)
T: 3.7 (9.5)
BR: 1.2 (3.1)
BE: 1.7 (4.2)
RU: 2.4 (6.2)

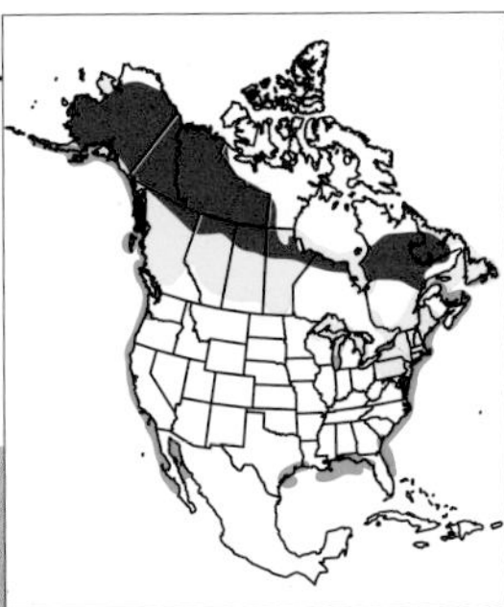

White-winged Scoter *(Melanitta fusca)*

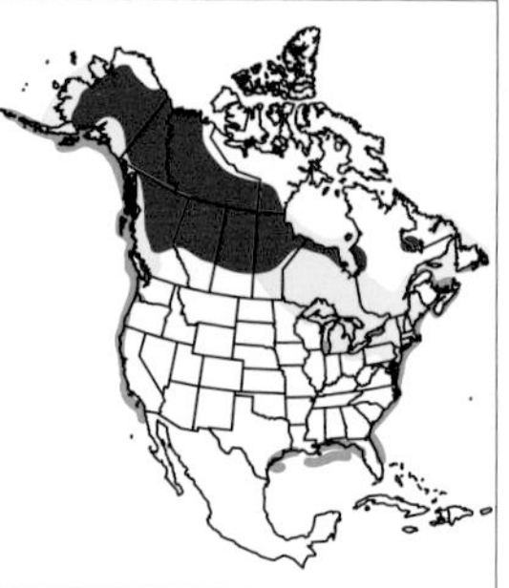

M>F

male
P: 7.8 (19.8)
P: 8.1 (20.5)
P: 6.6 (16.7)
S: 5.0 (12.7)
S: 4.9 (12.4)
TS: 5.0 (12.6)
T: 3.5 (8.9)
T: 3.9 (9.9)
T: 3.7 (9.4)
MA: 2.0 (5.0)
BR: 1.7 (4.2)
WC: 3.3 (8.4)
UNWC: 3.4 (8.6)
SC: 4.8 (12.1)

Black Scoter *(Melanitta nigra)*

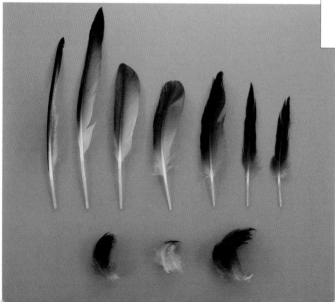

M>F

male
P: 5.5 (14.0)
P: 6.5 (16.5)
P: 4.7 (11.9)
S: 4.2 (10.6)
TS: 4.4 (11.2)
T: 4.0 (10.2)
T: 3.5 (9.0)
MA: 2.0 (5.1)
BR: 1.4 (3.6)
FL: 2.4 (6.2)

Long-tailed Duck *(Clangula hyemalis)*

M>F

male	**S:** 3.5 (9.0)	**MA:** 1.7 (4.4)
P: 6.3 (16.0)	**S:** 3.7 (9.5)	**T:** 3.1 (7.9)
P: 6.7 (16.9)	**TS:** 4.2 (10.7)	**T:** 3.7 (9.4)
P: 5.4 (13.7)	**NA:** 1.2 (3.1)	**T:** 4.8 (12.1)
P: 4.2 (10.7)	**BR:** 1.5 (3.7)	**T:** 7.8 (19.9)

Bufflehead *(Bucephala albeola)*

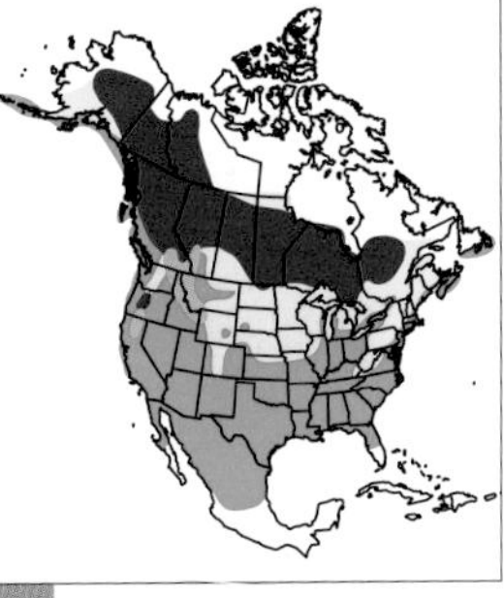

M>F

male	**S:** 3.1 (8.0)	**CR:** 0.9 (2.3)
P: 5.0 (12.8)	**S:** 3.1 (7.8)	**NA:** 1.1 (2.9)
P: 5.2 (13.3)	**T:** 3.5 (9.0)	**BR:** 1.6 (4.0)
P: 4.7 (11.9)	**T:** 3.6 (9.1)	
S: 3.5 (8.8)	**NE:** 1.3 (3.4)	

Common Goldeneye *(Bucephala clangula)*

M>F

male
P: 6.3 (16.0)
P: 6.5 (16.5)
P: 5.5 (14.0)
S: 3.9 (10.0)
S: 3.9 (10.0)
TS: 4.3 (11.0)
T: 3.6 (9.2)
T: 4.3 (10.8)
unknown:
4.3 (10.9)
NA: 1.7 (4.4)
BR: 1.3 (3.4)
MA: 1.6 (4.0)

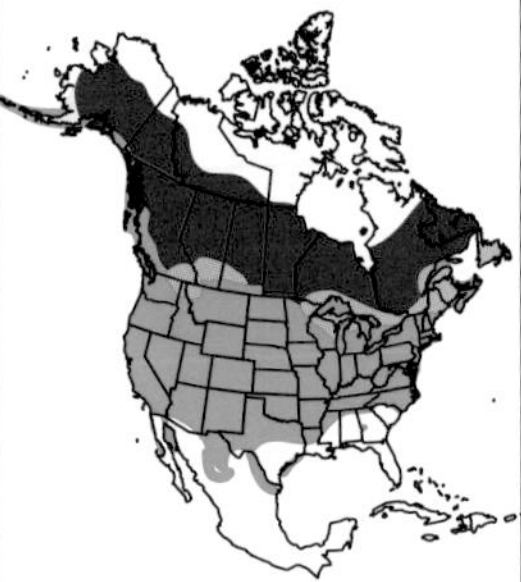

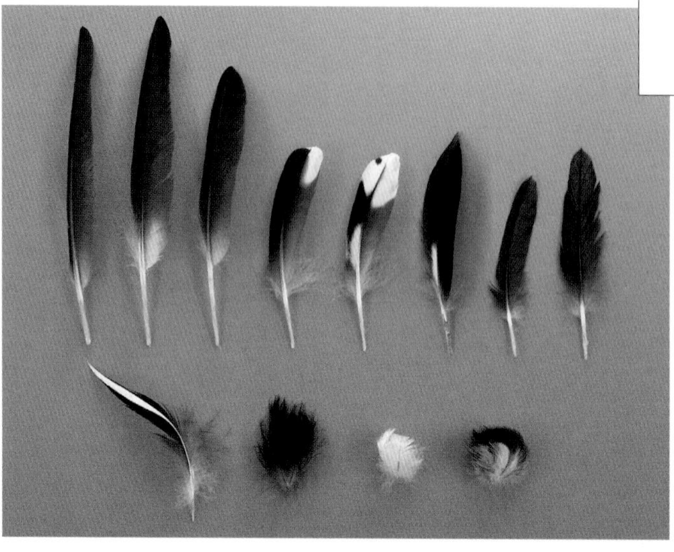

Barrow's Goldeneye *(Bucephala islandica)*

M>F

male
P: 6.3 (16.1)
P: 6.5 (16.6)
P: 4.8 (12.3)
S: 4.2 (10.7)
TS: 4.1 (10.4)
TS: 4.4 (11.3)
T: 2.9 (7.4)
[broken]
NA: 0.7 (1.9)
MA: 1.9 (4.9)
BR: 0.8 (2.0)
FL: 2.3 (5.9)

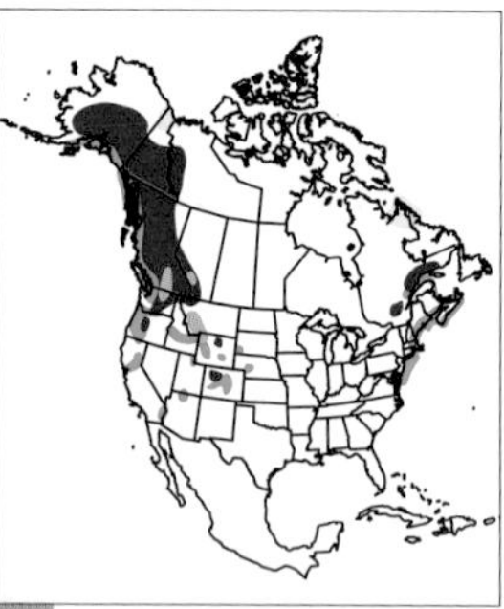

Hooded Merganser *(Lophodytes cucullatus)*

M>F

male
P: 5.4 (13.8)
P: 5.7 (14.4)
P: 3.9 (10.0)
S: 3.2 (8.1)
S: 3.4 (8.6)
T: 3.1 (8.0)
T: 4.0 (10.2)
SC: 4.3 (10.9)
BR: 1.3 (3.3)
MA: 1.4 (3.5)
RU: 2.6 (6.6)
FL: 3.2 (8.2)

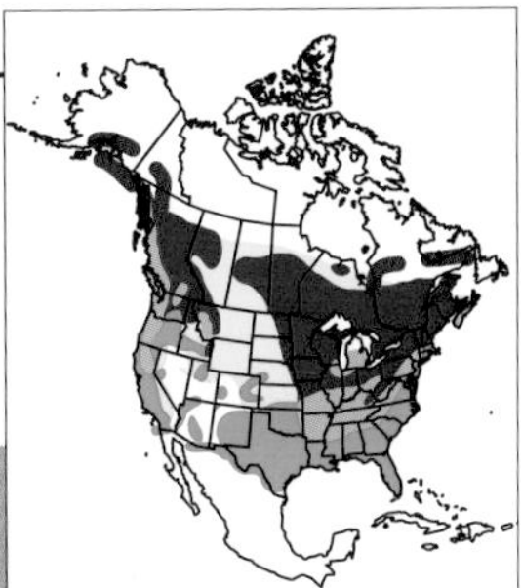

Common Merganser *(Mergus merganser)*

M>F

male
P: 7.9 (20.0)
P: 6.3 (15.9)
P: 5.4 (13.6)
S: 5.2 (13.1)
S: 4.9 (12.4)
S: 4.4 (11.3)
TS: 5.7 (14.5)
T: 4.6 (11.6)
T: 4.6 (11.8)
T: 4.5 (11.5)
MA: 2.0 (5.2)
BR: 1.7 (4.2)
RU: 1.5 (3.9)
SC: 4.5 (11.4)

Red-breasted Merganser *(Mergus serrator)*

M>F

male
P: 7.2 (18.2)
P: 6.2 (15.8)
P: 4.7 (12.0)

S: 4.2 (10.7)
S: 3.8 (9.6)
TS: 4.7 (11.9)
T: 4.0 (10.2)
T: 4.0 (10.2)

MA: 1.9 (4.8)
RU: 1.4 (3.6)
FL: 4.3 (10.8)
SC: 2.7 (6.9)

Ruddy Duck *(Oxyura jamaicensis)*

M>F

male
P: 3.9 (9.8)
P: 4.2 (10.6)
P: 3.9 (9.8)
P: 3.2 (8.1)

S: 2.8 (7.1)
S: 2.9 (7.4)
T: 2.4 (6.0)
T: 3.4 (8.7)

BR: 0.9 (2.2)
BE: 0.9 (2.3)
RU: 1.3 (3.4)
SC: 2.4 (6.2)

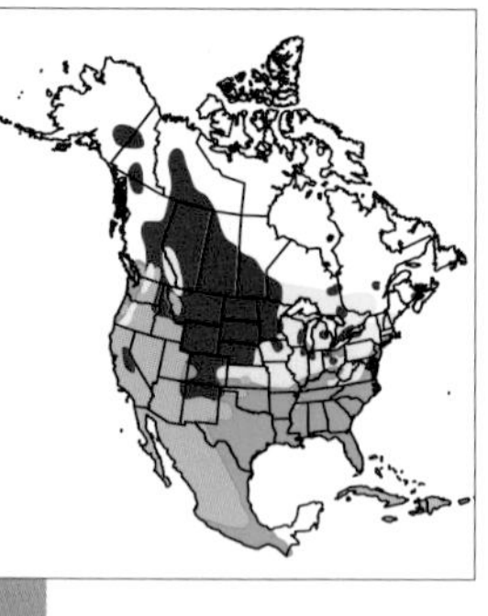

Partridges, Grouse, Turkeys, and Old World Quail *(Phasianidae)*

Camber is always present to varying degrees. Sickle-shaped leading primaries with slim waist. Feathers are often mottled for camouflage.

Chukar *(Alectoris chukar)*

M>F

male
P: 3.8 (9.7)
P: 4.6 (11.6)
P: 5.23 (13.3)
P: 5.3 (13.5)
P: 4.4 (11.1)
S: 4.4 (11.1)
S: 4.3 (11.0)
T: 3.8 (9.6)
T: 3.9 (10.0)
BR: 1.7 (4.2)
MA: 2.7 (6.9)
RU: 2.5 (6.4)
FL: 2.6 (6.7)

Gray Partridge *(Perdix perdix)*

M>F

male	S: 3.9 (9.8)	BR: 1.6 (4.1)
P: 4.4 (11.1)	S: 3.9 (9.9)	BE: 1.5 (3.8)
P: 5.4 (13.7)	T: 3.2 (8.1)	MA: 1.5 (3.8)
P: 5.2 (13.1)	T: 3.5 (8.9)	RU: 1.9 (4.7)
P: 4.1 (10.5)	T: 3.5 (9.0)	CR: 0.7 (1.9)

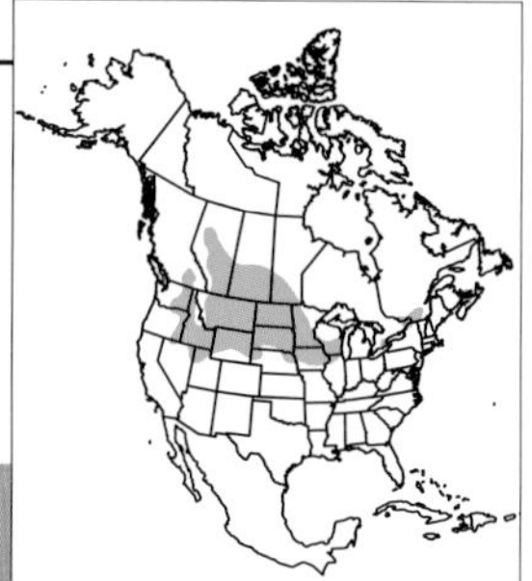

Ring-necked Pheasant *(Phasianus colchicus)*

M>F

male	S: 6.1 (15.6)	MA: 2.2 (5.6)
P: 6.5 (16.6)	S: 6.6 (16.7)	BR: 2.0 (5.0)
P: 7.8 (19.8)	TS: 4.8 (12.1)	NA: 1.0 (2.6)
P: 8.0 (20.2)	SC: 2.3 (5.9)	NA: 1.1 (2.8)
P: 6.5 (16.5)	RU: 2.4 (6.0)	

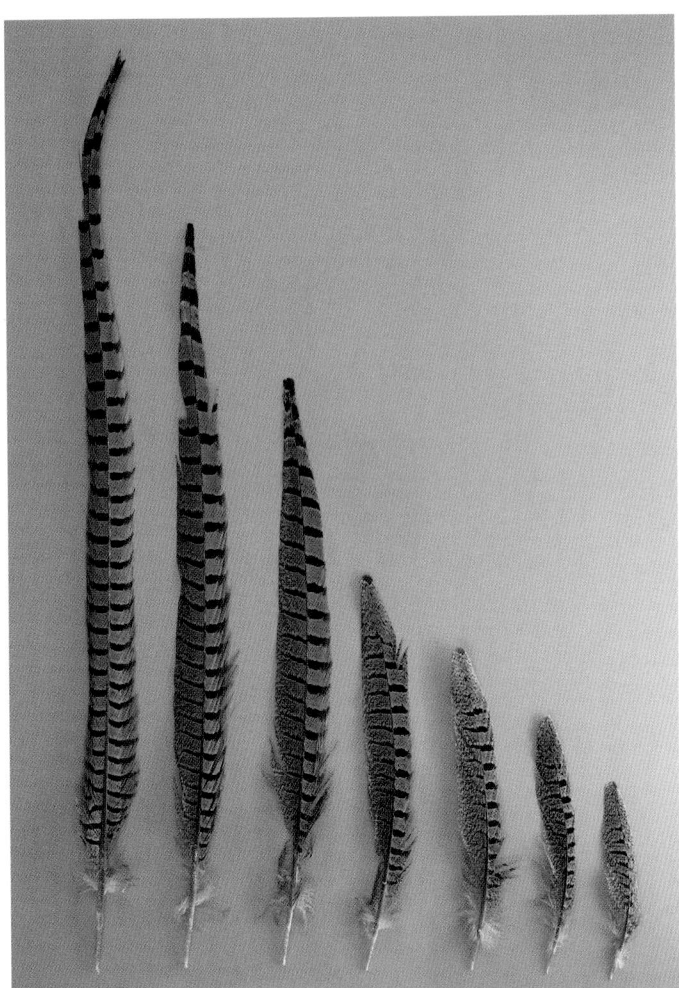

T: 21.6 (54.8)
T: 17.6 (44.8)
T: 13.7 (34.9)
T: 9.3 (23.5)
T: 7.6 (19.4)
T: 6.1 (15.6)
T: 4.7 (11.9)

Ruffed Grouse *(Bonasa umbellus)*

M>F

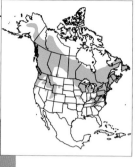

male	**S:** 4.2 (10.6)	**BR:** 2.6 (6.7)
P: 4.1 (10.4)	**S:** 4.0 (10.2)	**BE:** 2.6 (6.7)
P: 5.7 (14.4)	**T:** 5.2 (13.1)	**RU:** 1.5 (3.9)
P: 5.6 (14.3)	**T:** 5.2 (13.1)	
P: 5.2 (13.3)	**T:** 5.1 (12.9)	

gray phase
T: 6.3 (16.0)
T: 6.4 (16.2)
T: 6.6 (16.8)

Spruce Grouse *(Falcipennis canadensis)*

M>F

male
P: 4.1 (10.5)
P: 5.4 (13.6)
P: 5.9 (15.1)
P: 4.8 (12.2)
P: 4.5 (11.5)
S: 4.2 (10.6)
S: 4.0 (10.2)
TS: 3.7 (9.4)
T: 5.3 (13.4)
T: 5.4 (13.8)
UNTC: 3.2 (8.1)
BR: 1.7 (4.2)
MA: 1.88 (4.8)
RU: 2.8 (7.2)

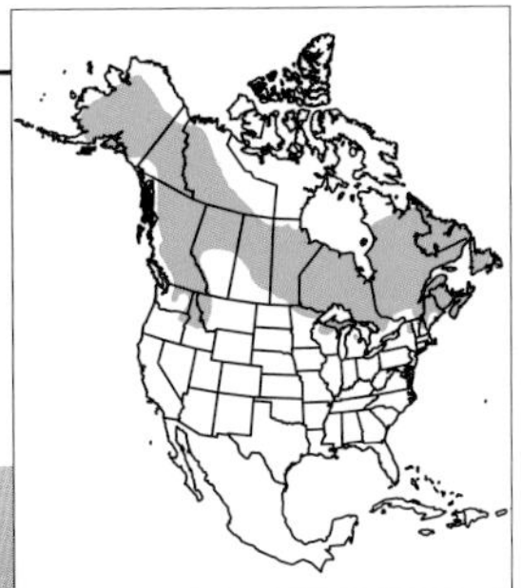

Greater Prairie-Chicken *(Tympanuchus cupido)*

(Attwater's subspecies)
M>F

male
P: 5.0 (12.8)
P: 6.8 (17.2)
P: 6.8 (17.3)
P: 5.8 (14.8)
S: 4.6 (11.6)
S: 4.4 (11.2)
S: 3.9 (9.9)
T: 3.9 (9.8)
T: 3.7 (9.3)
T: 2.8 (7.2)
neck plume: 2.6 (6.7)
MA: 2.4 (6.0)
RU: 2.6 (6.6)
UPTC: 3.0 (7.6)
UNTC: 3.7 (9.5)
FL: 3.3 (8.5)

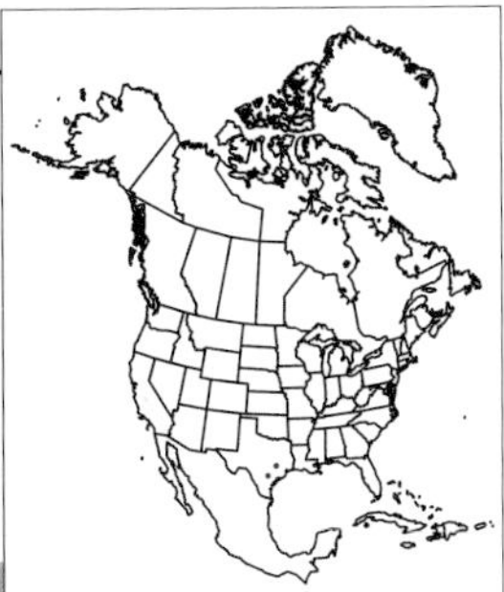

Wild Turkey *(Meleagris gallopavo)*

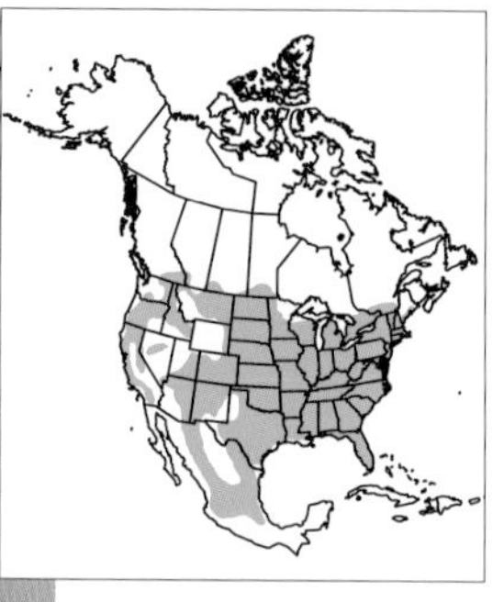

M>F

male
P: 10.0 (25.4)
P: 15.6 (39.8)
P: 14.5 (36.8)
P: 13.0 (33.1)
S: 13.1 (33.3)

T: 12.6 (31.9)
T: 15.5 (39.4)
BR: 3.3 (8.4)
BE: 2.7 (6.8)
NA: 1.7 (4.2)

New World Quail *(Odontophoridae)*

Camber is always present to varying degrees. Sickle-shaped leading primaries with slim waist. Feathers are often mottled for camouflage.

Mountain Quail *(Oreortyx pictus)*

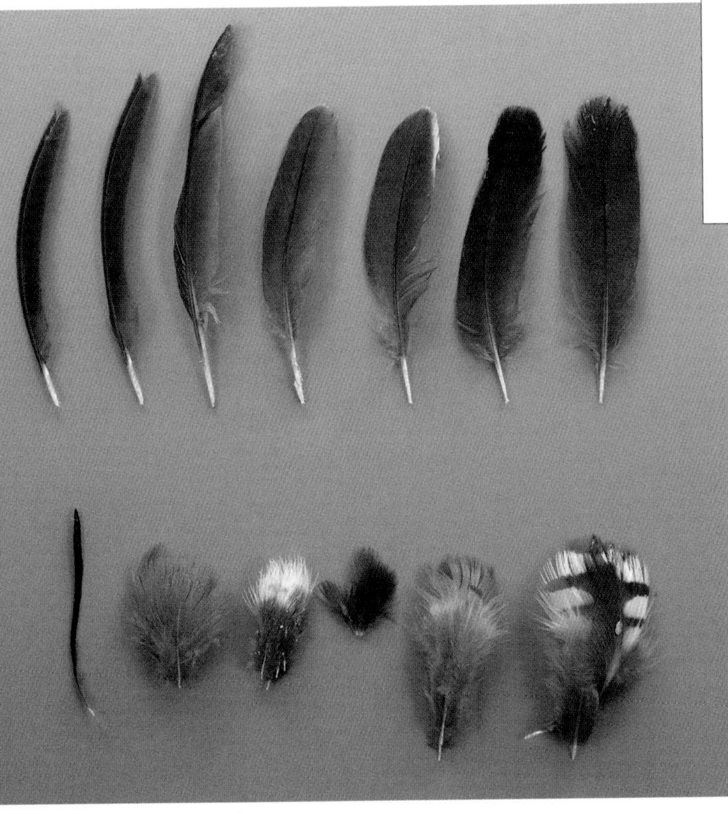

male
P: 3.7 (9.4)
P: 4.1 (10.4)
P: 4.6 (11.8)
S: 3.6 (9.2)
S: 3.6 (9.1)
T: 3.6 (9.1)
T: 3.7 (9.3)
head plume: 2.6 (6.7)
BR: 1.9 (4.7)
BE: 1.6 (4.0)
TH: 1.1 (2.7)
FL: 2.3 (5.9)
FL: 2.5 (6.4)

Scaled Quail *(Callipepla squamata)*

male
P: 3.2 (8.1)
P: 3.6 (9.2)
P: 4.1 (10.4)
P: 4.0 (10.2)
P: 3.5 (8.8)
S: 3.3 (8.3)
TS: 3.2 (8.2)
T: 3.0 (7.7)
T: 3.6 (9.2)
BR: 1.3 (3.3)
BE: 1.1 (2.8)
MA: 1.2 (3.1)
FL: 3.2 (8.2)
UNTC: 2.4 (6.2)

California Quail *(Callipepla californica)*

male
P: 2.8 (7.1)
P: 3.6 (9.1)
P: 3.9 (9.8)
P: 3.9 (9.9)
P: 3.1 (7.8)
S: 3.0 (7.6)
S: 2.8 (7.1)
T: 3.0 (7.5)
T: 3.5 (9.0)
BR: 1.1 (2.9)
BE: 1.3 (3.2)
MA: 1.4 (3.5)
UPTC: 2.9 (7.3)
head plume: 1.8 (4.5)

Gambel's Quail *(Callipepla gambelii)*

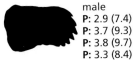

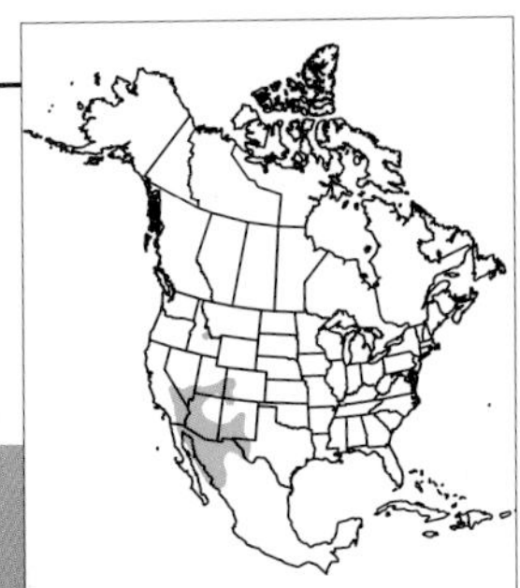

male	**S:** 3.1 (7.8)	**BR:** 1.1 (2.8)
P: 2.9 (7.4)	**S:** 3.2 (8.1)	**BE:** 1.3 (3.4)
P: 3.7 (9.3)	**T:** 3.4 (8.7)	**MA:** 1.8 (4.5)
P: 3.8 (9.7)	**T:** 4.1 (10.4)	**UNTC:** 2.8 (7.1)
P: 3.3 (8.4)		**FL:** 2.1 (5.3)

Northern Bobwhite *(Colinus virginianus)*

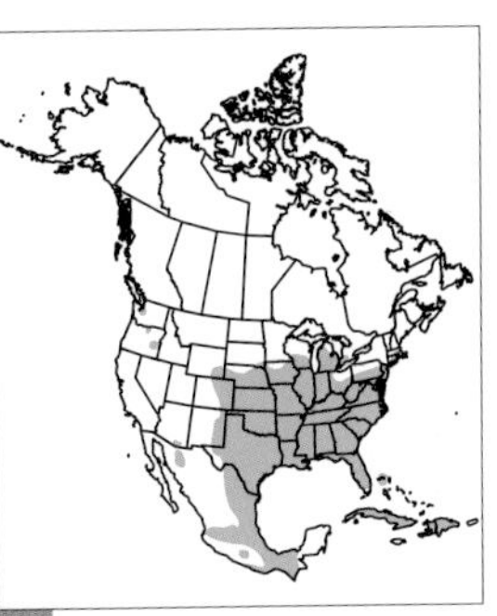

male	**S:** 2.8 (7.1)	**NA:** 0.7 (1.9)
P: 2.7 (6.8)	**TS:** 2.7 (6.8)	**BR:** 1.0 (2.5)
P: 3.5 (8.8)	**T:** 2.6 (6.7)	**BE:** 1.1 (2.9)
P: 3.5 (9.0)	**T:** 2.5 (6.4)	**MA:** 1.4 (3.5)
P: 2.9 (7.3)		**FL:** 2.6 (6.5)

Northern Bobwhite (Masked) *(Colinus virginianus)*

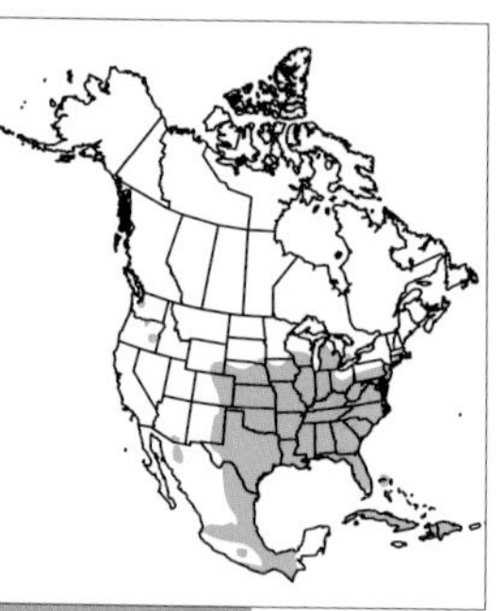

male
P: 3.1 (7.9)
P: 3.5 (8.9)
P: 3.7 (9.4)
S: 3.3 (8.4)
S: 3.3 (8.5)
TS: 3.0 (7.7)
T: 2.4 (6.0) [broken]
T: 2.2 (5.5) [broken]
BR: 1.3 (3.4)
MA: 1.4 (3.5)
NE: 1.2 (3.0)

Montezuma Quail *(Cyrtonyx montezumae)*

M>F

male
P: 3.6 (9.2)
P: 4.1 (10.4)
P: 3.7 (9.5)
P: 3.5 (9.0)
S: 3.5 (8.9)
S: 3.4 (8.7)
T: 1.8 (4.6)
T: 2.0 (5.0)
SC: 2.9 (7.3)
BR: 1.0 (2.5)
BE: 1.0 (2.5)
MA: 1.6 (4.0)
WC: 2.2 (5.5)

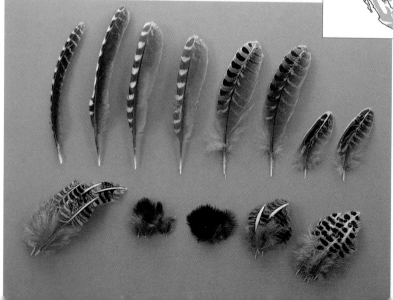

Loons (*Gaviidae*)

Tails very short. Primary feather camber is slight. Extremely long quills on the flight feathers of the wing. Secondaries vary in morphology by species.

Red-throated Loon (*Gavia stellata*)

M>F

unknown, non-breeding
P: 7.0 (17.9)
P: 6.3 (16.1)
P: 5.2 (13.3)
S: 4.2 (10.7)
S: 3.7 (9.5)
T: 2.6 (6.7)
T: 2.6 (6.5)
MA: 1.2 (3.0)
BR: 1.2 (3.1)
RU: 1.5 (3.7)

Pacific Loon *(Gavia pacifica)*

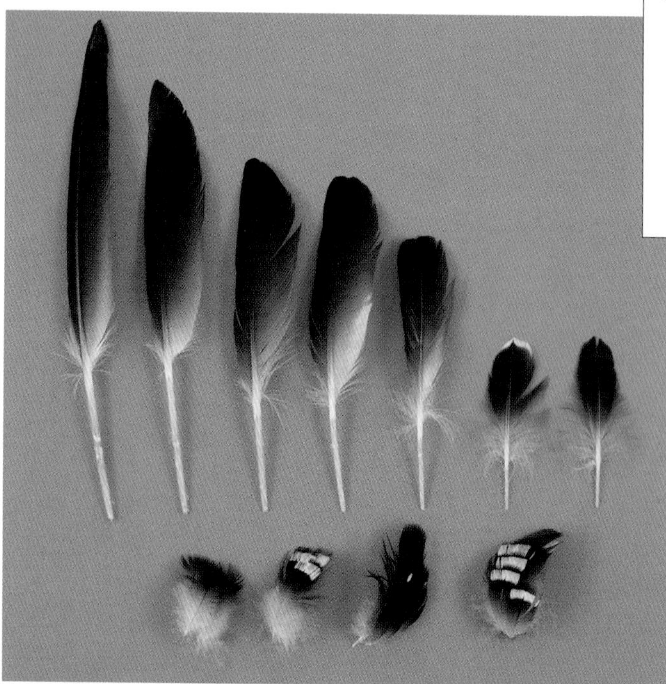

M>F

unknown,
breeding

P: 8.2 (20.9)
P: 7.2 (18.3)
P: 5.9 (15.0)
S: 5.5 (14.0)
S: 4.6 (11.6)
T: 2.9 (7.3)
T: 2.9 (7.3)
RU: 1.3 (3.4)
MA: 1.7 (4.3)
WC: 2.2 (5.5)
SC: 2.3 (5.9)

Common Loon *(Gavia immer)*

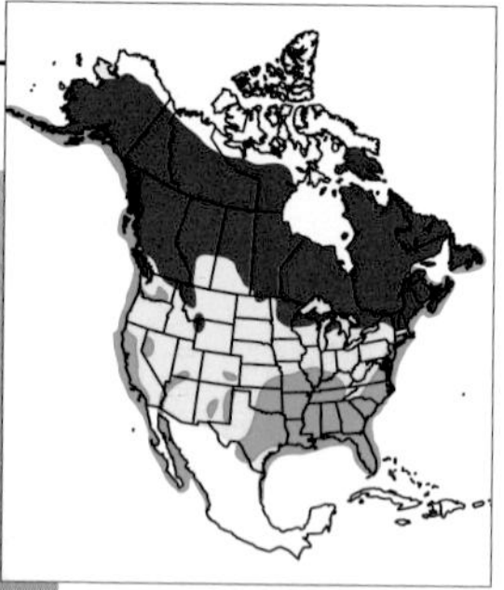

M>F

unknown

P: 9.6 (24.3)
P: 9.3 (23.6)
P: 7.2 (18.4)
S: 6.2 (15.7)
S: 6.0 (15.2)
T: 3.1 (7.8)
T: 3.5 (8.9)
T: 2.7 (6.8)
RU: 1.9 (4.8)
WC: 1.6 (4.0)
NA: 0.8 (2.1)
UNWC: 5.3 (13.4)

Grebes *(Podicipedidae)*

Vestigial tail. Camber is always present to varying degrees. Emargination present in the leading two or three primaries. Long, narrow, evenly vaned secondaries.

Pied-billed Grebe *(Podilymbus podiceps)*

unknown	**P:** 3.3 (8.4)	**BR:** 0.8 (2.1)
P: 3.3 (8.5)	**S:** 3.1 (8.0)	**MA:** 2.2 (5.5)
P: 3.8 (9.6)	**S:** 3.1 (7.8)	**RU:** 1.4 (3.5)
P: 3.8 (9.7)	**T:** 1.6 (4.0)	

M>F

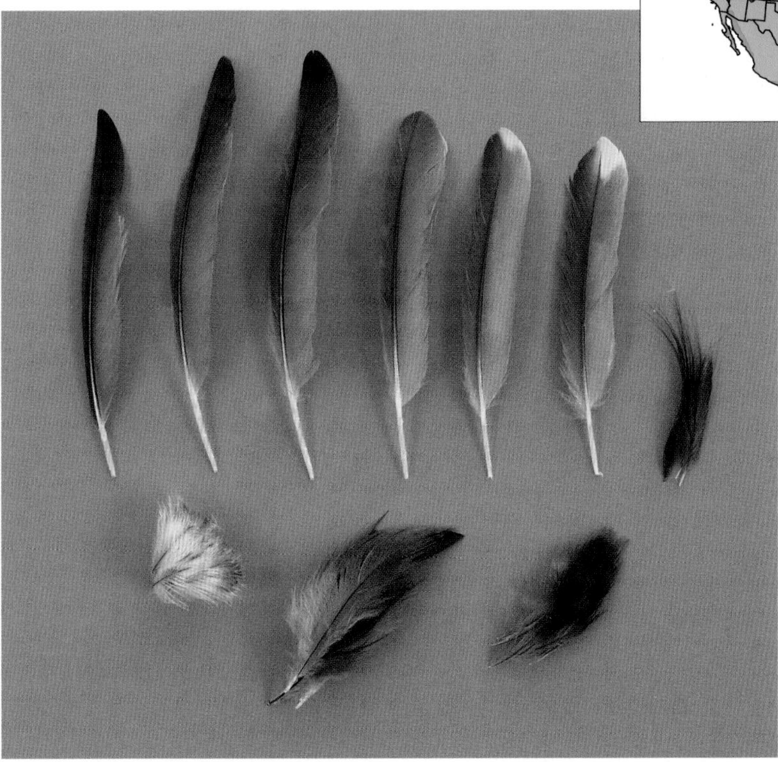

Horned Grebe *(Podiceps auritus)*

unknown,
non-breeding
P: 3.8 (9.6)
P: 4.0 (10.2)
P: 3.5 (9.0)
S: 3.1 (8.0)
S: 3.1 (7.8)
T: 1.6 (4.1)
MA: 1.3 (3.3)
NA: 0.8 (2.0)

Red-necked Grebe *(Podiceps grisegena)*

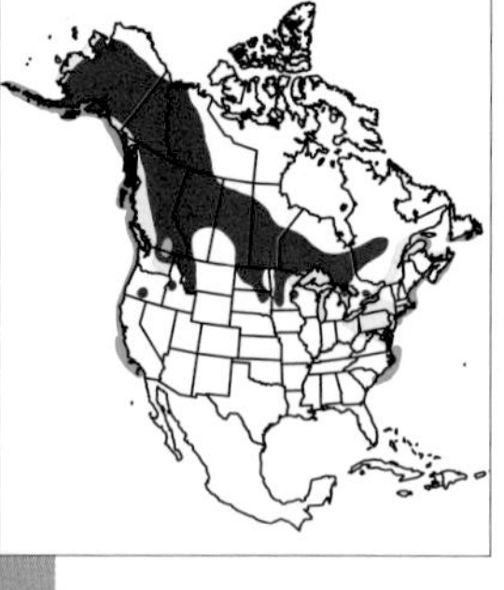

unknown,
breeding
P: 4.9 (12.5)
P: 5.5 (14.0)
P: 4.8 (12.3)
S: 4.5 (11.5)
S: 4.2 (10.7)
TS: 3.8 (9.7)
NE: 0.7 (1.9)
BR: 1.3 (3.3)
BE: 1.7 (4.3)
MA: 1.4 (3.5)

Western Grebe *(Aechmophorus occidentalis)*

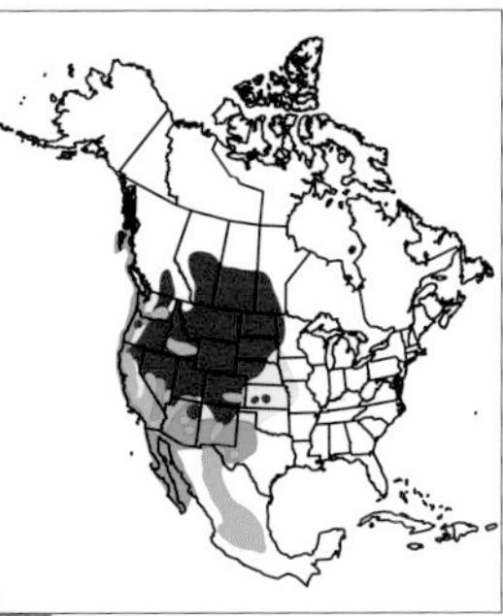

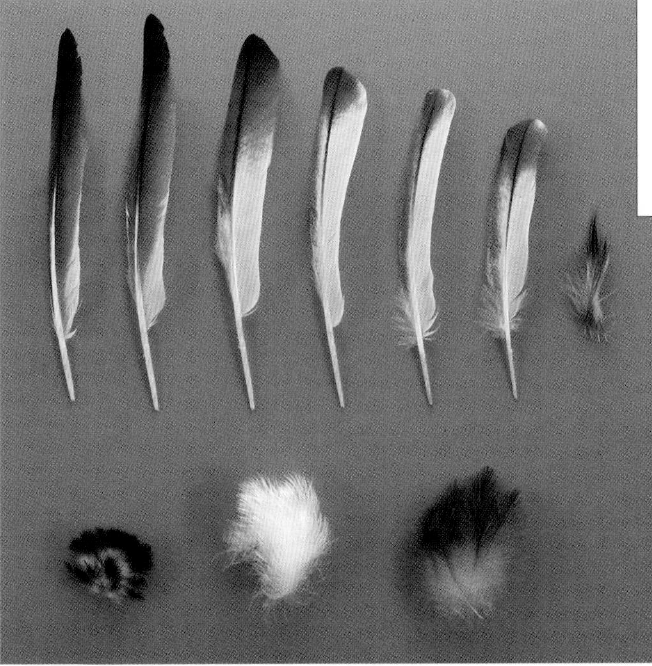

unknown
P: 4.9 (12.5)
P: 5.2 (13.2)
P: 5.0 (12.6)
P: 4.4 (11.3)
S: 4.2 (10.6)
S: 3.7 (9.3)
T: 1.7 (4.3)
NA: 0.6 (1.4)
BR: 1.2 (3.1)
MA: 2.0 (5.1)

Clark's Grebe *(Aechmophorus clarkii)*

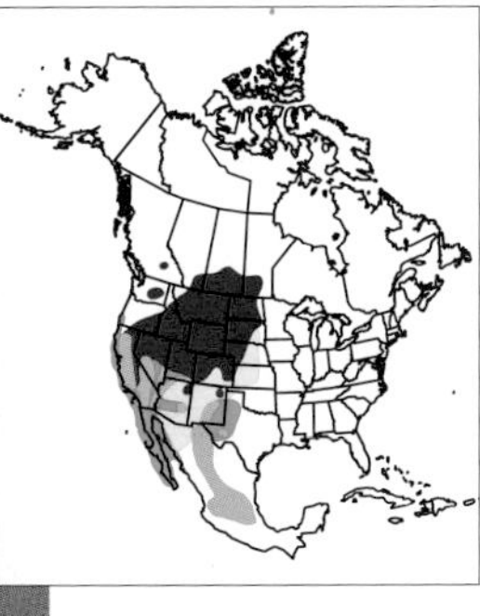

unknown
P: 4.7 (11.9)
P: 5.0 (12.7)
P: 4.8 (12.2)
P: 4.3 (11.0)
S: 3.7 (9.4)
TS: 3.3 (8.5)
T: 1.6 (4.1)
BR: 1.1 (2.9)
BE: 1.0 (2.6)
MA: 1.3 (3.3)

Shearwaters and Petrels *(Procellariidae)*, Storm-Petrels *(Hydrobatidae)*

Northern Fulmar *(Fulmarus glacialis)*

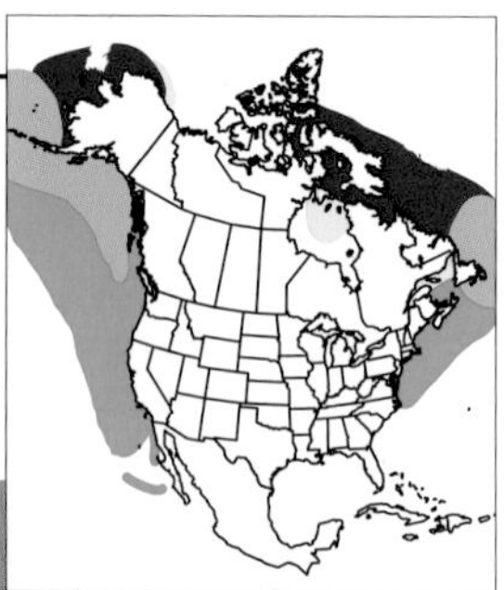

M>F

juvenile, leading primaries and tail severely worn

P: 8.3 (21.1)
P: 7.9 (20.1)
P: 7.0 (17.7)
P: 5.6 (14.3)

S: 3.3 (8.3)
S: 3.8 (9.7)
TS: 4.1 (10.4)
T: 4.5 (11.4)
T: 4.6 (11.6)

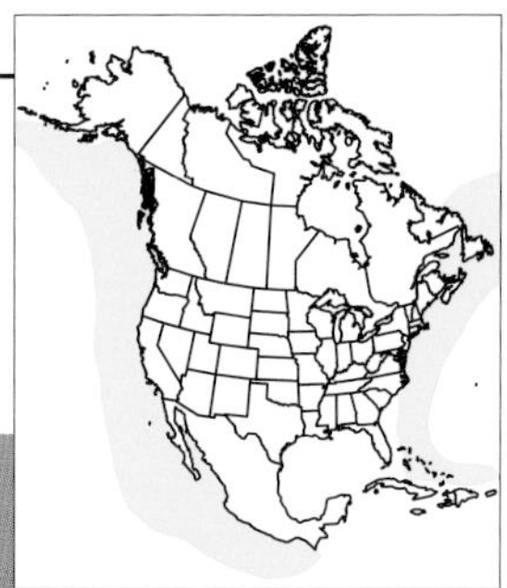

Sooty Shearwater *(Puffinus griseus)*

unknown
P: 8.8 (22.4)
P: 7.8 (19.9)
P: 6.6 (16.8)

S: 3.4 (8.7)
S: 3.4 (8.7)
T: 4.1 (10.4)
T: 3.6 (9.1)

MA: 1.6 (4.1)
BR: 1.6 (4.1)
BE: 2.0 (5.0)
UNWC: 2.4 (6.1)

Wilson's Storm-Petrel *(Oceanites oceanicus)*

unknown
P: 4.5 (11.4)
P: 3.9 (9.8)
P: 3.4 (8.7)

S: 2.7 (6.8)
S: 2.7 (6.9)
T: 3.6 (9.1)
T: 3.8 (9.6)

MA: 1.3 (3.2)
BE: 1.8 (4.5)
RU: 1.5 (3.9)

Tropicbirds *(Phaethontidae)*

Some species have long quills similar to waterfowl.

White-tailed Tropicbird *(Phaethon lepturus)*

juvenile
P: 9.2 (23.4)
P: 9.3 (23.5)
P: 7.5 (19.1)
P: 6.4 (16.3)
S: 3.5 (8.9)
TS: 4.3 (10.8)
T: 3.6 (9.2)
T: 5.8 (14.7)
BR: 1.1 (3.0)
BE: 1.1 (2.8)
RU: 1.2 (3.1)
UPTC: 3.7 (9.5)

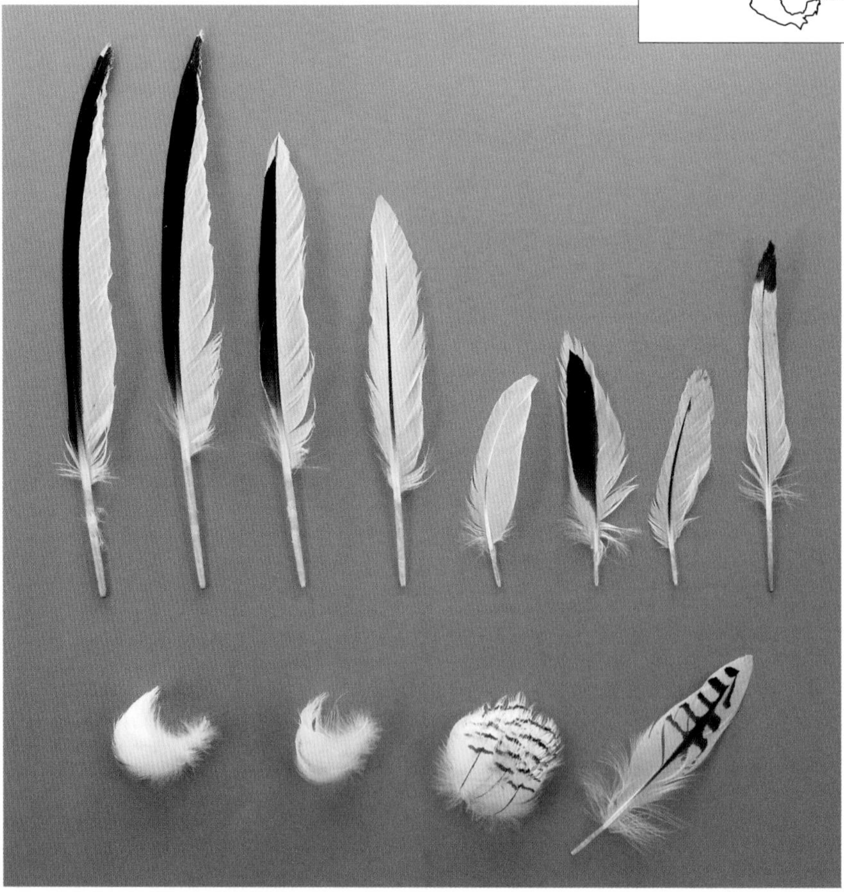

Boobies and Gannets *(Sulidae)*

Masked Booby *(Sula dactylatra)*

F>M

unknown, juvenile, molting to full adult plumage

P: 11.1 (28.3)
P: 9.8 (25.0)
P: 8.0 (20.4)
S: 5.0 (12.6)
TS: 4.6 (11.6)

T: 6.1 (15.5)
BR: 1.2 (3.1)
BE: 1.5 (3.8)
MA: 0.9 (2.4)

Brown Booby *(Sula leucogaster)*

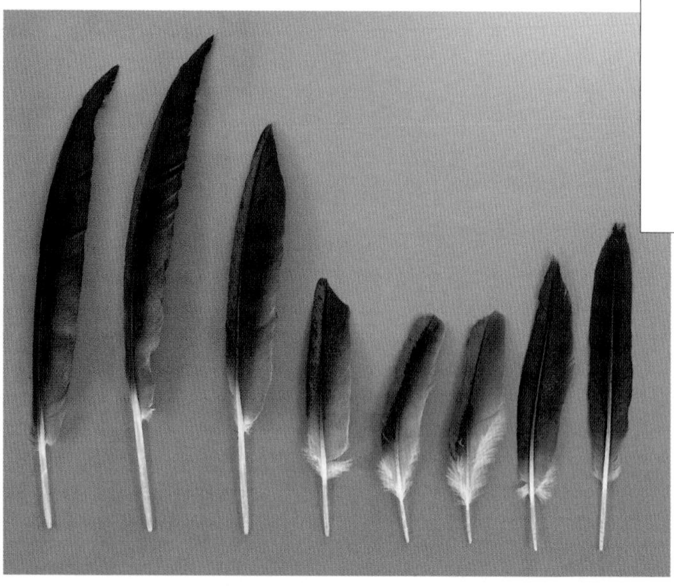

F>M

unknown
P: 12.3 (31.3)
P: 11.6 (29.4)
P: 10.1 (25.7)
P: 6.3 (16.1)
S: 5.6 (14.3)
S: 5.8 (14.7)
T: 7.3 (18.6)
T: 8.0 (20.4)

Northern Gannet *(Morus bassanus)*

unknown
P: 14.2 (36.1)
P: 14.1 (35.9)
P: 10.9 (27.7)
P: 9.9 (25.1)
S: 10.3 (26.2)
S: 5.7 (14.6)
T: 6.4 (16.2)
T: 7.8 (19.9)
T: 10.3 (26.2)
BR: 1.9 (4.9)
BE: 1.5 (3.7)
RU: 2.7 (6.9)

Pelicans *(Pelecanidae)*

Camber is always present to varying degrees. Long, narrow, heavily emarginated leading primaries, similar to those of vultures, cranes, eagles, and storks. Bottom half of the feather shaft is pale to white. Long quills similar to waterfowl.

American White Pelican *(Pelecanus erythrorhynchos)*

male
P: 15.0 (38.2)
P: 17.7 (44.9)
P: 18.1 (46.0)
P: 13.3 (33.7)
P: 13.3 (33.7)

M>F

continued on page 112

American White Pelican continued

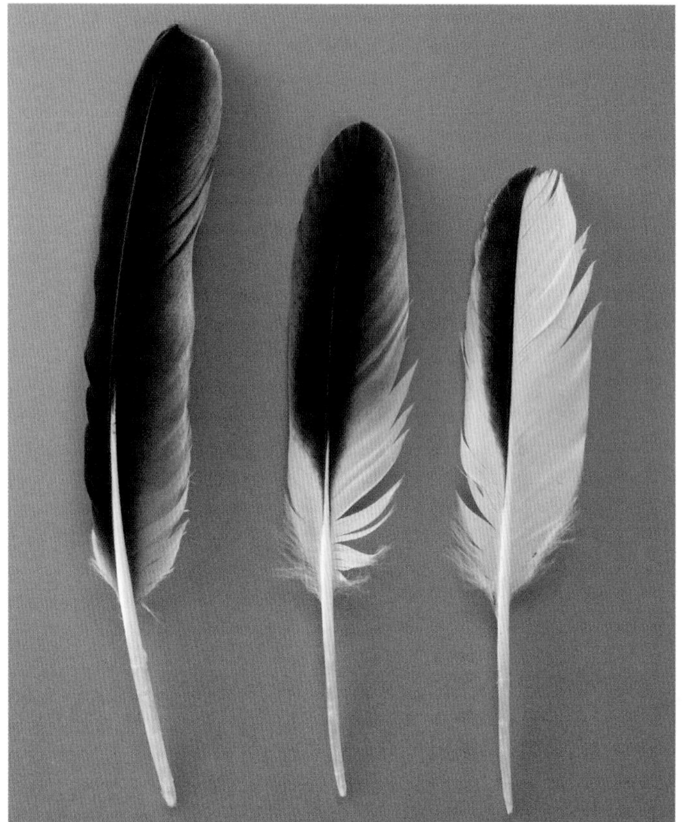

S: 12.5 (31.7)
S: 10.7 (27.2)
S: 10.3 (26.1)

T: 11.9 (30.2)
T: 11.4 (29.0)
WC: 6.3 (15.9)
CR: 3.8 (9.7)
BR: 2.3 (6.0)
MA: 5.8 (14.9)

Brown Pelican *(Pelecanus occidentalis)*

M>F

unknown,	**P:** 11.6 (29.4)	**S:** 9.0 (22.8)
primaries	**P:** 12.2 (31.0)	**TS:** 8.4 (21.4)
are worn,	**P:** 12.8 (32.6)	**T:** 6.1 (15.4)
tips broken	**P:** 9.4 (23.9)	**T:** 5.7 (14.4)

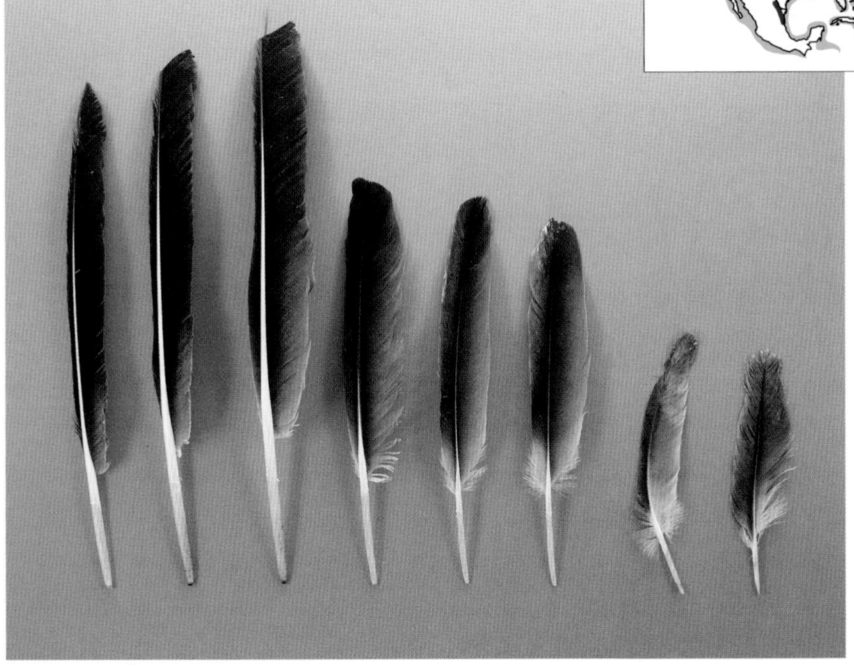

Frigatebirds *(Fregatidae)*

Magnificent Frigatebird *(Fregata magnificens)*

F>M

juvenile
P: 17.2 (43.6)
P: 17.2 (43.6)
P: 14.2 (36.1)
P: 9.6 (24.5)
P: 8.0 (20.2)

S: 7.9 (20.0)
S: 7.7 (19.5)
TS: 8.1 (20.7)
T: 15.5 (36.1)
T: 11.8 (29.9)

Cormorants *(Phalacrocoracidae)*

Unique long, narrow, stiff tail feathers. Stiff, narrow, evenly vaned secondaries.

Brandt's Cormorant *(Phalacrocorax penicillatus)*

juvenile	**S:** 7.3 (18.5)	**NA:** 0.8 (2.1)
P: 7.2 (18.2)	**TS:** 5.3 (13.5)	**BR:** 2.1 (5.4)
P: 8.5 (21.7)	**T:** 5.7 (14.4)	**BE:** 1.7 (4.3)
P: 7.0 (17.9)	**T:** 5.4 (13.7)	**RU:** 1.5 (3.7)

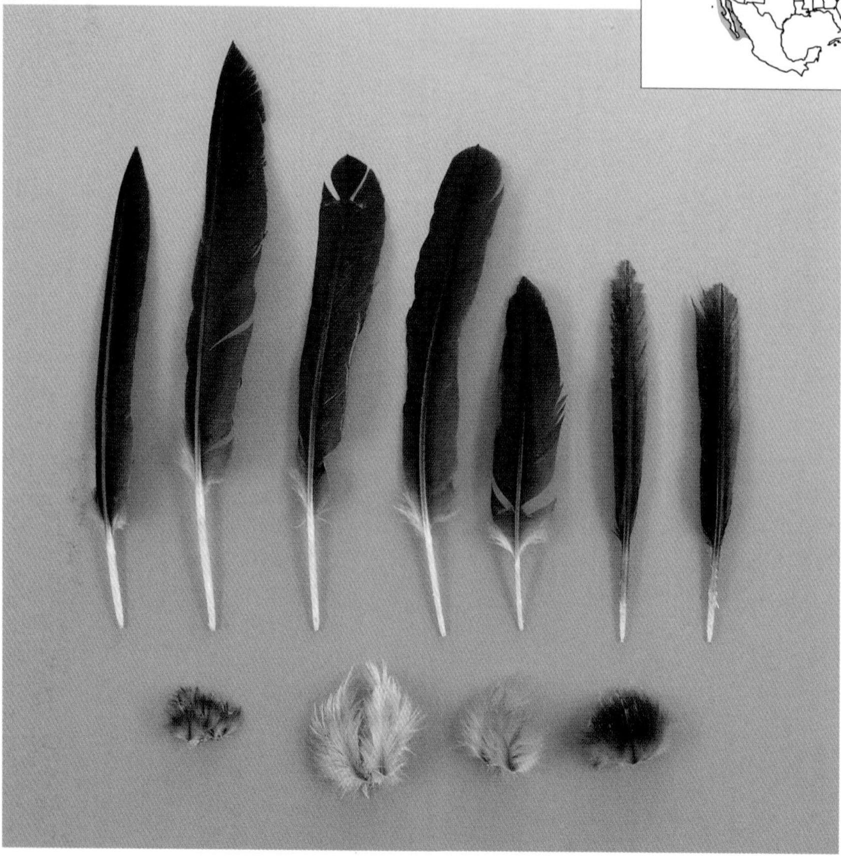

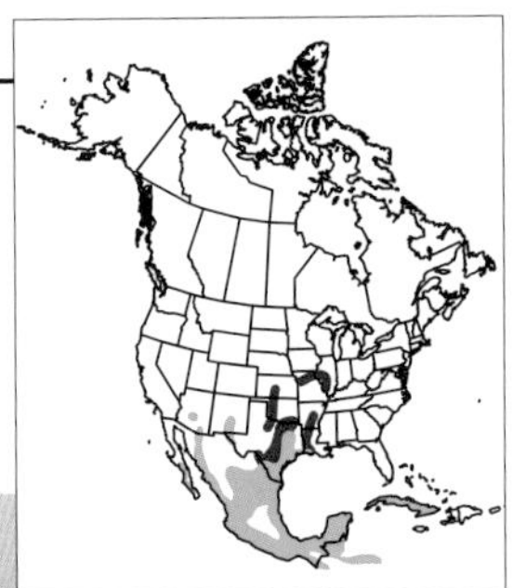

Neotropic Cormorant *(Phalacrocorax brasilianus)*

■■■■

M>F

unknown
P: 6.6 (16.7)
P: 7.6 (19.2)
P: 6.2 (15.8)
S: 5.6 (14.3)
S: 5.0 (12.8)
T: 5.3 (13.4)
T: 6.2 (15.7)
T: 6.8 (17.3)
MC: 2.2 (5.5)
BR: 1.5 (3.7)
MA: 1.3 (3.3)
RU: 1.0 (2.6)

Double-crested Cormorant *(Phalacrocorax auritus)*

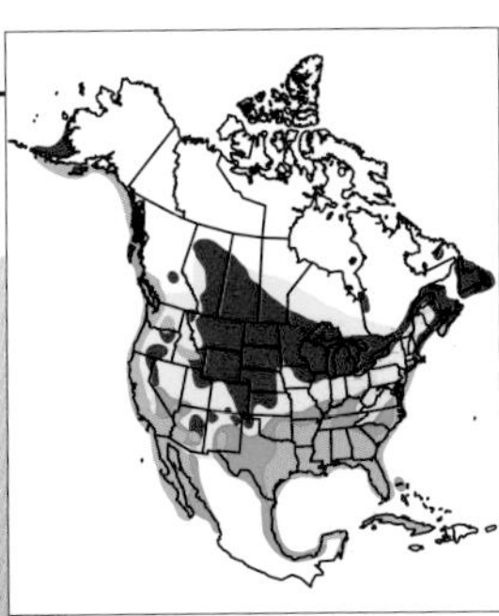

■■■■

M>F

juvenile
P: 6.4 (16.3)
P: 8.0 (20.2)
P: 8.3 (21.1)
P: 7.3 (18.5)
S: 6.5 (16.5)
S: 5.6 (14.1)
T: 6.4 (16.3)
T: 5.4 (13.8)
BR: 1.5 (3.7)
MA: 1.3 (3.2)
RU: 1.1 (2.7)

Pelagic Cormorant *(Phalacrocorax pelagicus)*

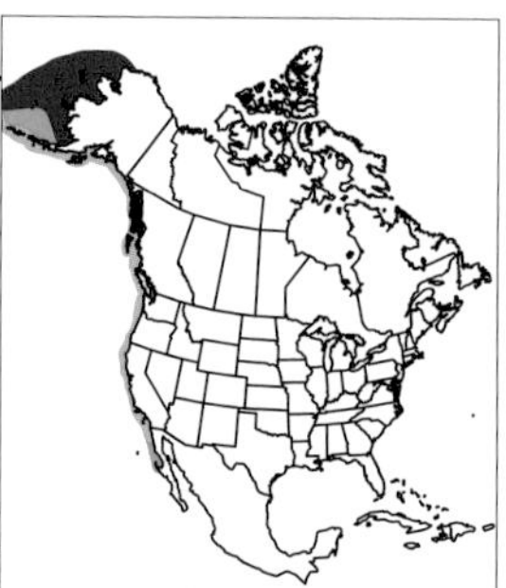

unknown
P: 6.0 (15.3)
P: 6.8 (17.3)
P: 6.9 (17.5)
P: 5.9 (15.1)
S: 5.4 (13.8)
S: 5.6 (14.1)
TS: 5.4 (13.6)
T: 5.2 (13.2)
T: 6.8 (17.2)
T: 7.0 (17.7)
BR: 0.9 (2.4)
MA: 1.2 (3.1)
FL: 1.6 (4.1)

Darters *(Anhingidae)*

Unique long, stiff tail feathers. Stiff, narrow, evenly vaned secondaries.

Anhinga *(Anhinga anhinga)*

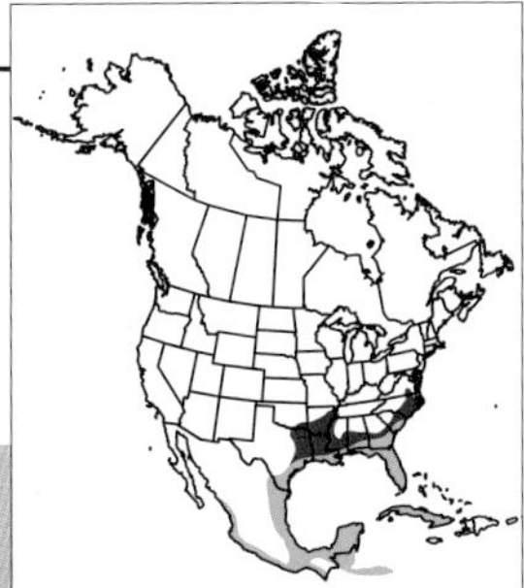

male
P: 9.3 (23.6)
P: 10.6 (26.8)
P: 9.1 (23.2)
P: 8.0 (20.4)
S: 7.6 (19.2)
S: 6.9 (17.5)
T: 9.8 (25.0)
T: 11.5 (29.1)

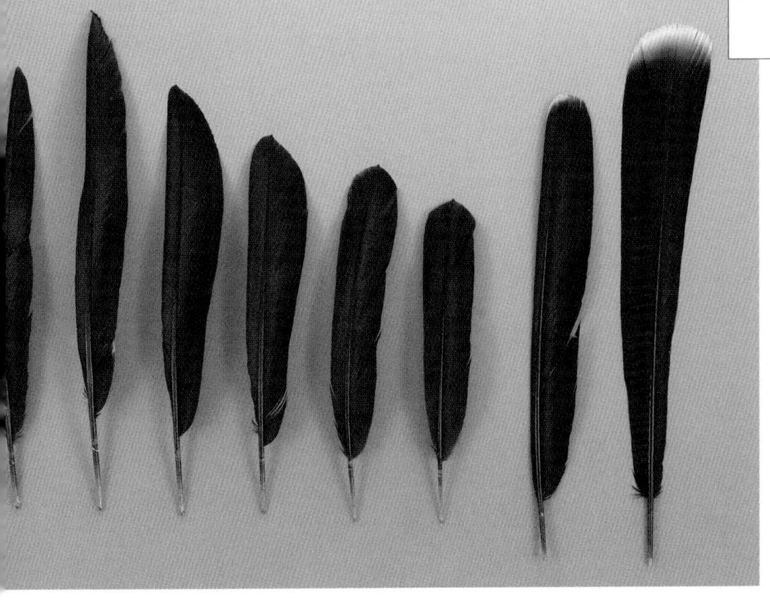

Bitterns, Herons, and Allies *(Ardeidae)*

American Bittern *(Botaurus lentiginosus)*

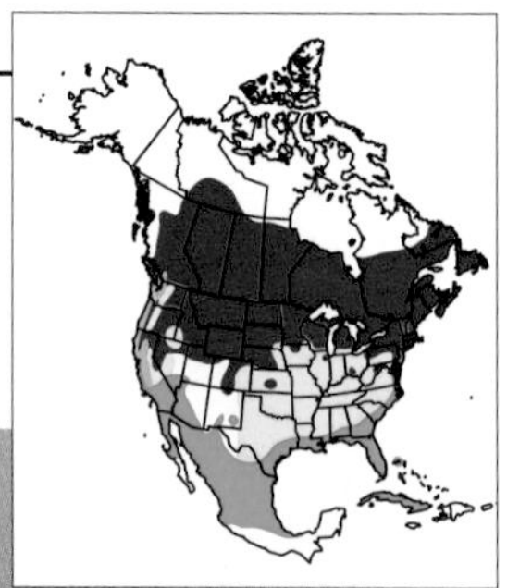

unknown	**S:** 6.2 (15.7)	**NA:** 3.4 (8.7)
P: 6.9 (17.5)	**TS:** 5.8 (14.8)	**BR:** 2.5 (6.4)
P: 7.3 (18.6)	**T:** 3.3 (8.4)	**MA:** 1.8 (4.6)
P: 7.7 (19.6)	**T:** 3.3 (8.5)	**TH:** 2.7 (6.9)
P: 7.1 (18.0)		**WC:** 3.3 (8.3)

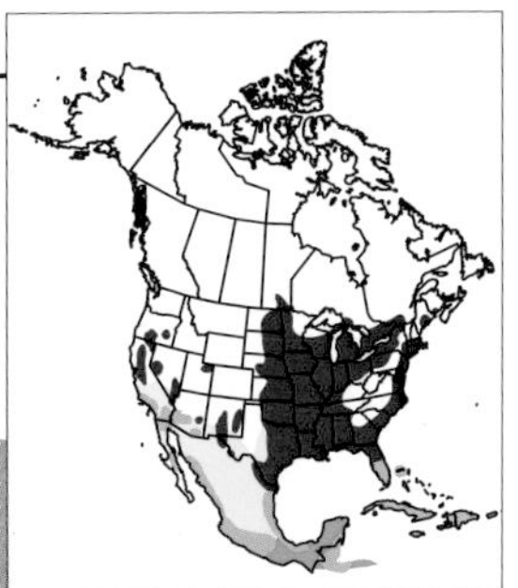

Least Bittern *(Ixobrychus exilis)*

■ ■ ■ ■

male
P: 3.4 (8.6)
P: 3.6 (9.2)
P: 3.4 (8.6)
P: 3.1 (7.9)
S: 3.0 (7.7)
S: 2.8 (7.2)
T: 1.9 (4.8)
T: 1.9 (4.9)
NA: 0.7 (1.9)
BR: 1.6 (4.0)
MA: 1.5 (3.7)
WC: 2.1 (5.4)

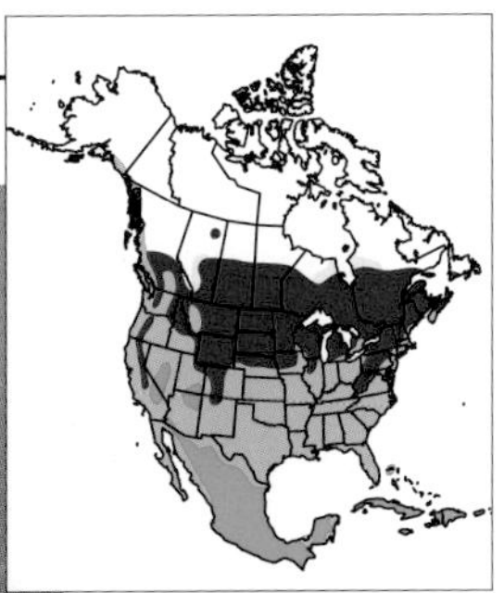

Great Blue Heron *(Ardea herodias)*

M>F

unknown
T: 7.2 (18.4)
T: 4.0 (10.1)
BR: 3.9 (9.9)
BE: 3.9 (9.9)
RU: 2.5 (6.4)
MA: 1.8 (4.6)

continued on page 122

Great Blue Heron continued

P: 13.2 (33.6)
P: 14.2 (36.0)
P: 14.8 (37.5)

S: 11.3 (28.7)
S: 11.1 (28.3)

Great Egret *(Ardea alba)*

M>F

unknown, breeding
P: 11.6 (29.4)
P: 12.0 (30.5)
P: 11.1 (28.1)
S: 9.3 (23.7)
S: 10.4 (26.4)
T: 6.9 (17.4)
T: 6.7 (16.9)
MA: 3.2 (8.2)
BR: 3.1 (7.9)
BE: 4.1 (10.3)

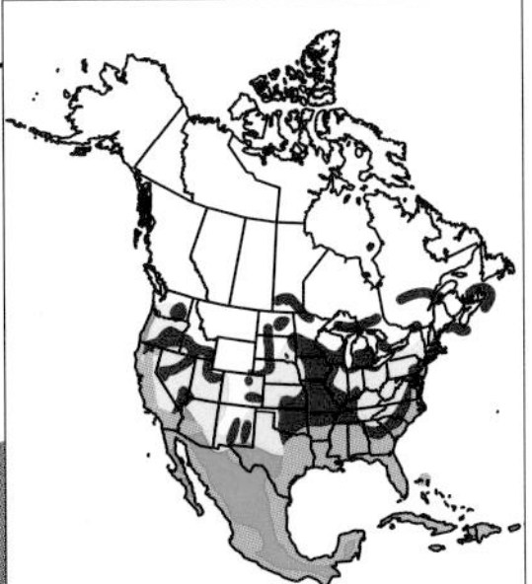

breeding plumes: 19.3 (48.9)

Cattle Egret *(Bubulcus ibis)*

unknown,
breeding
P: 7.0 (17.9)
P: 7.7 (19.5)
P: 7.8 (19.9)
P: 6.3 (16.0)
S: 5.9 (15.0)
S: 5.8 (14.7)
T: 3.9 (9.8)
T: 3.9 (9.9)
T: 3.9 (9.8)
BR: 3.1 (8.0)
CR: 2.0 (5.1)
TH: 3.0 (7.7)
MA: 3.1 (7.9)

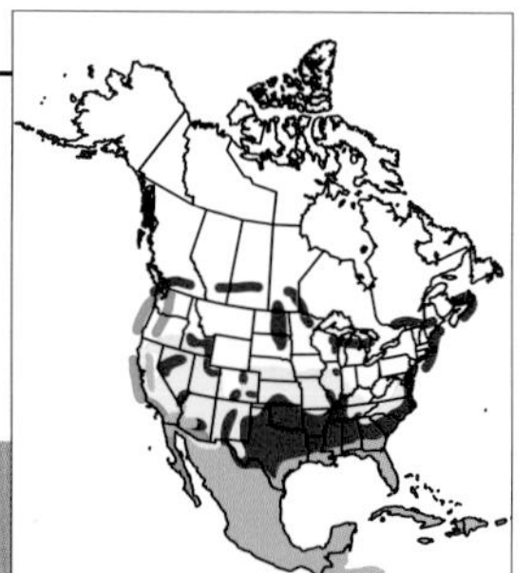

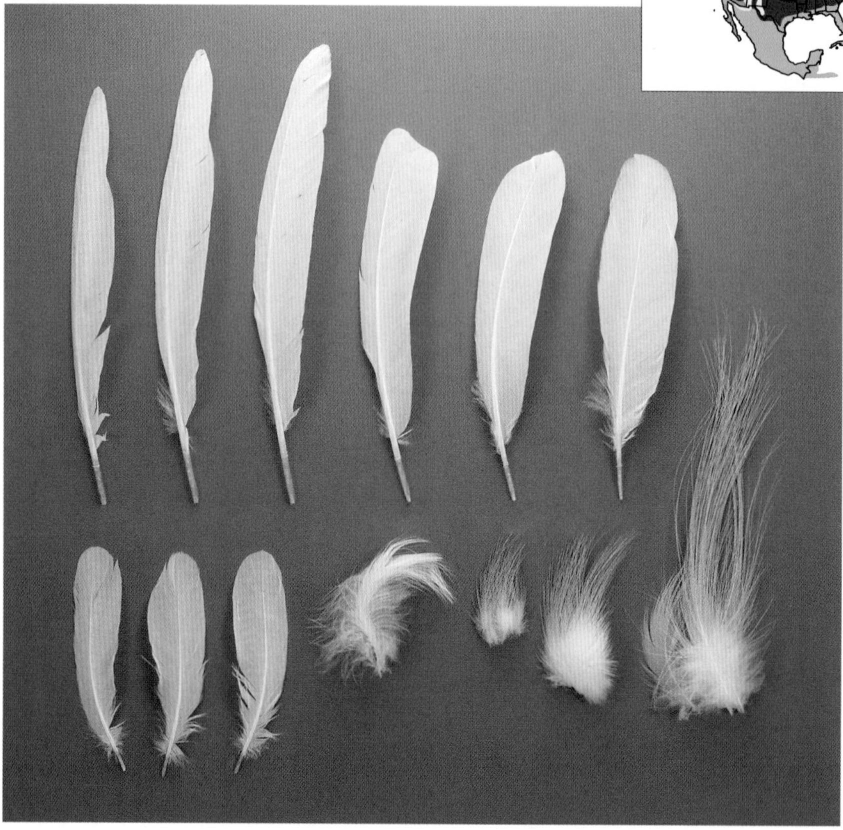

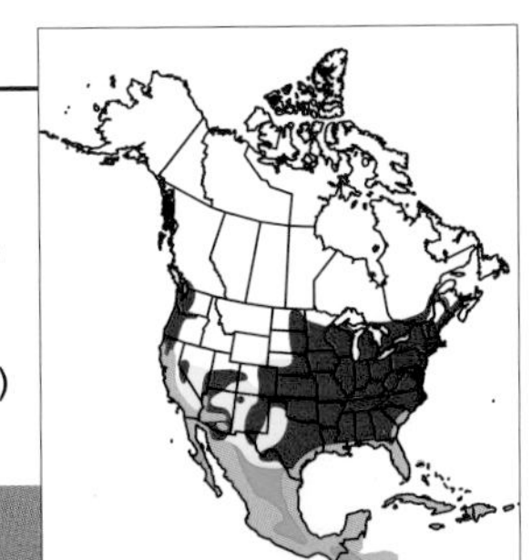

Green Heron *(Butorides virescens)*

■■■■

unknown, flight feathers iridescent green

P: 5.6 (14.3)	**T:** 3.1 (7.9)
P: 6.0 (15.3)	**NA:** 1.3 (3.2)
P: 5.3 (13.5)	**BR:** 1.5 (3.8)
S: 4.9 (12.4)	**BE:** 2.2 (5.5)
TS: 4.7 (12.0)	**MA:** 2.4 (6.2)
T: 3.0 (7.7)	**CR:** 1.9 (4.8)

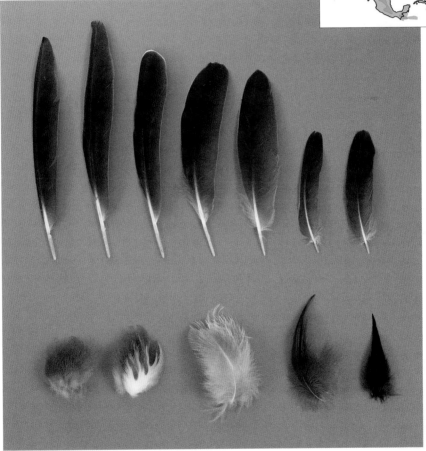

Yellow-crowned Night-Heron *(Nyctanassa violacea)*

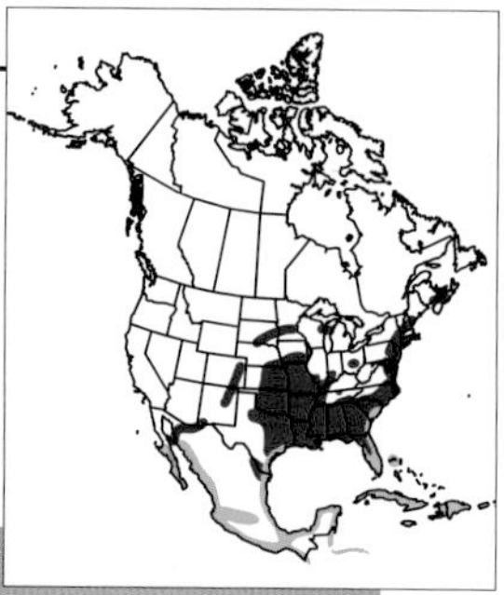

unknown, breeding
P: 8.5 (21.7)
P: 9.5 (24.2)
P: 8.7 (22.2)
P: 7.4 (18.8)
S: 7.4 (18.8)
TS: 7.1 (18.0)
T: 4.6 (11.8)
T: 4.5 (11.4)
CR: 4.1 (10.5)
WC: 3.8 (9.7)
BR: 2.8 (7.0)
MA: 2.6 (6.7)
plume: 8.0 (20.4)

Ibises and Spoonbills *(Threskiornithidae)*

White Ibis *(Eudocimus albus)*

M>F

unknown
P: 7.4 (18.8)
P: 8.3 (21.2)
P: 8.7 (22.2)
P: 8.9 (22.6)
P: 7.2 (18.2)
S: 6.6 (16.8)
S: 6.4 (16.2)
T: 4.4 (11.2)
T: 4.6 (11.7)

Glossy Ibis *(Plegadis falcinellus)*

M>F

unknown, breeding, flight and body feathers iridescent green

P: 8.2 (20.8)
P: 8.8 (22.4)
P: 8.4 (21.4)
S: 6.5 (16.4)
S: 6.4 (16.2)
T: 4.7 (12.0)
T: 4.5 (11.4)

T: 4.4 (11.3)
NA: 1.1 (2.7)
BR: 2.2 (5.5)
BE: 2.3 (5.9)
MA: 1.7 (4.4)
RU: 2.0 (5.2)
WC: 1.9 (4.8)

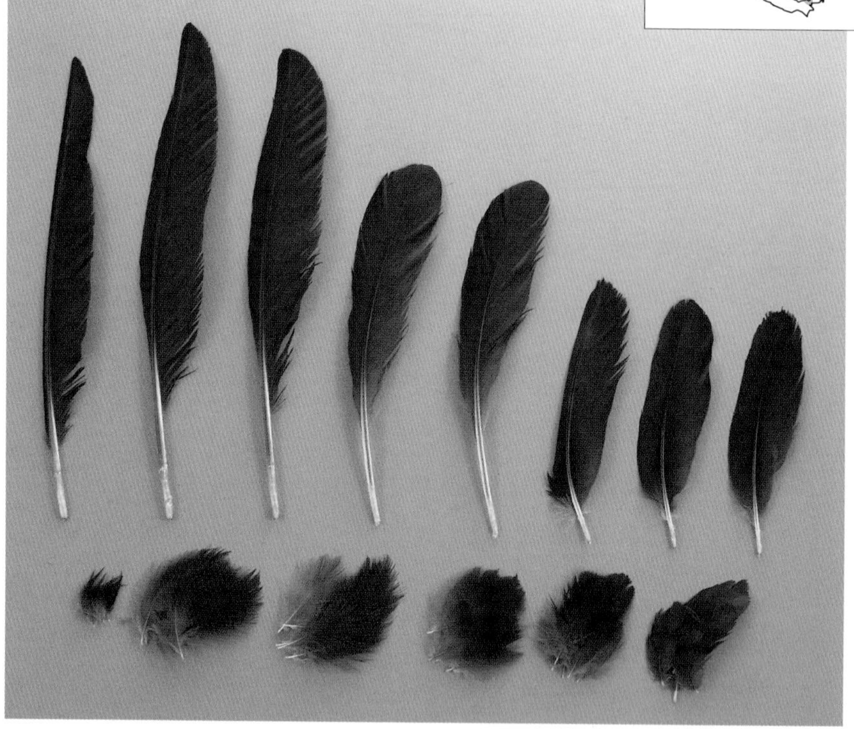

Storks (Ciconiidae)

Camber is always present to varying degrees. Long, narrow, heavily emarginated leading primaries, similar to those of vultures, cranes, eagles, and pelicans.

Wood Stork *(Mycteria americana)*

M>F

male, iridescent green in flight feathers	**P:** 12.0 (30.6)	**P:** 11.4 (28.9)
	P: 15.4 (39.0)	**S:** 10.8 (27.4)
	P: 14.2 (36.0)	**S:** 10.5 (26.6)

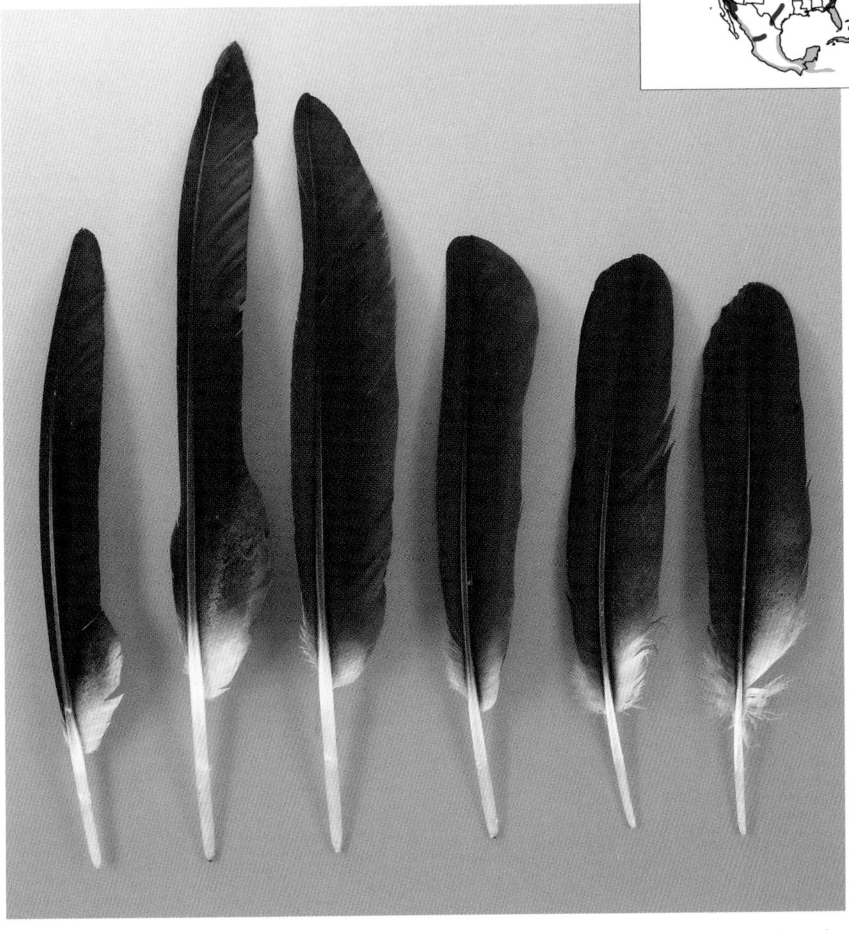

continued on page 130

Wood Stork continued

male, iridescent green and purple in tail feathers	**T:** 7.2 (18.4) **T:** 7.0 (17.7) **T:** 7.0 (17.8)

New World Vultures *(Cathartidae)*

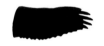

Camber is always present to varying degrees. Long, narrow, heavily emarginated leading primaries, similar to those of cranes, eagles, pelicans, and storks. Black Vulture primary feather shafts are white compared to the dingy brown color of Turkey Vulture primary feather shafts.

Black Vulture *(Coragyps atratus)*

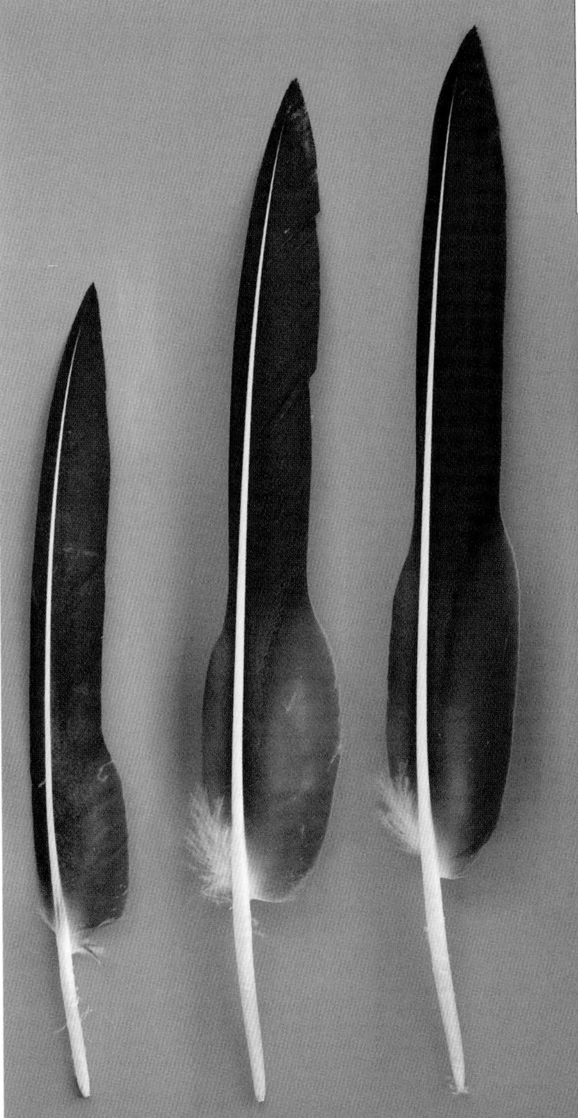

unknown
P: 10.8 (27.5)
P: 13.7 (34.9)
P: 14.3 (36.2)

continued on page 132

Black Vulture continued

S: 10.6 (26.9)
S: 10.6 (26.8)

T: 8.1 (20.6)
T: 8.1 (20.6)
T: 8.4 (21.4)
NA: 1.7 (4.4)
BR: 2.7 (6.9)
BE: 4.0 (10.1)
MA: 2.4 (6.2)

Turkey Vulture *(Cathartes aura)*

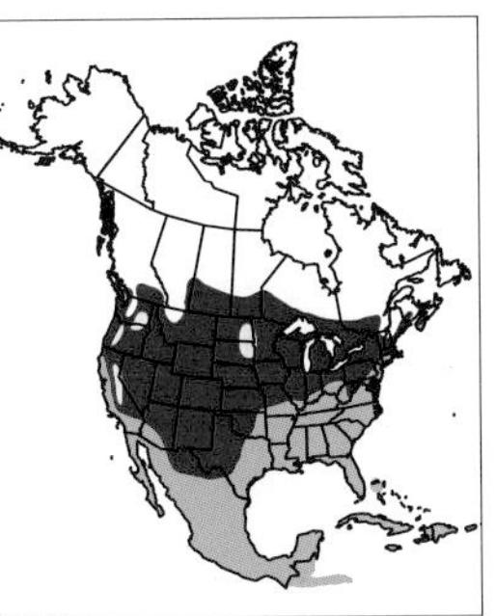

unknown
P: 13.5 (34.2)
P: 17.8 (45.3)
P: 17.4 (44.3)

S: 11.3 (28.7)
S: 11.3 (28.7)
TS: 10.3 (26.1)

continued on page 134

Turkey Vulture continued

T: 11.3 (28.6)
T: 11.8 (30.0)
T: 11.8 (29.6)
NA: 1.7 (4.4)
BR: 3.1 (8.0)
BE: 5.5 (13.9)
RU: 4.2 (10.7)

California Condor *(Gymnogyps californianus)*

unknown
P: 19.0 (48.3)
P: 24.8 (62.9)
P: 24.8 (62.9)
P: 21.7 (55.2)
P: 18.5 (47.0)
S: 16.5 (41.9)
TS: 17.0 (43.2)
T: 14.8 (37.5)
T: 14.0 (35.6)

Hawks, Kites, Eagles, and Allies *(Accipitridae)*

Camber is always present to varying degrees. Heavily emarginated leading primaries. Distinct barring is often present in flight feathers.

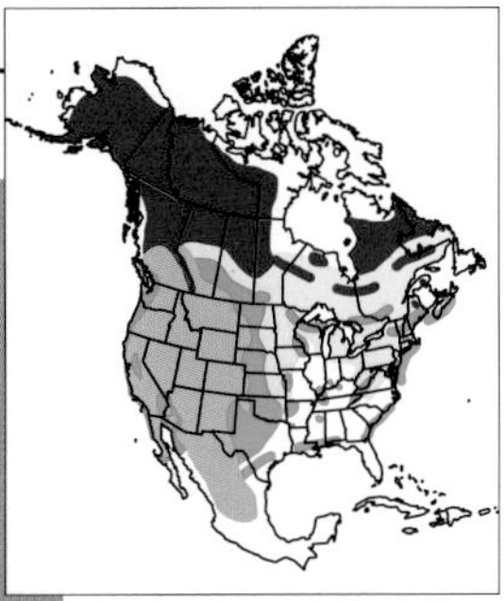

Golden Eagle *(Aquila chrysaetos)*

F>M

male
P: 18.0 (45.7)
P: 19.5 (49.6)
P: 16.0 (40.6)

continued on page 136

Golden Eagle continued

S: 12.8 (32.6)
S: 11.9 (30.1)
T: 13.9 (35.4)
T: 14.3 (36.2)

Bald Eagle *(Haliaeetus leucocephalus)*

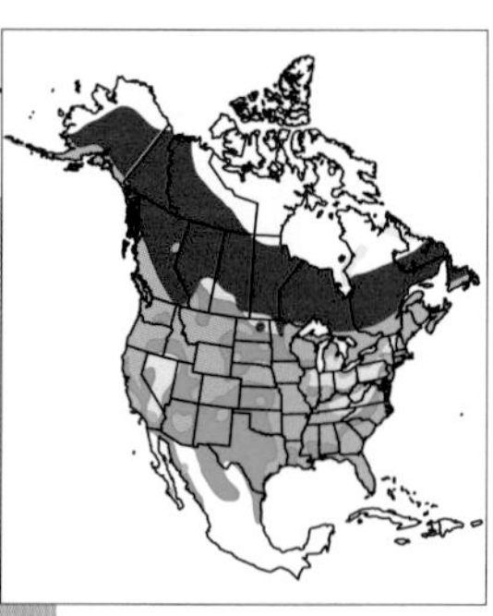

F>M

female
P: 16.5 (41.8)
P: 20.7 (52.6)
P: 18.8 (47.7)
P: 14.6 (37.0)

S: 14.1 (35.9)
S: 12.5 (31.7)
T: 13.0 (33.0)
T: 13.0 (32.9)
MA: 3.9 (9.8)
BR: 5.1 (12.9)
NA: 3.9 (9.8)
WC: 8.7 (22.2)

Osprey *(Pandion haliaetus)*

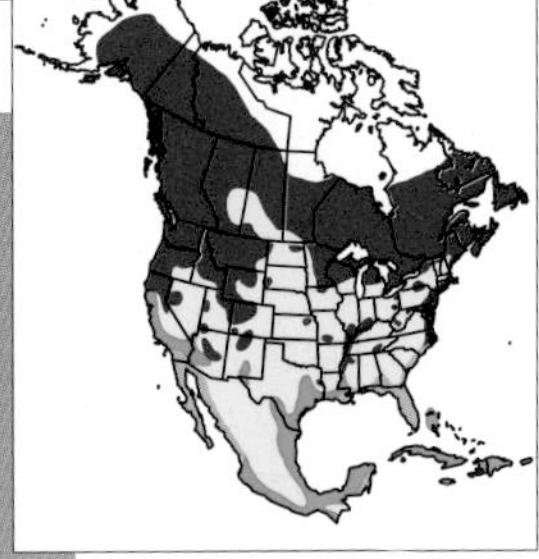

unknown
P: 12.4 (31.6)
P: 15.4 (39.1)
P: 13.8 (35.1)
S: 8.3 (21.0)
S: 8.2 (20.8)

continued on page 138

Osprey continued

T: 9.2 (23.3)
T: 9.0 (22.8)
T: 9.1 (23.2)
BR: 1.9 (4.9)
BE: 3.3 (8.5)
RU: 2.7 (6.8)
MA: 2.6 (6.7)

Swallow-tailed Kite *(Elanoides forficatus)*

unknown
P: 10.0 (25.4)
P: 12.6 (32.1)
P: 13.3 (33.9)
P: 10.9 (27.7)
P: 8.5 (21.6)
S: 5.5 (14.0)
S: 5.1 (13.0)

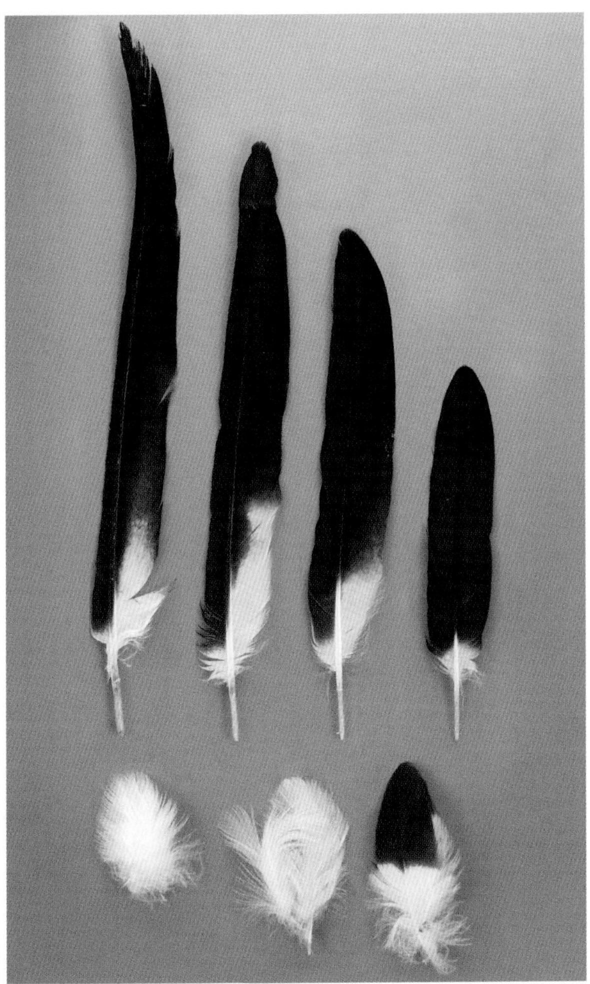

T: 11.4 (28.9)
T: 9.6 (24.4)
T: 8.3 (21.0)
T: 6.0 (15.3)
BR: 2.2 (5.5)
BE: 3.0 (7.7)
MA: 3.6 (9.1)

White-tailed Kite *(Elanus leucurus)*

unknown
P: 7.5 (19.0)
P: 7.1 (18.1)
P: 6.6 (16.7)
P: 5.7 (14.5)

S: 5.6 (14.2)
TS: 4.6 (11.6)
[no leading primaries]

Snail Kite *(Rostrhamus sociabilis)*

unknown
P: 8.1 (20.5)
P: 8.0 (20.2)
P: 7.4 (18.7)

T: 7.9 (20.0)
T: 7.4 (18.8)
T: 7.2 (18.3)

Mississippi Kite *(Ictinia mississippiensis)*

unknown
P: 6.6 (16.7)
P: 8.3 (21.1)
P: 9.0 (23.0)
P: 6.0 (15.3)

S: 5.1 (13.0)
TS: 4.3 (11.0)
T: 6.6 (16.7)
T: 6.5 (16.5)
T: 6.4 (16.3)

NA: 1.0 (2.6)
BR: 2.1 (5.4)
BE: 2.8 (7.0)
MA: 2.6 (6.7)

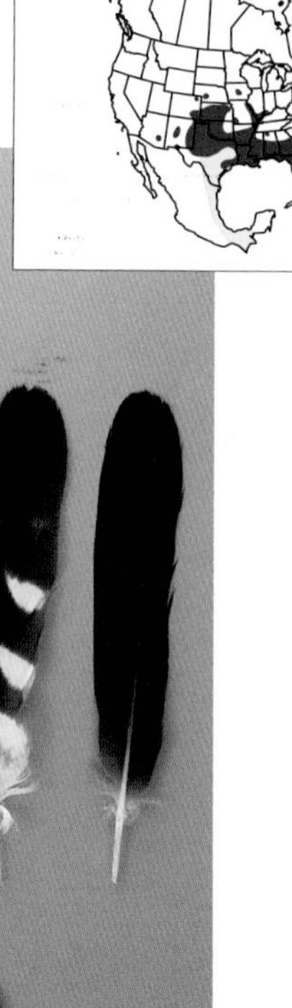

Northern Harrier *(Circus cyaneus)*

F>M

female
P: 8.1 (20.6)
P: 10.8 (27.4)
P: 11.5 (29.2)
P: 8.9 (22.7)
S: 7.4 (18.9)
S: 6.4 (16.2)
T: 10.0 (25.4)
T: 9.7 (24.7)
MA: 2.0 (5.2)
BR: 2.0 (5.1)
BE: 3.7 (9.4)

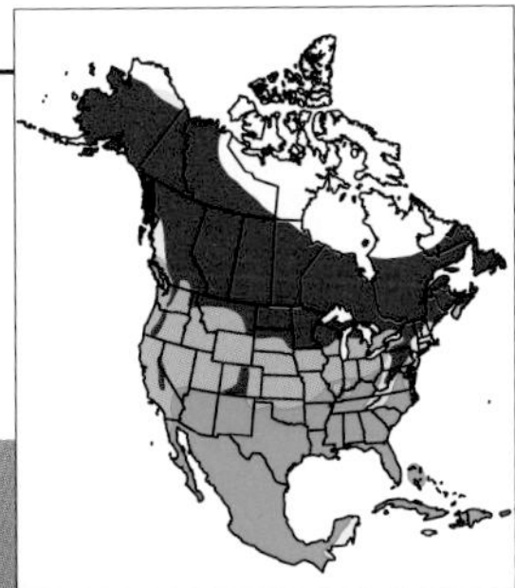

male
P: 7.8 (19.7)
P: 7.4 (18.7)
P: 7.0 (17.9)
S: 6.9 (17.5)

Sharp-shinned Hawk *(Accipiter striatus)*

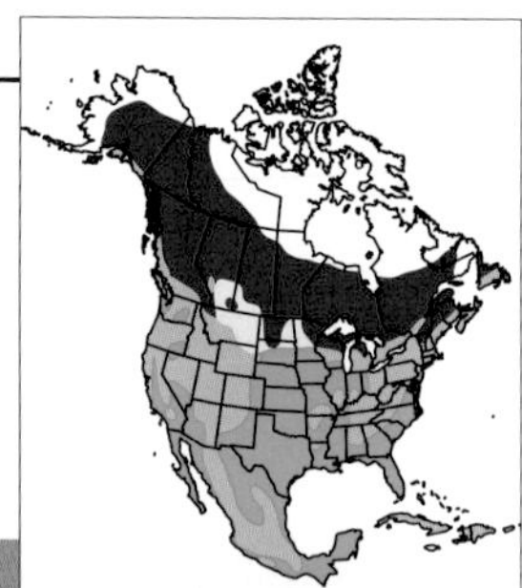

F>M

female	
P: 3.8 (9.6)	**T:** 6.7 (17.1)
P: 6.2 (15.8)	**T:** 6.8 (17.3)
P: 6.5 (16.5)	**T:** 6.5 (16.6)
P: 5.4 (13.6)	**FL:** 3.1 (7.8)
S: 5.0 (12.8)	**RU:** 1.9 (4.9)
TS: 3.5 (9.0)	**BE:** 2.6 (6.5)
	BR: 1.8 (4.5)

Cooper's Hawk *(Accipiter cooperii)*

F>M

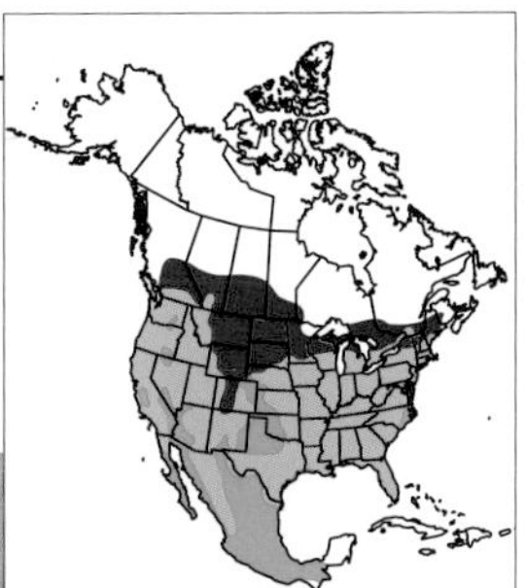

female
P: 4.2 (10.6)
P: 7.1 (18.0)
P: 7.4 (18.9)
P: 6.2 (15.7)
S: 5.6 (14.2)
S: 5.2 (13.3)
T: 7.3 (18.6)
T: 7.7 (19.6)
T: 7.7 (19.6)
NA: 2.0 (5.2)
BR: 2.2 (5.5)
BE: 3.0 (7.6)

Northern Goshawk *(Accipiter gentilis)*

F>M

female	**S:** 7.7 (19.5)	**NA:** 2.0 (5.1)
P: 10.3 (26.1)	**TS:** 5.6 (14.3)	**BR:** 2.0 (5.1)
P: 10.7 (27.1)	**T:** 9.4 (23.8)	**BE:** 4.0 (10.1)
P: 8.7 (22.2)	**T:** 9.9 (25.1)	**FL:** 4.8 (12.2)

Harris's Hawk *(Parabuteo unicinctus)*

F>M

male	**P:** 6.7 (16.9)	**S:** 8.1 (20.5)
primaries and	**P:** 10.6 (26.9)	**TS:** 6.6 (16.8)
secondaries,	**P:** 11.2 (28.4)	**T:** 11.3 (28.8)
female tail	**P:** 11.0 (27.9)	**T:** 10.8 (27.4)
feathers	**P:** 9.5 (24.1)	

Red-shouldered Hawk *(Buteo lineatus)*

F>M

female
P: 5.9 (15.0)
P: 8.9 (22.5)
P: 10.9 (27.6)
P: 9.0 (22.8)
S: 7.3 (18.5)
S: 6.7 (17.1)

T: 8.1 (20.6)
T: 8.1 (20.6)
T: 7.9 (20.0)
NA: 0.8 (2.0)
BR: 3.0 (7.6)
BE: 4.8 (12.1)
MA: 3.1 (8.0)

Broad-winged Hawk *(Buteo platypterus)*

F>M

male	**S:** 5.1 (13.0)	**NA:** 1.9 (4.7)
P: 5.0 (12.6)	**S:** 5.3 (13.4)	**BR:** 1.5 (3.9)
P: 8.1 (20.5)	**T:** 6.1 (15.6)	**BE:** 3.2 (8.2)
P: 7.1 (18.0)	**T:** 6.0 (15.3)	**WC:** 3.3 (8.3)
P: 6.2 (15.7)	**T:** 6.1 (15.6)	

Short-tailed Hawk *(Buteo brachyurus)*

F>M

unknown	**S:** 6.5 (16.5)	**MA:** 2.4 (6.2)
P: 6.1 (15.6)	**TS:** 5.4 (13.6)	**BR:** 2.2 (5.6)
P: 10.0 (25.3)	**T:** 6.7 (17.0)	**BE:** 3.8 (9.6)
P: 9.1 (23.1)	**T:** 7.0 (17.8)	
P: 7.8 (19.9)	**T:** 7.3 (18.6)	

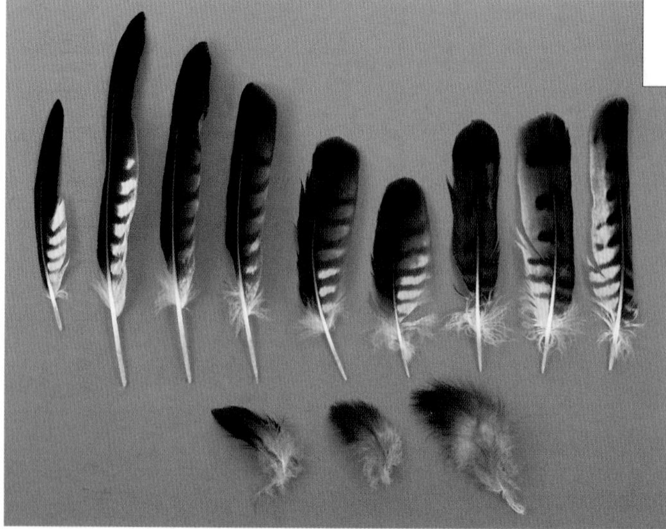

Swainson's Hawk *(Buteo swainsoni)*

F>M

unknown
P: 12.3 (31.2)
P: 14.2 (36.0)
P: 12.2 (31.0)
P: 10.0 (25.4)
S: 8.4 (21.3)
TS: 7.3 (18.5)

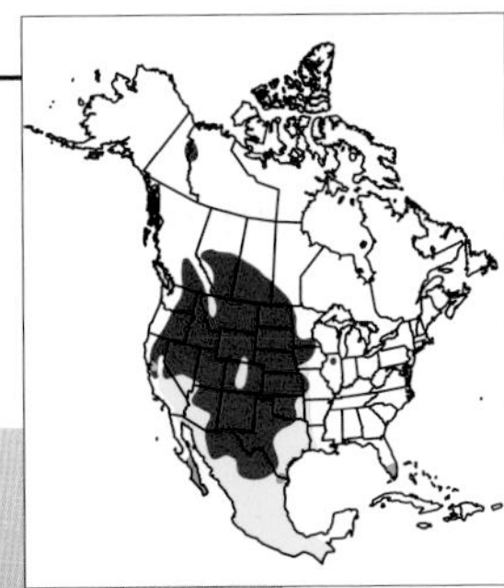

T: 9.3 (23.5)
T: 9.3 (23.6)
T: 9.1 (23.2)
UPTC: 3.5 (8.9)
MA: 2.3 (5.9)
BR: 2.4 (6.1)
FL: 3.9 (10.0)

Red-tailed Hawk *(Buteo jamaicensis)*

F>M

female
P: 8.7 (22.1)
P: 11.4 (29.0)
P: 12.9 (32.8)
P: 13.7 (34.7)
P: 10.9 (27.8)
S: 9.0 (22.8)
S: 9.3 (23.5)

T: 10.0 (25.3)
T: 10.0 (25.5)
T: 10.1 (25.7)
TS: 7.5 (19.0)
NA: 1.1 (2.9)
BR: 3.1 (8.0)
BE: 5.0 (12.6)
MA: 3.3 (8.3)

Ferruginous Hawk *(Buteo regalis)*

F>M

unknown
P: 8.7 (22.0)
P: 13.5 (34.2)
P: 13.9 (35.4)
P: 10.9 (27.8)
S: 9.2 (23.3)
TS: 7.7 (19.6)
TS: 6.4 (16.3)

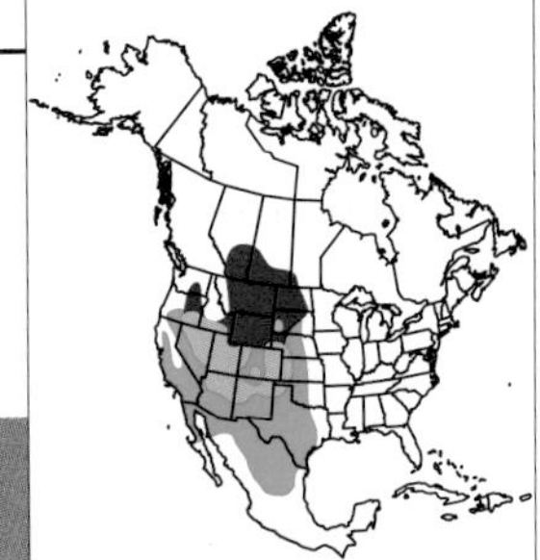

T: 9.8 (25.0)
T: 9.4 (24.0)
BR: 2.4 (6.2)
BE: 3.6 (9.2)
FL: 5.2 (13.2)
NA: 1.7 (4.2)
MA: 3.3 (8.5)
MC: 3.9 (10.0)

Rough-legged Hawk *(Buteo lagopus)*

F>M

unknown
P: 8.9 (22.7)
P: 13.8 (35.0)
P: 12.4 (31.5)
P: 10.6 (26.8)
S: 8.5 (21.5)
S: 7.6 (19.3)

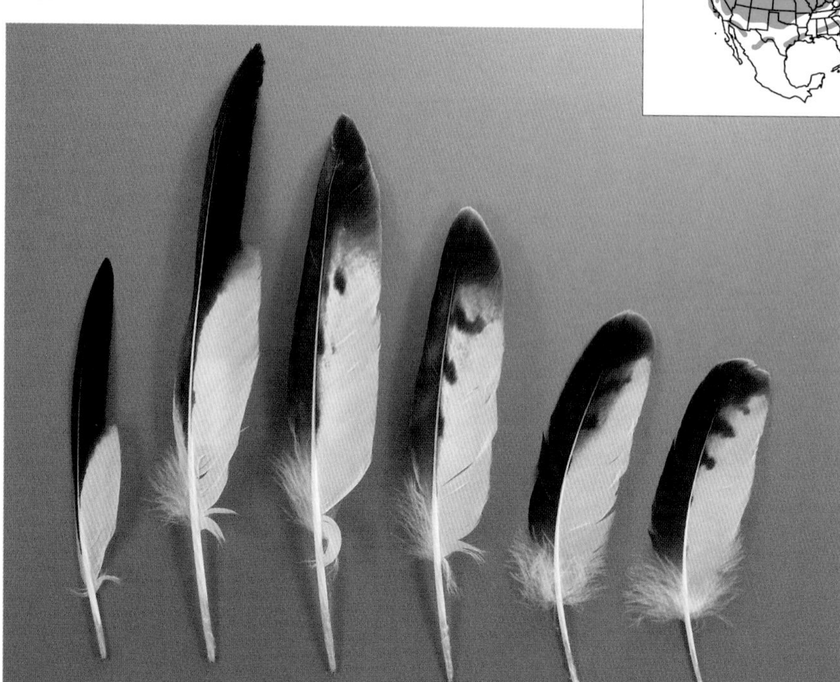

T: 9.4 (24.0)
T: 9.7 (24.6)
CR: 2.5 (6.4)
RU: 4.8 (12.2)
MA: 3.5 (9.0)

Caracaras and Falcons *(Falconidae)*

Emargination is slight and occurs primarily on the upper trailing vane of the first two to three leading primaries (exception: Crested Caracara). Barring usually appears as elongated spots on trailing vane as opposed to stripes (exception: Crested Caracara).

Crested Caracara *(Caracara cheriway)*

unknown	**P:** 11.9 (30.1)
P: 8.5 (21.7)	**P:** 8.6 (21.8)
P: 11.2 (28.5)	**S:** 7.7 (19.5)
P: 12.2 (31.1)	**S:** 7.1 (18.0)

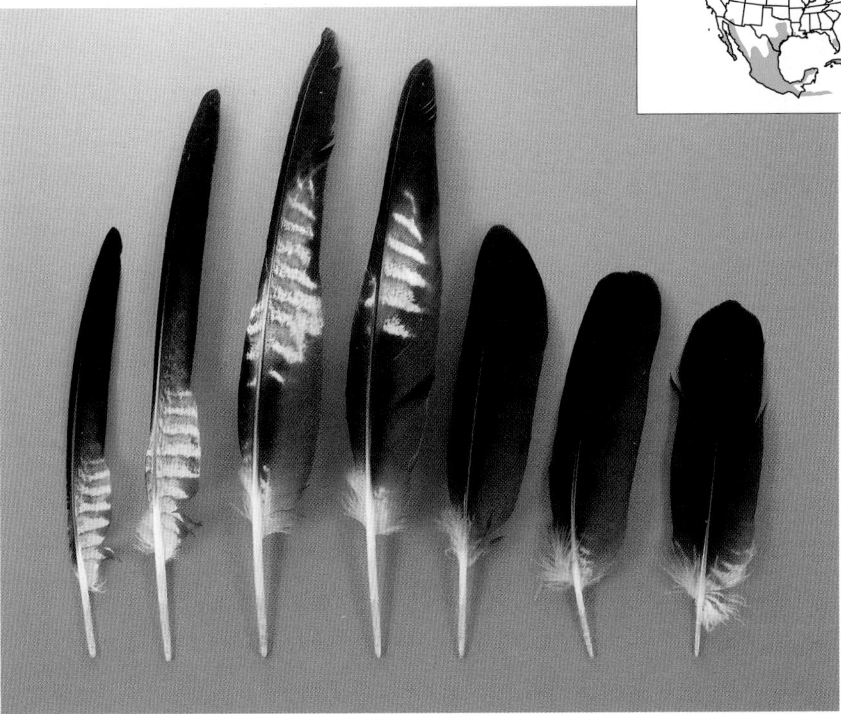

continued on page 152

Crested Caracara continued

T: 8.5 (21.5)
T: 8.4 (21.3)
T: 8.5 (21.5)
NA: 1.6 (4.0)
BR: 2.4 (6.0)
BE: 2.9 (7.3)
MA: 2.8 (7.0)

American Kestrel *(Falco sparverius)*

F>M

male
P: 5.2 (13.1)
P: 5.9 (14.9)
P: 4.6 (11.7)
P: 3.7 (9.3)
S: 3.1 (8.0)
S: 3.1 (8.0)
T: 4.8 (12.3)
T: 4.9 (12.4)
T: 4.7 (11.9)
T: 4.3 (10.8)
BR: 2.0 (5.2)
BE: 1.6 (4.0)
MA: 1.9 (4.7)
UPTC: 1.5 (3.7)
MC: 1.9 (4.7)

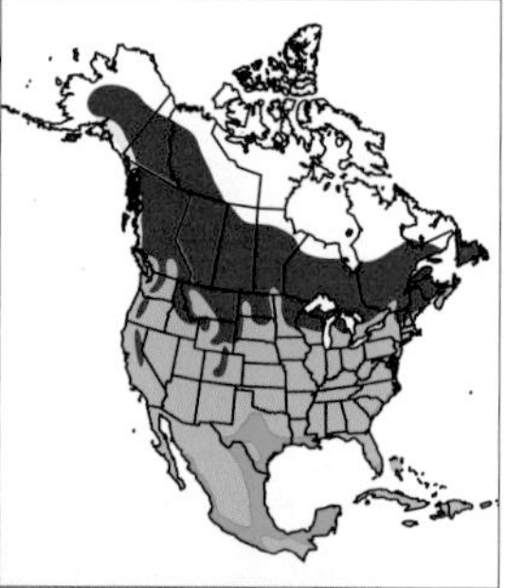

female
P: 6.2 (15.8)
P: 5.7 (14.6)
P: 5.0 (12.7)
P: 4.3 (10.9)
S: 3.3 (8.5)
TS: 3.3 (8.5)
T: 4.6 (11.7)
T: 5.1 (12.9)
T: 5.0 (12.6)
BR: 1.1 (2.8)
MA: 1.7 (4.4)
RU: 1.3 (3.2)
WC: 1.6 (4.1)

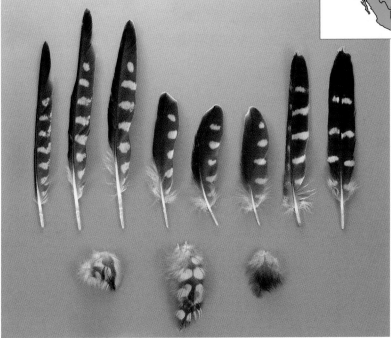

Merlin *(Falco columbarius)*

F>M

female
P: 5.7 (14.5)
P: 6.8 (17.2)
P: 6.4 (16.2)
P: 4.1 (10.5)
S: 3.7 (9.5)
TS: 3.6 (9.2)
T: 5.4 (13.6)
T: 5.6 (14.2)
BR: 1.0 (2.6)
BE: 2.4 (6.0)
RU: 1.2 (3.0)

Aplomado Falcon *(Falco femoralis)*

F>M

unknown
P: 6.7 (17.1)
P: 8.1 (20.5)
P: 8.0 (20.3)
P: 6.3 (16.1)
P: 4.9 (12.5)

S: 4.6 (11.7)
S: 4.2 (10.7)
T: 8.5 (21.5)
T: 8.5 (21.5)
T: 7.9 (20.0)
BR: 2.3 (5.8)

BE: 2.0 (5.0)
MA: 2.0 (5.0)
UPTC: 1.6 (4.0)
FL: 1.7 (4.4)

very rare in the United States, occurring only in desert and coastal grasslands in western Texas and southern New Mexico

Peregrine Falcon *(Falco peregrinus)*

F>M

juvenile, molting to adult plumage (note tail)
P: 9.0 (23.0)
P: 9.6 (24.4)
P: 8.7 (22.0)
P: 6.8 (17.3)
S: 5.0 (12.8)
TS: 4.6 (11.6)
T: 6.6 (16.8)
T: 6.5 (16.6)
T: 6.6 (16.8)
MA: 2.1 (5.4)
FL: 2.6 (6.5)
SC: 3.4 (8.6)
UPTC: 3.3 (8.4)
alula: 3.5 (8.9)

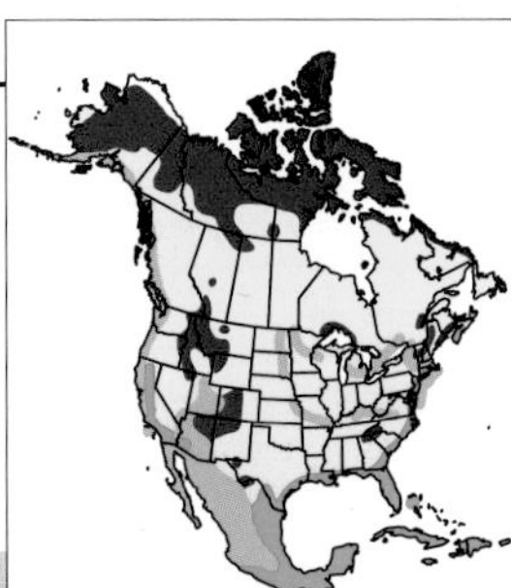

Prairie Falcon *(Falco mexicanus)*

F>M

unknown
P: 9.4 (23.9)
P: 11.1 (28.2)
P: 10.5 (26.6)
P: 7.0 (17.9)

S: 6.0 (15.2)
TS: 5.8 (14.8)
T: 7.7 (19.6)
T: 8.3 (21.0)

BR: 2.7 (6.8)
BE: 1.5 (3.9)
RU: 2.0 (5.0)
FL: 2.3 (5.8)

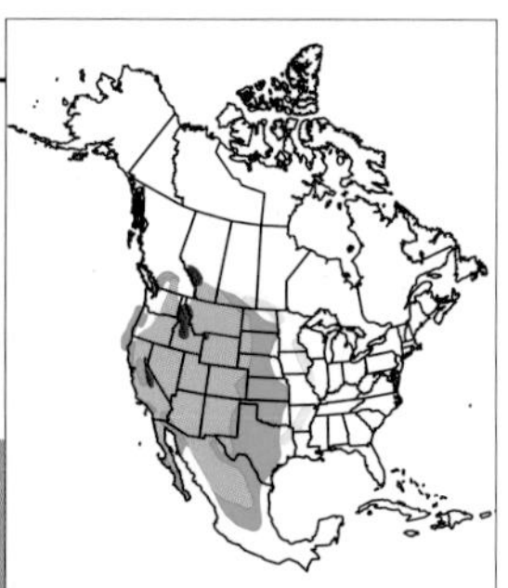

Rails, Gallinules, and Coots *(Rallidae)*

Short, compact primaries and secondaries. Secondaries are relatively wide and evenly vaned.

Clapper Rail *(Rallus longirostris)*

(Atlantic)
M>F

unknown
P: 3.9 (10.0)
P: 5.1 (13.0)
P: 5.0 (12.8)
P: 4.8 (12.3)

S: 4.4 (11.3)
S: 4.3 (10.8)
T: 2.8 (7.0)
T: 2.6 (6.5)
UPTC: 2.3 (5.9)

BR: 1.3 (3.2)
BE: 1.5 (3.7)
RU: 2.3 (5.8)
MA: 1.5 (3.7)
FL: 2.5 (6.3)

Virginia Rail *(Rallus limicola)*

■ ■ ■ ■

M>F

unknown
P: 3.2 (8.1)
P: 3.3 (8.3)
S: 2.8 (7.0)

S: 2.8 (7.0)
T: 1.8 (4.6)
T: 1.9 (4.9)
T: 1.7 (4.4)

BR: 1.2 (3.1)
BE: 1.8 (4.6)
MA: 1.7 (4.2)

Sora *(Porzana carolina)*

■ ■ ■ ■

M>F

unknown
P: 2.4 (6.2)
P: 2.7 (6.8)
P: 3.3 (8.5)
P: 3.0 (7.6)

S: 2.5 (6.4)
S: 2.7 (6.8)
T: 1.5 (3.9)
T: 2.1 (5.3)

TS: 2.8 (7.0)
NA: 0.8 (2.0)
BR: 0.7 (1.9)
FL: 1.9 (4.9)
MA: 1.5 (3.8)

Purple Gallinule *(Porphyrio martinica)*

■ ■ ■ ■

M>F

unknown
P: 4.8 (12.2)
P: 5.7 (14.4)
P: 5.8 (14.8)
P: 4.1 (10.5)

S: 3.7 (9.4)
TS: 3.8 (9.6)
T: 2.7 (6.8)
T: 3.1 (7.9)

WC: 2.1 (5.3)
MA: 2.7 (6.8)
BR: 1.1 (2.8)
NA: 1.0 (2.5)

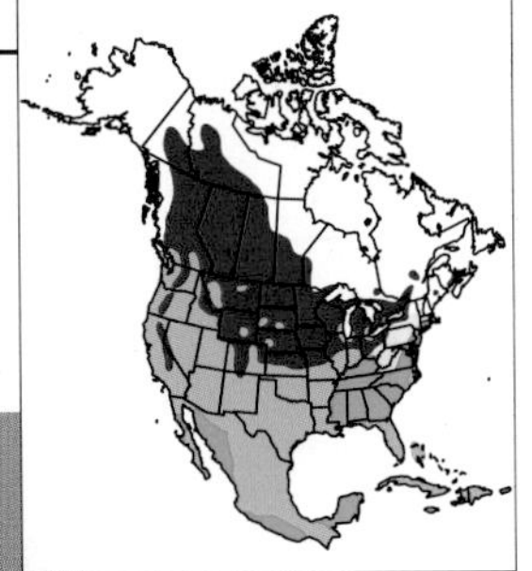

American Coot *(Fulica americana)*

■ ■ ■ ■

M>F

unknown
P: 4.4 (11.3)
P: 5.4 (13.6)
P: 4.9 (12.5)
S: 3.9 (10.0)

S: 4.0 (10.2)
TS: 4.9 (12.5)
T: 2.4 (6.0)
T: 2.3 (5.8)

BR: 1.3 (3.4)
BE: 1.5 (3.9)
NA: 0.8 (2.1)
UNTC: 2.1 (5.3)

Cranes *(Gruidae)*

Camber is always present to varying degrees. Long, narrow, heavily emarginated leading primaries, similar to those of vultures, eagles, pelicans, and storks.

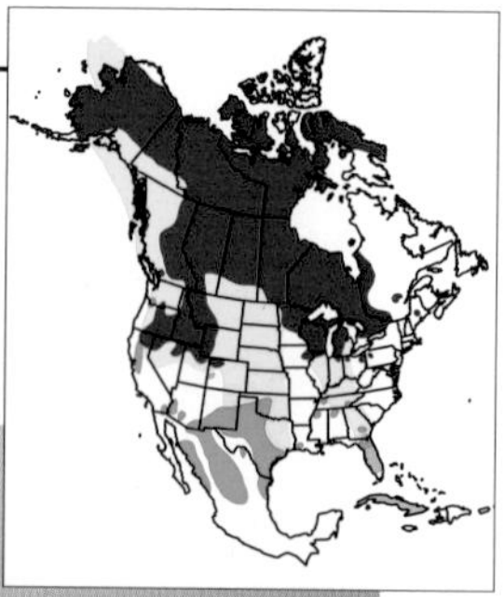

Sandhill Crane *(Grus canadensis)*

M>F

unknown,
breeding
P: 15.0 (38.2)
P: 15.6 (40.3)
P: 13.6 (34.5)
P: 11.8 (30.0)
S: 11.6 (29.4)
S: 11.9 (30.1)

T: 8.0 (20.4)
T: 7.9 (20.1)
BR: 2.3 (5.8)
NE: 2.4 (6.1)
MA: 6.5 (16.4)
RU: 2.4 (6.0)

Lapwings and Plovers *(Charadriidae)*

Black-bellied Plover *(Pluvialis squatarola)*

male, breeding	**P:** 3.5 (8.9)	**T:** 3.4 (8.7)
	S: 2.8 (7.1)	**BR:** 1.1 (2.9)
P: 6.0 (15.2)	**TS:** 4.1 (10.4)	**BE:** 1.4 (3.6)
P: 5.8 (14.7)	**T:** 3.3 (8.4)	**MA:** 1.3 (3.4)

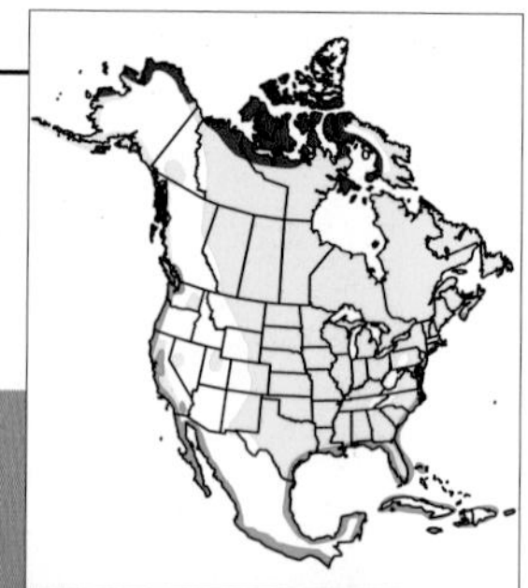

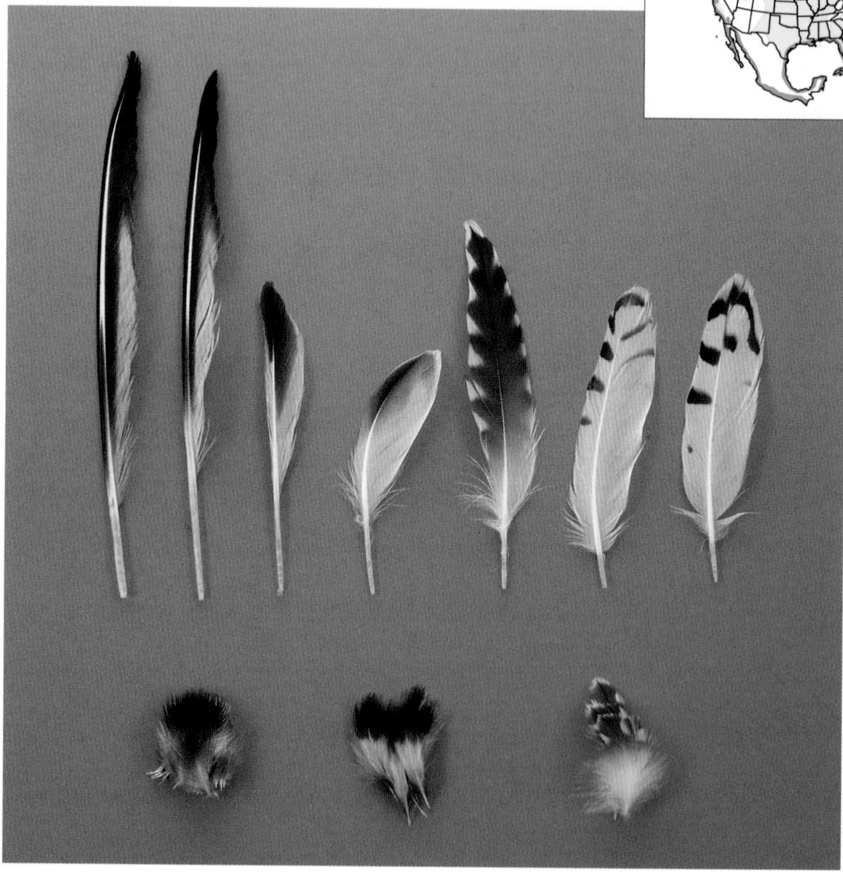

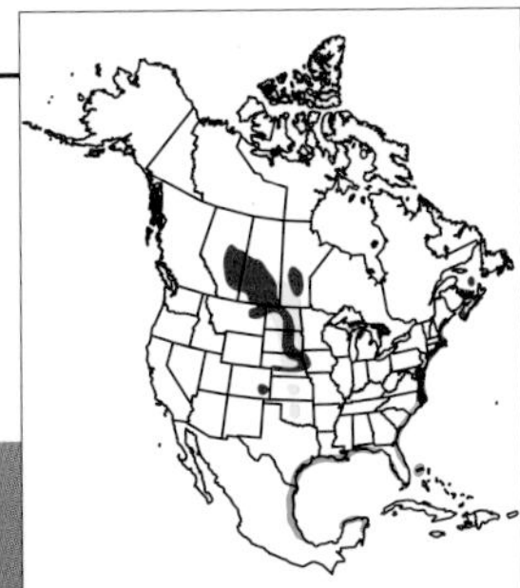

Piping Plover *(Charadrius melodus)*

unknown
P: 3.7 (9.3)
P: 3.5 (8.8)
P: 2.9 (7.4)
TS: 2.7 (6.8)
T: 2.2 (5.5)
T: 2.2 (5.7)
T: 2.2 (5.7)
T: 2.2 (5.6)

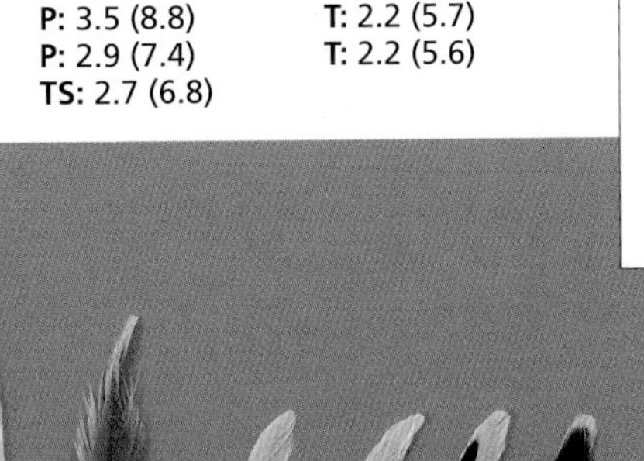

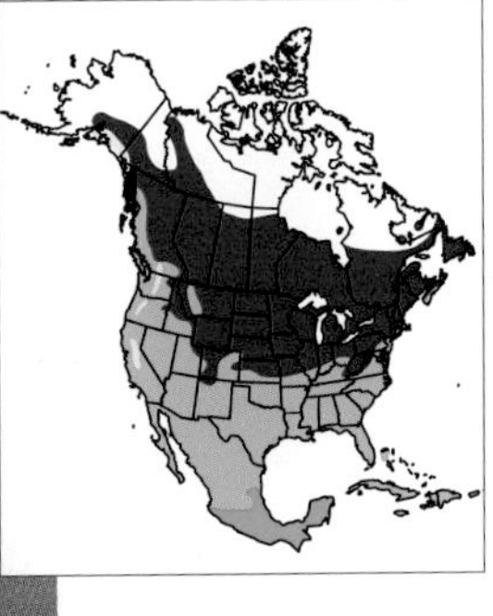

Killdeer *(Charadrius vociferus)*

F>M

unknown
P: 5.3 (13.5)
P: 5.3 (13.5)
P: 4.3 (11.0)
P: 3.9 (10.0)
P: 3.2 (8.1)
T: 4.1 (10.4)
T: 4.1 (10.4)
UPTC: 2.5 (6.4)
MA: 1.1 (2.7)
TH: 0.7 (1.9)
BR: 1.3 (3.4)

Stilts and Avocets *(Recurvirostridae)*

American Avocet *(Recurvirostra americana)*

unknown,
non-breeding
P: 6.8 (17.2)
P: 6.9 (17.6)
S: 4.3 (10.8)
S: 3.7 (9.3)

T: 3.6 (9.2)
WC: 2.6 (6.6)
BR: 1.6 (4.0)
RU: 2.3 (5.8)
FL: 3.1 (7.9)

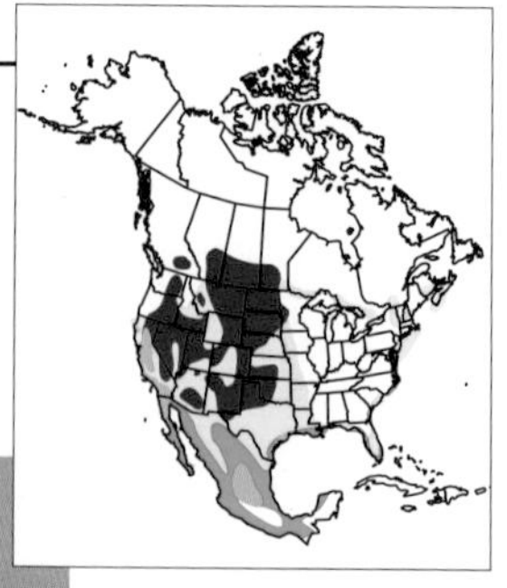

Sandpipers and Allies *(Scolopacidae)*

Greater Yellowlegs *(Tringa melanoleuca)*

unknown
P: 3.5 (8.9)
S: 3.3 (8.3)
TS: 4.1 (10.5)
T: 3.4 (8.6)

T: 3.4 (8.6)
RU: 2.0 (5.0)
NA: 1.0 (2.5)
BR: 1.0 (2.6)
UPTC: 2.1 (5.4)

Willet *(Tringa semipalmata)*

unknown, breeding	**S:** 3.6 (9.1)	**BR:** 1.3 (3.4)
P: 5.9 (15.0)	**S:** 3.5 (8.8)	**BE:** 1.3 (3.4)
P: 6.0 (15.3)	**TS:** 3.6 (9.2)	**MA:** 1.5 (3.9)
P: 5.1 (13.0)	**T:** 3.2 (8.2)	**RU:** 2.0 (5.2)
P: 3.8 (9.6)	**T:** 3.1 (7.8)	
	T: 3.1 (7.9)	

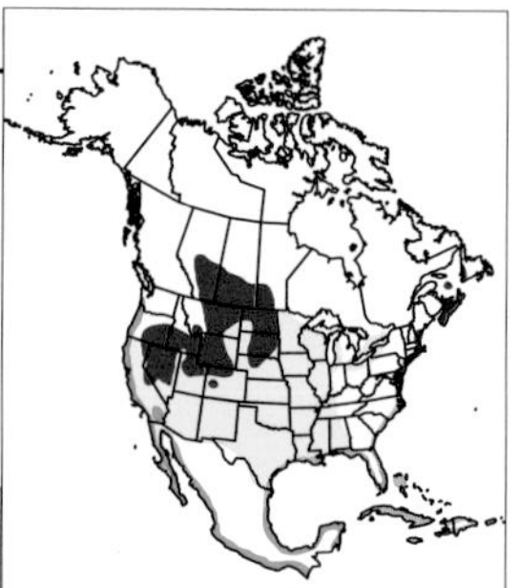

Long-billed Curlew *(Numenius americanus)*

F>M

unknown	**S:** 5.0 (12.8)	**MA:** 1.1 (2.8)
P: 8.2 (20.8)	**S:** 4.7 (11.9)	**TH:** 0.9 (2.2)
P: 7.9 (20.0)	**T:** 4.4 (11.1)	**BR:** 1.4 (3.6)
P: 6.6 (16.8)	**T:** 4.4 (11.3)	**RU:** 1.2 (3.1)
P: 5.4 (13.7)		

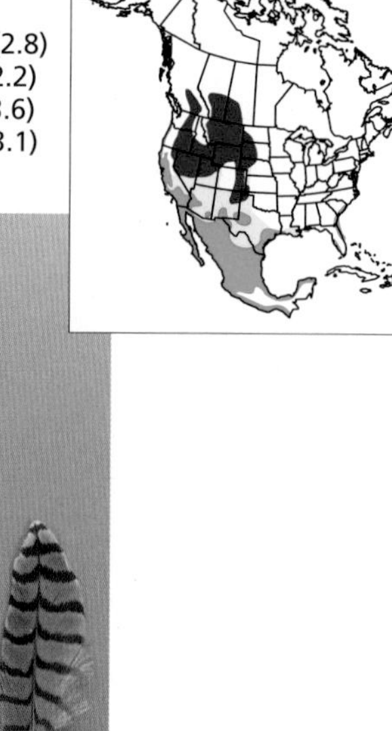

Marbled Godwit *(Limosa fedoa)*

F>M

unknown, non-breeding
P: 6.6 (16.8)
P: 5.9 (15.1)
P: 4.3 (11.0)
P: 4.0 (10.2)
S: 3.8 (9.7)
S: 3.8 (9.6)
T: 3.5 (8.9)
T: 3.4 (8.7)
T: 3.0 (7.5)
WC: 3.0 (7.6)
BR: 1.5 (3.9)
BE: 1.4 (3.6)
UPTC: 2.2 (5.5)
axillar: 3.7 (9.3)

Ruddy Turnstone *(Arenaria interpres)*

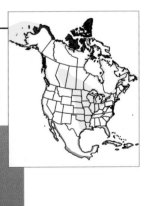

F>M

male, breeding
P: 4.8 (12.1)
P: 4.5 (11.5)
P: 4.0 (10.1)
P: 2.7 (6.9)
S: 2.3 (5.8)
S: 2.4 (6.0)
TS: 3.5 (9.0)
T: 2.6 (6.6)
T: 2.6 (6.7)
BR: 0.9 (2.2)
BE: 1.0 (2.6)
MA: 1.1 (2.9)
WC: 1.8 (4.5)

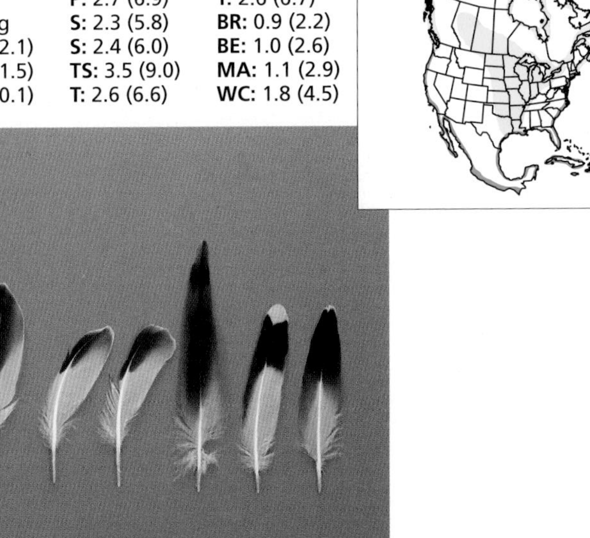

Red Knot *(Calidris canutus)*

F>M

unknown, breeding
P: 5.5 (14.0)
P: 5.0 (12.6)
P: 3.9 (9.9)
P: 3.5 (8.9)
P: 2.9 (7.3)
S: 2.7 (6.9)
S: 2.7 (6.8)
T: 2.9 (7.4)
T: 3.0 (7.7)
SC: 1.6 (4.1)
RU: 1.1 (2.8)
BE: 1.3 (3.2)
BR: 1.2 (3.1)

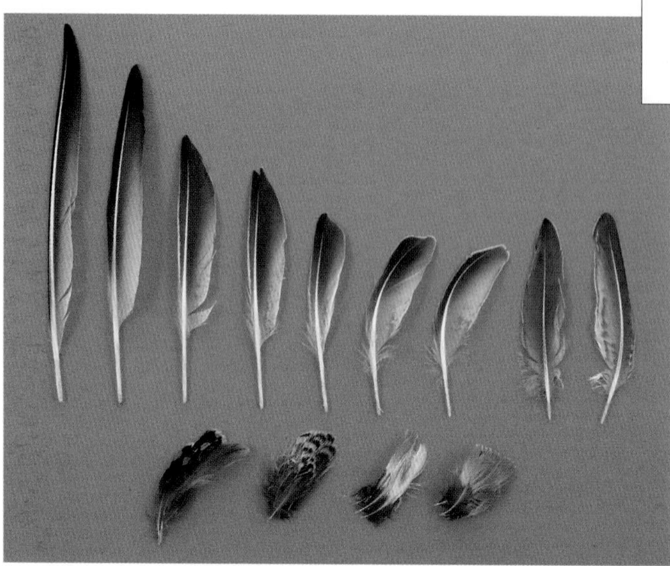

Sanderling *(Calidris alba)*

F>M

unknown, non-breeding
P: 3.2 (8.2)
P: 2.9 (7.3)
P: 2.4 (6.0)
S: 1.6 (4.1)
S: 1.6 (4.1)
T: 1.8 (4.6)
T: 1.7 (4.4)
UNTC: 1.2 (3.0)
RU: 0.9 (2.2)
MA: 0.6 (1.4)
BR: 0.8 (2.0)

Semipalmated Sandpiper *(Calidris pusilla)*

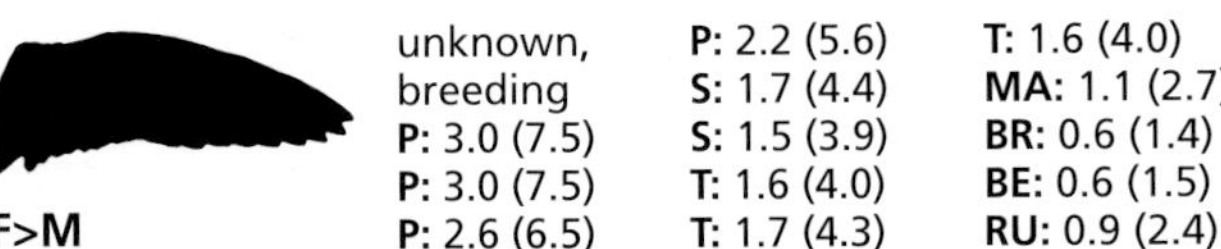

F>M

unknown, breeding

P: 3.0 (7.5)
P: 3.0 (7.5)
P: 2.6 (6.5)
P: 2.2 (5.6)
S: 1.7 (4.4)
S: 1.5 (3.9)
T: 1.6 (4.0)
T: 1.7 (4.3)
T: 1.6 (4.0)
MA: 1.1 (2.7)
BR: 0.6 (1.4)
BE: 0.6 (1.5)
RU: 0.9 (2.4)

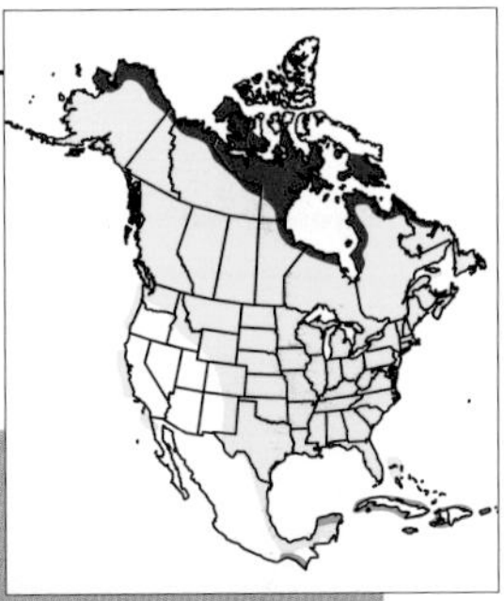

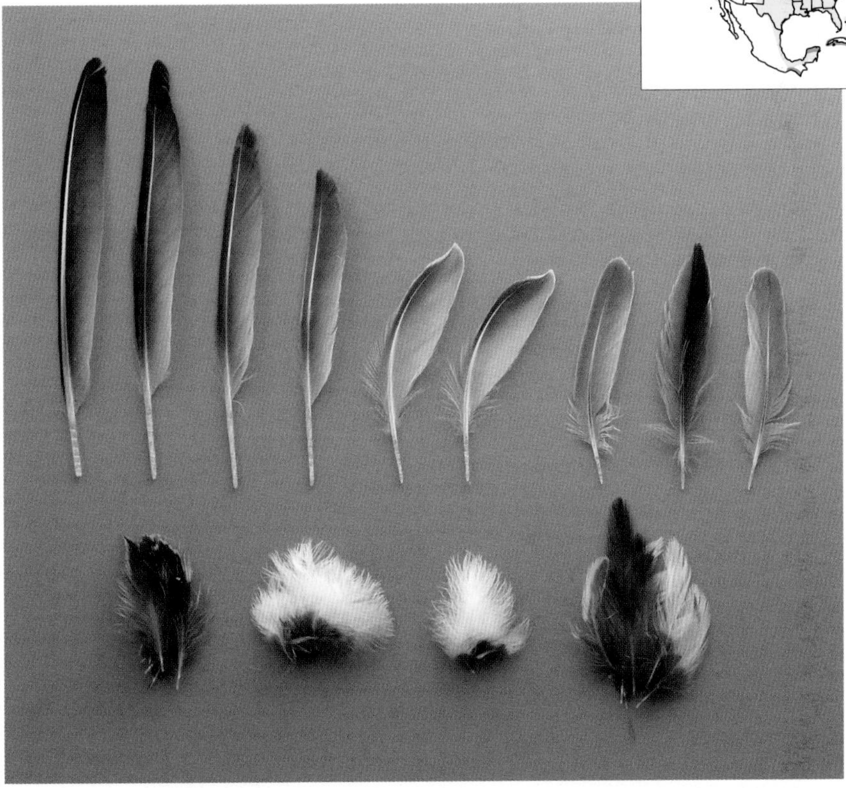

Western Sandpiper *(Calidris mauri)*

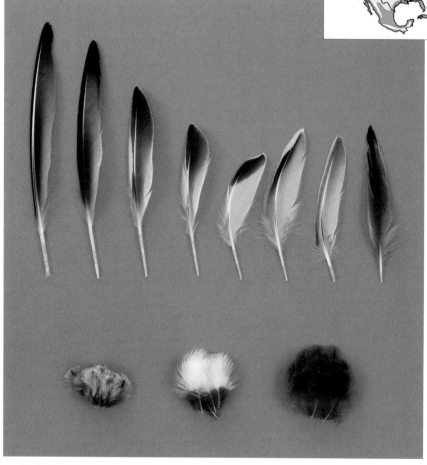

unknown, breeding
P: 3.8 (9.7)
P: 3.6 (9.2)
P: 2.9 (7.4)
P: 2.3 (5.9)
S: 1.9 (4.9)
TS: 2.3 (5.8)
T: 2.2 (5.6)
T: 2.4 (6.2)
BR: 0.6 (1.5)
BE: 0.9 (2.3)
MA: 1.0 (2.6)

Least Sandpiper *(Calidris minutilla)*

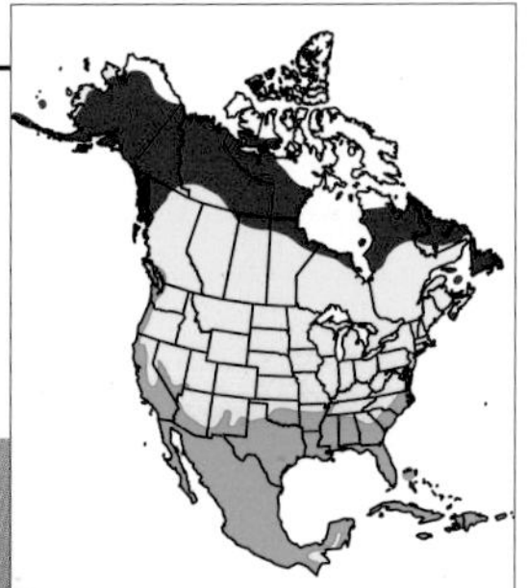

unknown	**S:** 1.6 (4.1)	**BE:** 0.6 (1.5)
P: 2.8 (7.1)	**S:** 1.6 (4.1)	**BE:** 0.7 (1.8)
P: 2.8 (7.0)	**T:** 1.6 (4.1)	**MA:** 1.1 (2.7)
P: 2.5 (6.3)	**T:** 1.7 (4.3)	**RU:** 0.8 (2.1)
P: 2.0 (5.0)	**T:** 1.7 (4.2)	

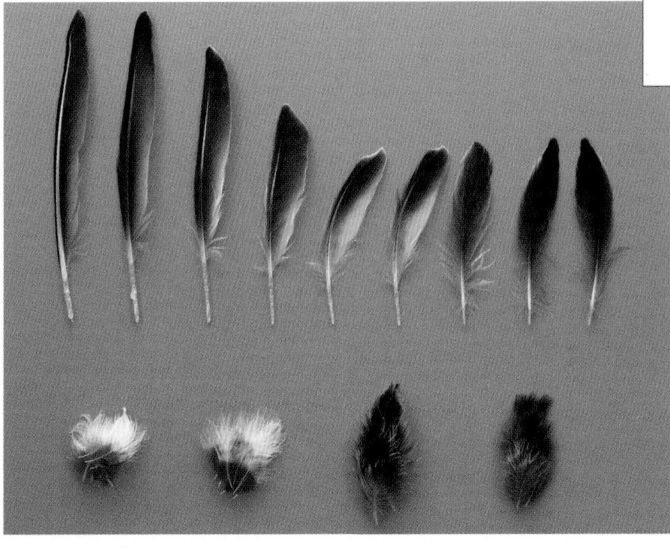

White-rumped Sandpiper *(Calidris fuscicollis)*

F>M

unknown	**S:** 2.0 (5.0)	**BR:** 0.8 (2.1)
P: 3.8 (9.6)	**S:** 2.0 (5.2)	**BE:** 0.7 (1.9)
P: 3.9 (9.9)	**T:** 2.0 (5.2)	**MA:** 1.1 (2.7)
P: 2.6 (6.6)	**T:** 2.0 (5.2)	**RU:** 1.3 (3.4)
P: 2.4 (6.0)	**T:** 2.2 (5.6)	

Dunlin *(Calidris alpina)*

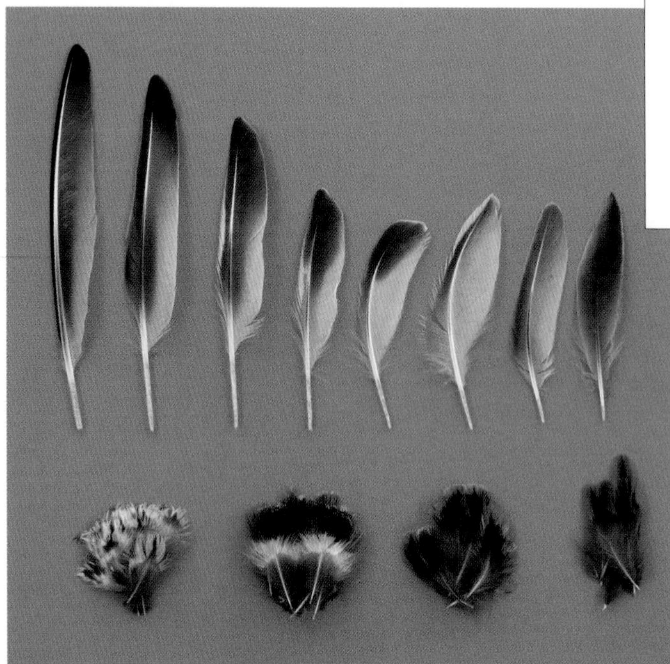

unknown, breeding
P: 3.6 (9.2)
P: 3.3 (8.5)
P: 3.0 (7.5)
P: 2.3 (5.8)
S: 2.0 (5.0)
S: 2.2 (5.6)
T: 2.1 (5.3)
T: 2.1 (5.4)
BR: 0.8 (2.0)
BE: 0.9 (2.3)
MA: 1.0 (2.6)
RU: 1.1 (2.8)

Stilt Sandpiper *(Calidris himantopus)*

F>M

unknown
P: 4.0 (10.1)
P: 4.1 (10.4)
P: 3.7 (9.3)
P: 2.8 (7.2)
S: 2.4 (6.0)
TS: 2.8 (7.2)
T: 2.2 (5.5)
T: 2.1 (5.3)
MA: 0.9 (2.2)
BR: 1.0 (2.5)
BE: 0.9 (2.4)
MA: 1.0 (2.6)

Short-billed Dowitcher *(Limnodromus griseus)*

F>M

unknown
P: 4.4 (11.1)
P: 4.2 (10.6)
P: 3.4 (8.6)
P: 2.7 (6.9)

S: 2.4 (6.0)
S: 2.2 (5.7)
T: 2.2 (5.7)
T: 2.2 (5.7)

MA: 0.9 (2.3)
BR: 0.7 (1.8)
RU: 0.9 (2.2)
SC: 2.0 (5.2)

Wilson's Snipe *(Gallinago delicata)*

F>M

unknown
P: 4.1 (10.3)
P: 3.8 (9.7)
P: 2.7 (6.9)
S: 2.6 (6.5)

S: 2.5 (6.4)
TS: 3.1 (8.0)
T: 2.5 (6.3)
T: 2.6 (6.5)

NA: 0.9 (2.3)
BR: 0.8 (2.0)
BE: 1.1 (2.7)
RU: 1.1 (2.7)

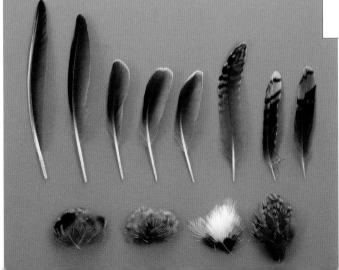

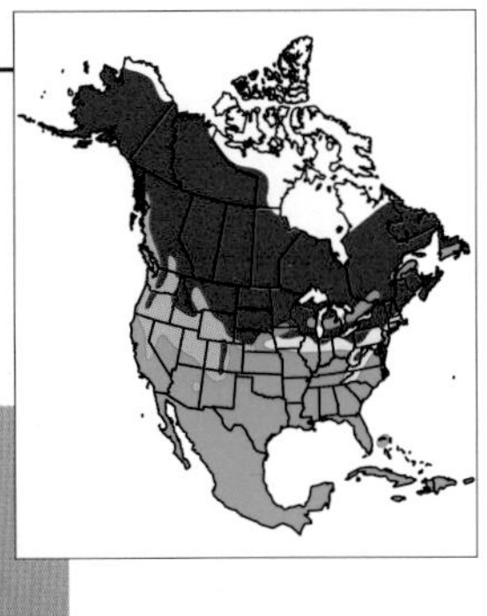

American Woodcock *(Scolopax minor)*

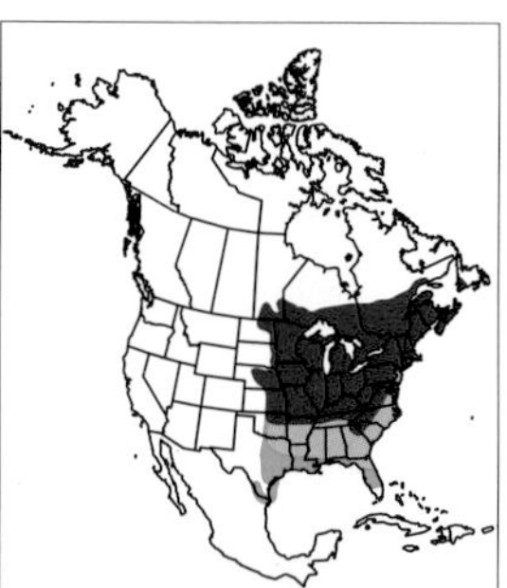

■■■■

F>M

unknown, note leading "whistling" primary
P: 3.3 (8.5)
P: 4.0 (10.2)

P: 4.1 (10.5)
S: 3.6 (9.2)
S: 3.5 (8.8)
TS: 3.3 (8.3)
T: 2.0 (5.2)
T: 2.6 (6.6)

WC: 1.9 (4.9)
BE: 1.2 (3.0)
BR: 1.0 (2.6)
MA: 1.2 (3.1)
CR: 0.6 (1.6)

Gulls, Terns, and Skimmers *(Laridae)*

Gulls often have boldly marked primary feather tips.

Laughing Gull *(Larus atricilla)*

M>F

unknown	**S:** 5.0 (12.7)	**NA:** 1.1 (2.8)
P: 9.8 (25.0)	**S:** 4.9 (12.5)	**BE:** 1.6 (4.0)
P: 9.6 (24.4)	**T:** 5.5 (14.0)	**RU:** 3.7 (9.4)
P: 8.9 (22.5)	**T:** 5.4 (13.8)	**MA:** 1.6 (4.0)
P: 7.8 (19.8)		

Bonaparte's Gull *(Chroicocephalus philadelphia)*

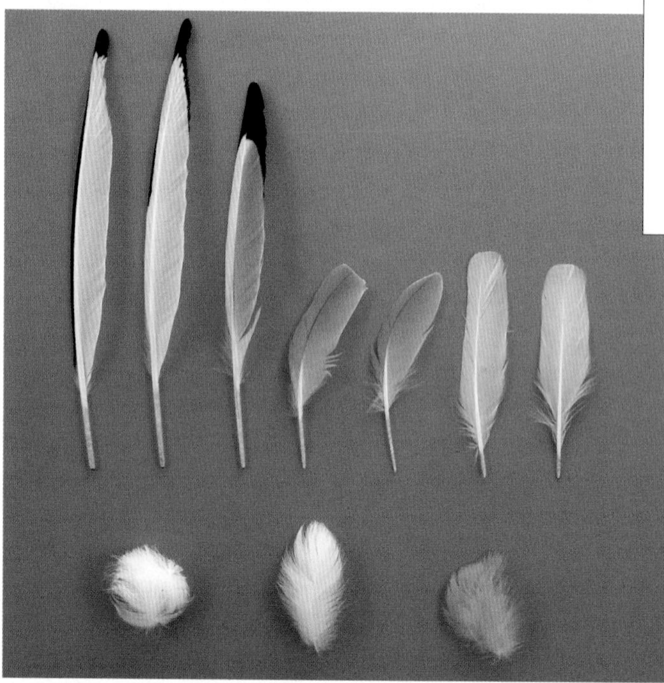

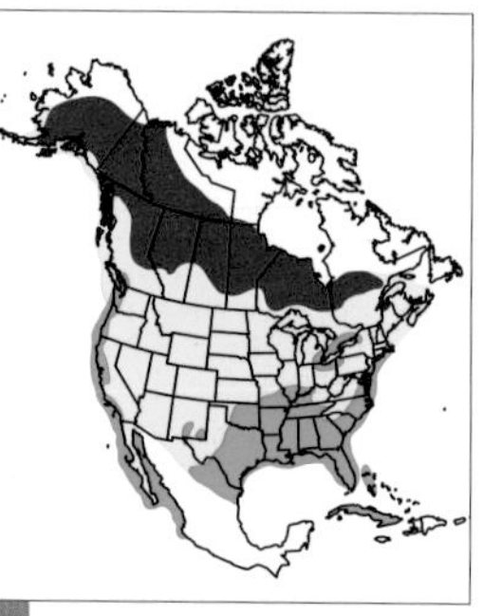

unknown,
non-breeding
P: 8.1 (20.6)
P: 8.3 (21.0)
P: 7.1 (18.0)
S: 3.9 (9.8)
S: 3.8 (9.6)
T: 4.2 (10.6)
T: 4.0 (10.2)
BR: 1.4 (3.6)
BE: 2.8 (7.0)
RU: 1.7 (4.3)

Heermann's Gull *(Larus heermanni)*

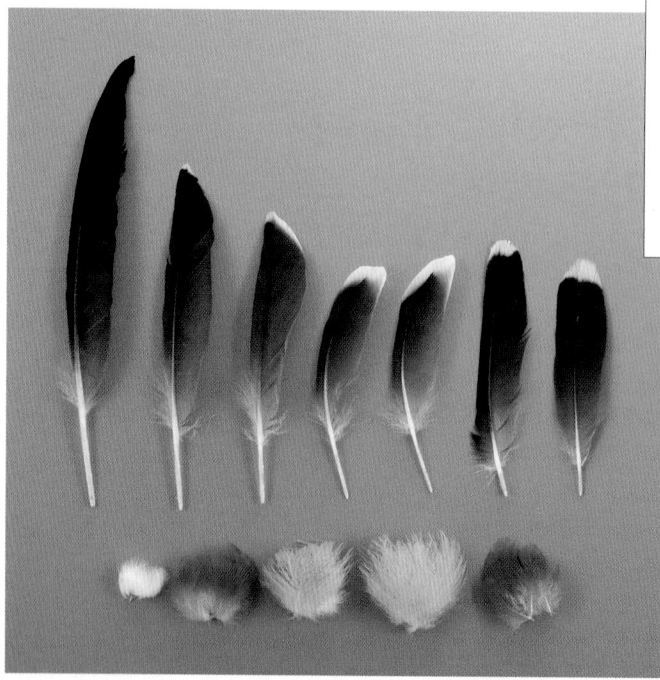

M>F

unknown
P: 10.7 (27.1)
P: 8.1 (20.7)
P: 6.9 (17.6)
S: 5.6 (14.2)
S: 5.6 (14.3)
T: 6.0 (15.3)
T: 5.6 (14.1)
CR: 0.8 (2.1)
NA: 2.0 (5.1)
BR: 2.3 (5.9)
BE: 2.6 (6.5)
RU: 2.2 (5.5)

Mew Gull *(Larus canus)*

M>F

unknown,
non-breeding
P: 10.7 (27.2)
P: 10.9 (27.7)
P: 9.3 (23.5)
P: 6.3 (16.1)
S: 5.1 (13.0)
S: 5.1 (12.9)
T: 6.1 (15.4)
T: 5.9 (14.9)
T: 5.9 (15.0)

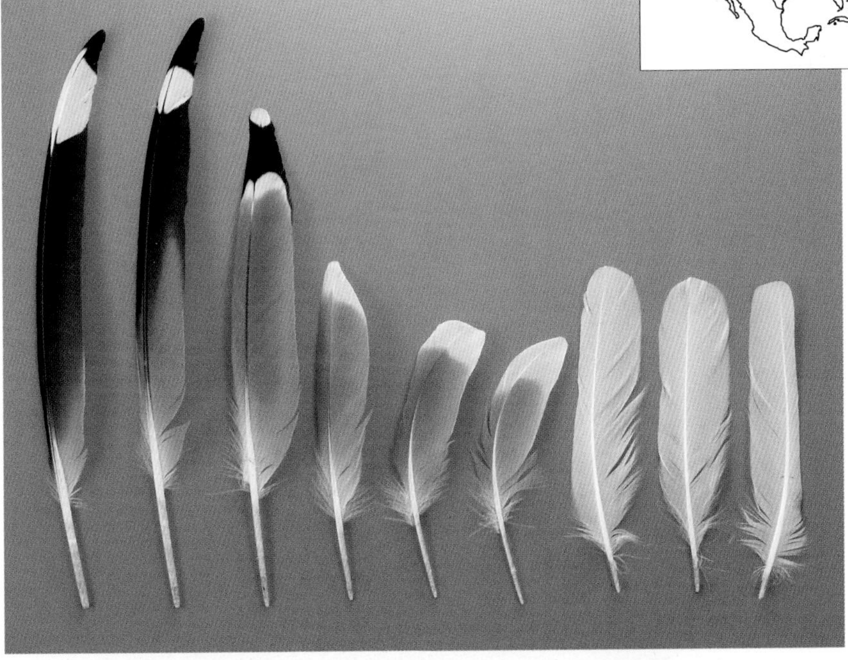

NA: 0.9 (2.4)
BR: 2.2 (5.5)
BE: 2.0 (5.0)
MA: 2.7 (6.9)

Ring-billed Gull *(Larus delawarensis)*

M>F

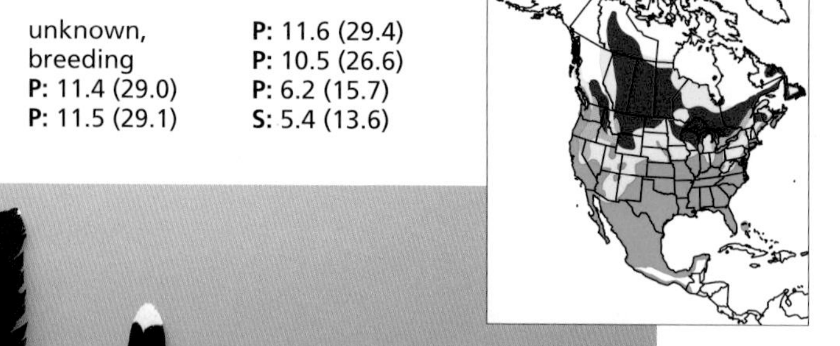

unknown,
breeding
P: 11.4 (29.0)
P: 11.5 (29.1)
P: 11.6 (29.4)
P: 10.5 (26.6)
P: 6.2 (15.7)
S: 5.4 (13.6)

T: 6.3 (16.1)
T: 6.1 (15.5)
BR: 1.7 (4.4)
RU: 2.6 (6.6)
MA: 2.1 (5.4)

Herring Gull *(Larus argentatus)*

M>F

unknown	**P:** 10.9 (27.8)	**T:** 7.5 (19.1)
P: 13.0 (32.9)	**S:** 7.3 (18.6)	**BR:** 2.3 (5.9)
P: 13.3 (33.9)	**S:** 6.6 (16.7)	**RU:** 3.3 (8.5)
P: 11.7 (29.7)	**T:** 7.6 (19.4)	**NE:** 2.0 (5.0)

Western Gull *(Larus occidentalis)*

M>F

unknown	**P:** 7.4 (18.9)	**T:** 6.2 (15.7)
P: 11.4 (28.9)	**S:** 5.9 (14.9)	**BR:** 1.9 (4.8)
P: 10.3 (26.2)	**S:** 5.8 (14.7)	**BE:** 2.7 (6.9)
P: 9.5 (24.1)	**T:** 6.4 (16.2)	**MA:** 2.5 (6.4)

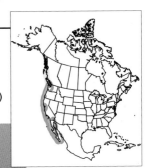

Glaucous-winged Gull *(Larus glaucescens)*

M>F

unknown,
breeding
P: 12.6 (31.9)
P: 10.9 (27.6)
P: 8.2 (20.8)

S: 6.7 (17.1)
S: 6.9 (17.5)
T: 7.7 (19.6)
T: 7.4 (18.9)

MA: 3.0 (7.6)
BR: 3.0 (7.5)
RU: 2.4 (6.1)

Great Black-backed Gull *(Larus marinus)*

M>F

unknown
P: 14.0 (35.6)
P: 14.6 (37.2)
P: 14.6 (37.0)
P: 12.0 (30.4)

S: 8.5 (21.6)
S: 7.6 (19.4)
T: 8.3 (21.1)
T: 7.7 (19.5)

RU: 2.7 (6.9)
MA: 3.0 (7.7)
BE: 2.6 (6.5)
BR: 2.2 (5.5)

Least Tern *(Sternula antillarum)*

unknown
P: 5.4 (13.6)
P: 4.8 (12.3)
P: 4.1 (10.4)
P: 2.0 (5.2)
S: 1.8 (4.5)
TS: 2.0 (5.1)
T: 3.3 (8.4)
T: 2.4 (6.0)
BR: 1.1 (2.8)
MA: 1.3 (3.4)

Caspian Tern *(Hydroprogne caspia)*

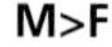

M>F

unknown
P: 11.3 (28.8)
P: 10.7 (27.2)
P: 8.8 (22.4)
S: 4.5 (11.5)
S: 4.5 (11.5)
T: 4.8 (12.2)
T: 5.6 (14.3)
BR: 2.0 (5.0)
RU: 1.7 (4.4)

Black Tern *(Chlidonias niger)*

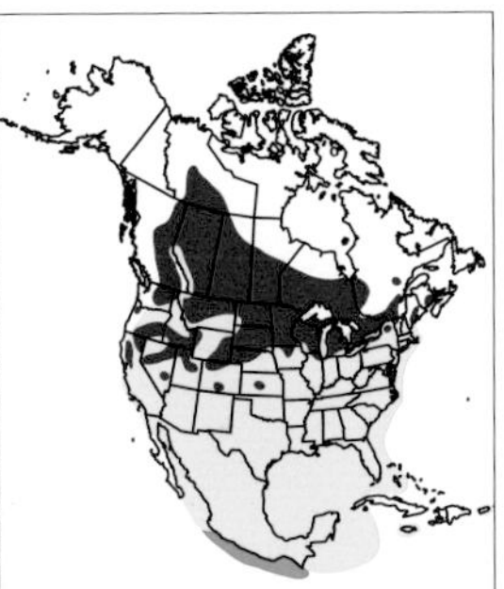

unknown,
non-breeding
P: 5.1 (13.0)
P: 3.1 (7.9)
S: 2.6 (6.6)
UPTC: 2.2 (5.5)
BR: 2.1 (5.3)
RU: 2.1 (5.4)

Common Tern *(Sterna hirundo)*

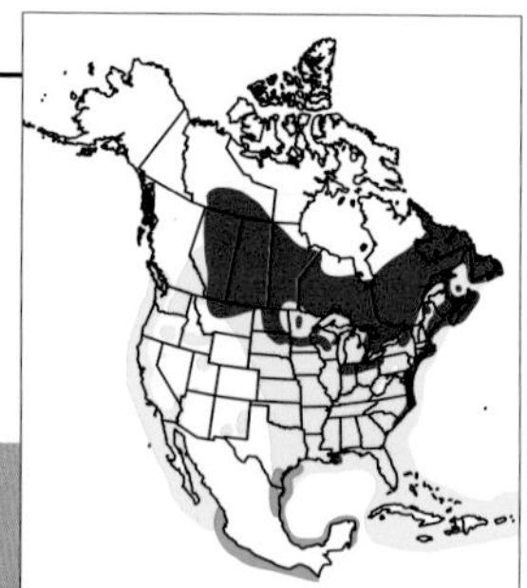

unknown
P: 8.5 (21.5)
P: 7.6 (19.2)
P: 6.9 (17.5)
P: 4.9 (12.4)

S: 3.0 (7.6)
S: 2.8 (7.2)
T: 6.3 (16.0)
T: 3.6 (9.2)
T: 4.8 (12.3)

NA: 0.7 (1.9)
BE: 1.4 (3.5)
BR: 1.2 (3.0)
MA: 1.3 (3.3)

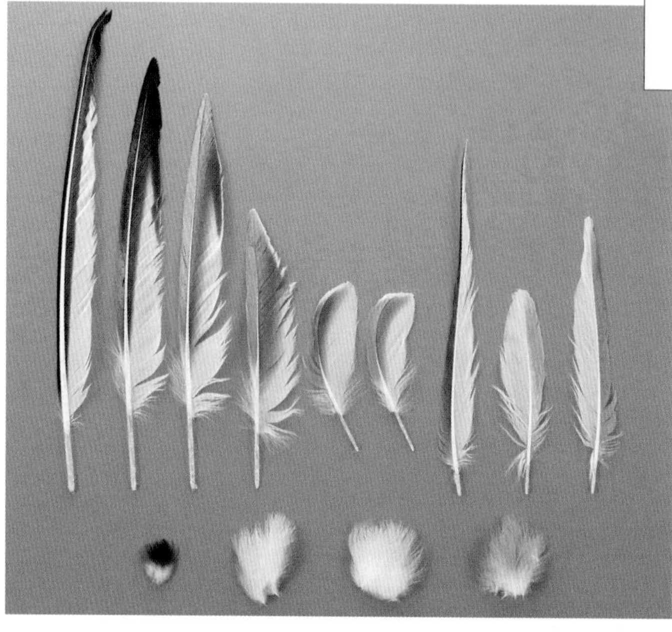

Forster's Tern *(Sterna forsteri)*

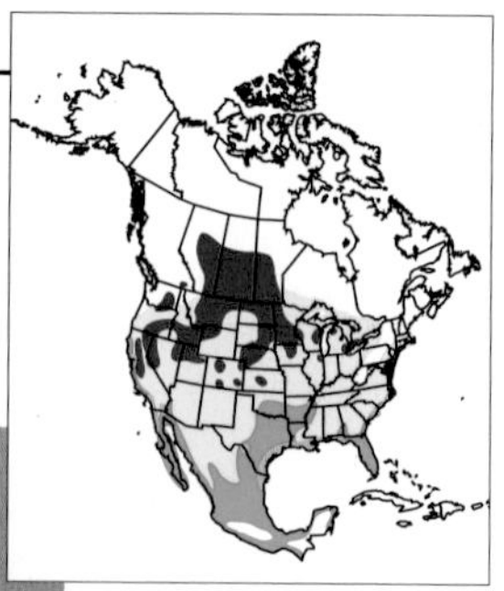

unknown
P: 7.6 (19.2)
P: 7.0 (17.8)
P: 6.5 (16.4)
P: 4.6 (11.8)

S: 2.9 (7.3)
S: 3.1 (7.8)
S: 3.0 (7.6)
TS: 3.3 (8.3)
TS: 3.1 (7.8)

BR: 1.1 (2.7)
BE: 1.0 (2.6)
MA: 1.1 (2.8)
RU: 1.4 (3.6)

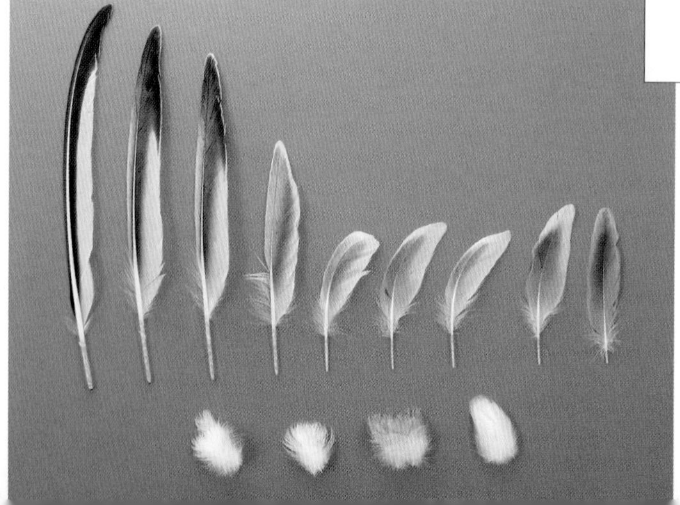

Elegant Tern *(Thalasseus elegans)*

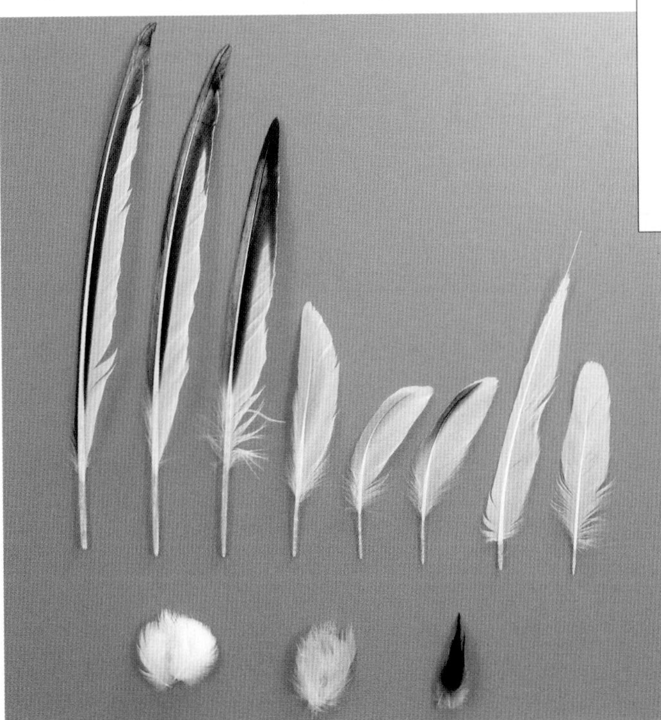

M>F

unknown
P: 9.9 (25.2)
P: 9.7 (24.7)
P: 8.3 (21.0)
P: 4.8 (12.1)
S: 3.4 (8.6)
S: 4.9 (12.4)
T: 6.3 (16.1)
T: 4.0 (10.2)
BR: 1.1 (2.7)
MA: 1.3 (3.3)
CR: 1.8 (4.5)

Sooty Tern *(Onychoprion fuscatus)*

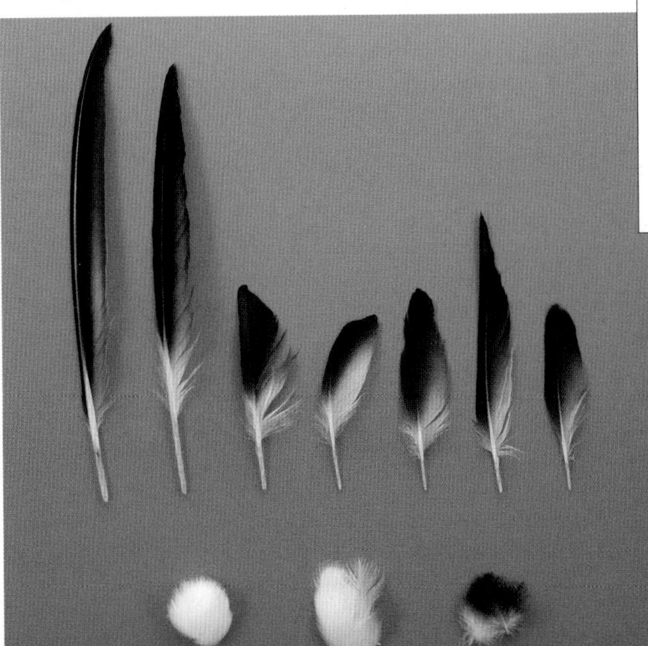

unknown
P: 9.1 (23.2)
P: 8.2 (20.8)
P: 4.0 (10.1)
S: 3.2 (8.1)
TS: 3.8 (9.7)
T: 5.2 (13.3)
T: 3.6 (9.1)
BR: 1.2 (3.1)
BE: 1.7 (4.2)
NA: 1.3 (3.3)

Black Skimmer *(Rynchops niger)*

M>F

unknown	**S:** 3.8 (9.6)	**BR:** 1.5 (3.7)
P: 11.5 (29.1)	**TS:** 4.5 (11.5)	**BE:** 2.2 (5.5)
P: 11.4 (28.9)	**T:** 3.4 (8.7)	**MA:** 1.4 (3.5)
P: 9.6 (24.3)	**T:** 2.9 (7.4)	**RU:** 1.3 (3.3)
P: 6.2 (15.8)	**T:** 2.5 (6.4)	

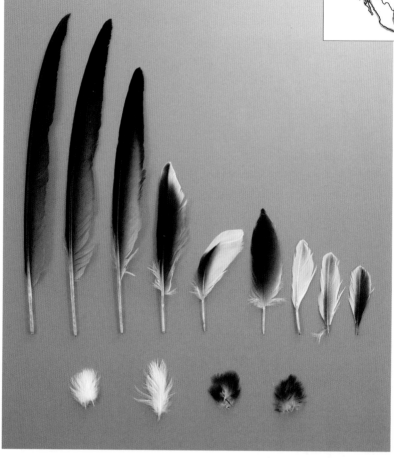

Alcids *(Alcidae)*

Short, compact primaries. Secondaries resemble those of high-speed wings.

Dovekie *(Alle alle)*

unknown
P: 3.1 (7.8)
P: 3.0 (7.6)
P: 2.7 (6.8)

S: 1.7 (4.4)
S: 1.8 (4.6)
T: 1.6 (4.0)
T: 1.6 (4.1)

BR: 0.9 (2.2)
MA: 1.1 (2.9)
RU: 1.2 (3.0)

Common Murre *(Uria aalge)*

unknown
P: 5.4 (13.6)
P: 5.1 (13.0)
P: 4.4 (11.3)
S: 3.0 (7.5)
S: 3.0 (7.7)
TS: 3.5 (9.0)
T: 2.3 (5.9)
T: 2.2 (5.5)
T: 1.7 (4.2)
BR: 1.6 (4.1)
RU: 1.9 (4.9)
FL: 2.5 (6.3)

Pigeon Guillemot *(Cepphus columba)*

juvenile
P: 5.0 (12.8)
P: 5.1 (12.9)
P: 4.6 (11.7)
P: 4.1 (10.5)
S: 3.9 (9.9)
S: 3.3 (8.5)
T: 2.4 (6.0)
T: 2.4 (6.2)
NA: 1.2 (3.0)
BR: 1.2 (3.0)
BE: 1.3 (3.3)

Cassin's Auklet *(Ptychoramphus aleuticus)*

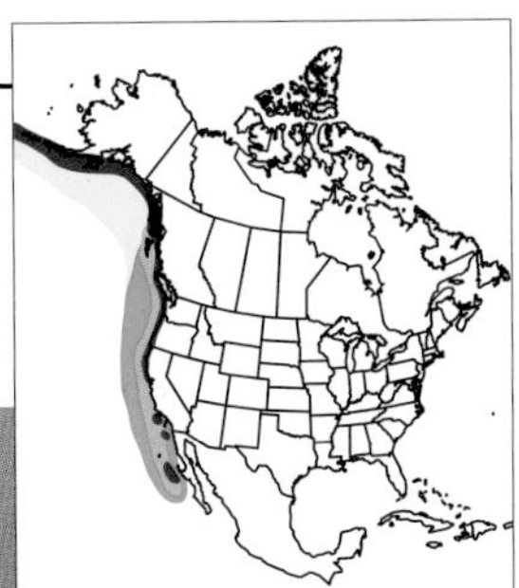

■■■■

unknown
P: 3.4 (8.7)
P: 3.6 (9.1)
P: 3.4 (8.6)
S: 2.1 (5.4)
S: 2.0 (5.2)
T: 1.6 (4.0)
T: 1.5 (3.9)
NA: 0.6 (1.5)
BE: 1.2 (3.0)
MA: 0.9 (2.3)

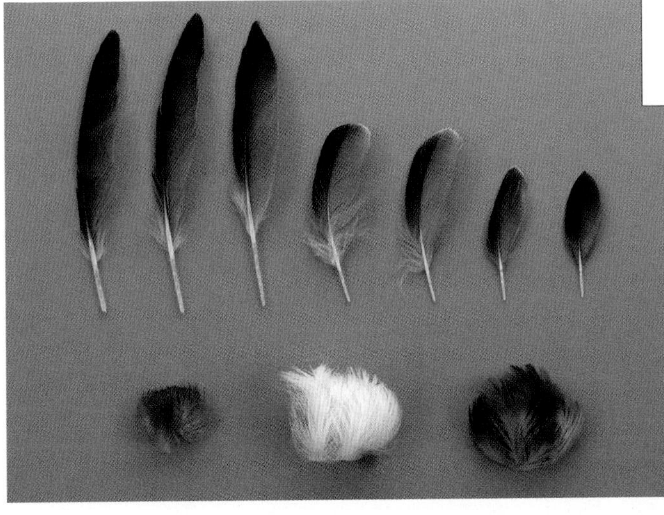

Crested Auklet *(Aethia cristatella)*

■■■■

unknown, breeding
P: 3.9 (9.8)
P: 4.0 (10.1)
P: 3.6 (9.1)
P: 2.9 (7.4)
S: 2.4 (6.0)
S: 2.2 (5.6)
T: 2.0 (5.2)
T: 2.0 (5.1)
BR: 1.1 (2.8)
BE: 1.4 (3.5)
FL: 2.0 (5.0)
CR: 1.8 (4.5)

Pigeons and Doves *(Columbidae)*

Tails often boldly tipped with white, gray, or black. Feathers often have a unique steel-blue–gray color that can be used to easily distinguish family.

Rock Pigeon *(Columba livia)*

(feral pigeon)

unknown	**S:** 4.8 (12.2)	**NA:** 1.0 (2.6)
P: 6.5 (16.4)	**TS:** 0.4 (1.1)	**BR:** 1.3 (3.4)
P: 7.2 (18.3)	**T:** 6.1 (15.5)	**BE:** 1.7 (4.4)
P: 6.5 (16.6)	**T:** 6.1 (15.4)	**MA:** 1.3 (3.2)
P: 5.2 (13.2)		

Rock Pigeon *(Columba livia)*

(fancy pigeon)

unknown	**S:** 4.3 (11.0)	**SC:** 2.4 (6.1)
P: 6.7 (17.0)	**S:** 4.1 (10.5)	**RU:** 1.3 (3.2)
P: 7.4 (18.8)	**T:** 5.1 (13.0)	**BE:** 1.4 (3.6)
P: 6.8 (17.2)	**T:** 4.9 (12.4)	**NA:** 1.0 (2.5)
P: 5.6 (14.2)		

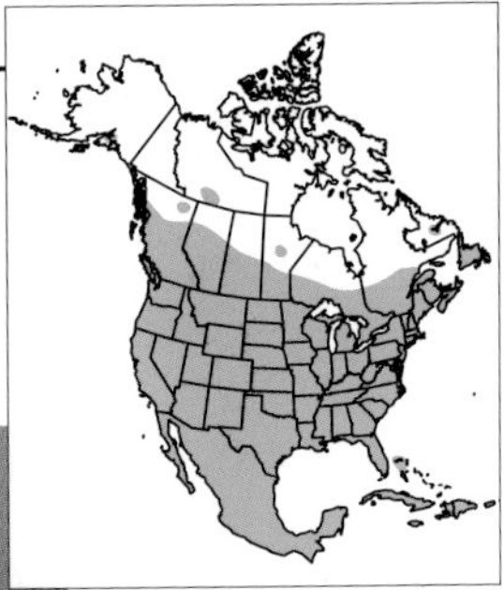

Band-tailed Pigeon *(Patagioenas fasciata)*

M>F

unknown	**S:** 5.0 (12.6)	**NA:** 1.0 (2.5)
P: 6.7 (16.9)	**TS:** 4.4 (11.2)	**BR:** 1.5 (3.8)
P: 7.2 (18.3)	**T:** 6.0 (15.3)	**BE:** 1.5 (3.7)
P: 6.6 (16.8)	**T:** 6.2 (15.7)	**MA:** 1.6 (4.1)
P: 5.6 (14.1)		**RU:** 2.2 (5.7)

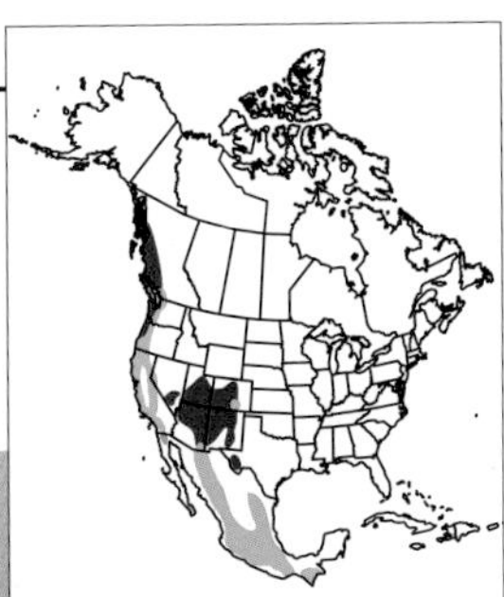

Eurasian Collared-Dove *(Streptopelia decaocto)*

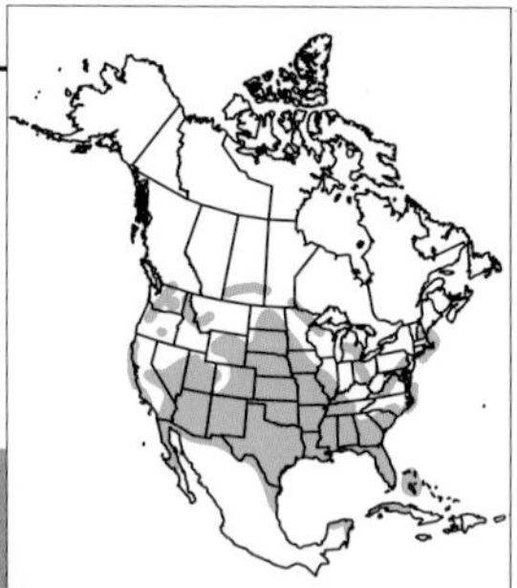

unknown	**S:** 4.1 (10.5)	**MA:** 1.4 (3.6)
P: 5.3 (13.4)	**TS:** 3.7 (9.3)	**BR:** 1.5 (3.7)
P: 5.3 (13.5)	**T:** 5.6 (14.3)	**BE:** 1.5 (3.7)
P: 4.3 (10.8)	**T:** 5.4 (13.6)	**RU:** 1.2 (3.0)

White-winged Dove *(Zenaida asiatica)*

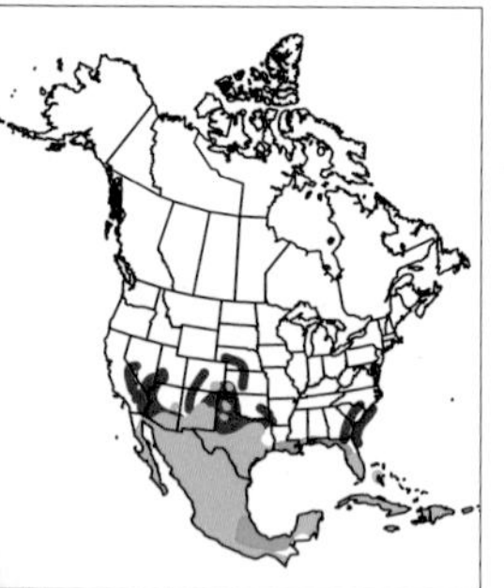

unknown	**S:** 3.9 (9.9)	**BR:** 1.2 (3.1)
P: 5.0 (12.7)	**TS:** 3.6 (9.2)	**BE:** 1.3 (3.4)
P: 5.4 (13.6)	**T:** 5.2 (13.2)	**RU:** 1.3 (3.2)
P: 5.3 (13.5)	**T:** 5.1 (13.0)	NE: 0.6 (1.4)
P: 4.5 (11.4)	**T:** 5.0 (12.7)	**UPTC:** 3.1 (8.0)

Mourning Dove *(Zenaida macroura)*

M>F

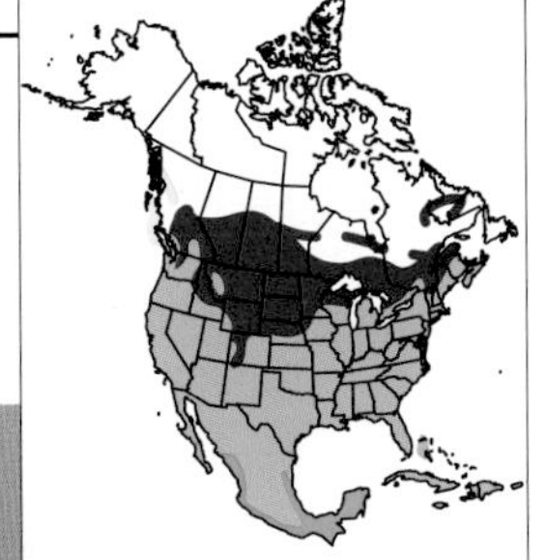

unknown	**S:** 3.2 (8.2)	**MA:** 1.3 (3.2)
P: 4.6 (11.7)	**S:** 3.3 (8.5)	**BR:** 1.1 (2.9)
P: 4.9 (12.4)	**T:** 3.4 (8.7)	**BE:** 1.6 (4.1)
P: 4.7 (11.9)	**T:** 5.3 (13.4)	**FL:** 1.4 (3.6)
P: 3.8 (9.7)	**T:** 6.3 (15.9)	**WC:** 2.7 (6.9)

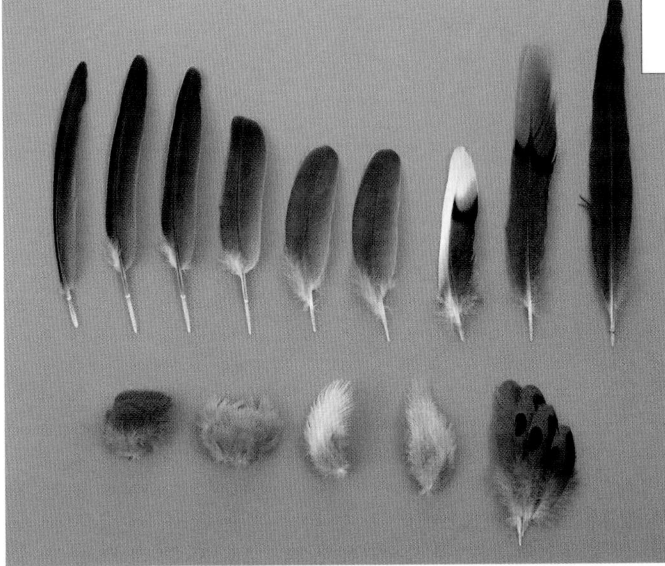

Inca Dove *(Columbina inca)*

unknown
P: 3.1 (7.8)
P: 3.0 (7.7)
S: 2.5 (6.3)
S: 2.5 (6.3)
T: 4.0 (10.1)
T: 4.0 (10.1)
T: 3.9 (9.8)
MA: 1.7 (4.4)
BE: 0.9 (2.3)
BR: 4.6 (11.7)

Common Ground-Dove *(Columbina passerina)*

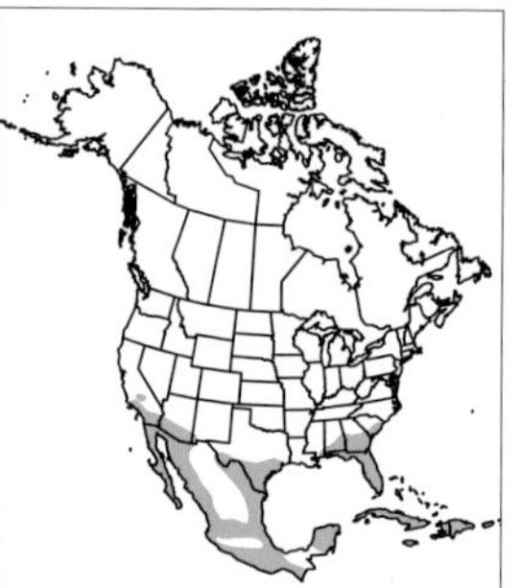

unknown	**S:** 2.5 (6.4)	**WC:** 1.2 (3.1)
P: 2.9 (7.3)	**T:** 2.7 (6.8)	**BR:** 0.9 (2.2)
P: 2.9 (7.4)	**T:** 2.7 (6.8)	**MA:** 1.0 (2.5)
P: 2.6 (6.5)	**WC:** 0.6 (1.4)	**UPTC:** 1.4 (3.6)

Cuckoos and Roadrunners *(Cuculidae)*

Tail feathers are long. White markings often present on outer tail feathers. Flight feathers similar to those of thrashers and towhees.

Yellow-billed Cuckoo *(Coccyzus americanus)*

unknown
P: 2.6 (6.7)
P: 4.8 (12.3)
P: 3.8 (9.7)
P: 3.3 (8.5)
S: 2.9 (7.4)
S: 2.6 (6.7)
T: 4.2 (10.6)
T: 5.8 (14.7)
T: 5.4 (13.6)
MA: 0.8 (2.0)
BR: 0.8 (2.1)
BE: 1.3 (3.3)

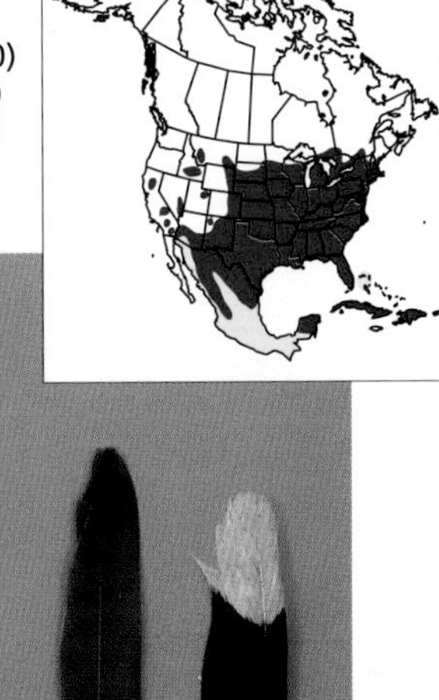

Black-billed Cuckoo *(Coccyzus erythropthalmus)*

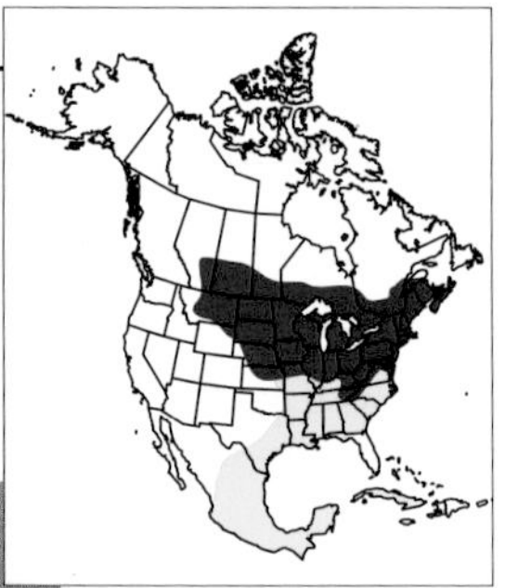

unknown
P: 5.0 (12.8)
P 4.4 (11.1)
P: 3.6 (9.2)
S: 3.3 (8.3)
S: 3.0 (7.7)
T: 4.4 (11.2)
T: 6.7 (16.9)
T: 6.1 (15.5)
MA: 1.3 (3.3)
BR: 1.1 (2.9)
BE: 1.6 (4.1)
RU: 1.3 (3.2)

Greater Roadrunner *(Geococcyx californianus)*

unknown
P: 3.2 (8.2)
P: 4.6 (11.6)
P: 6.2 (15.7)
P: 6.3 (16.1)
S: 5.9 (15.0)
TS: 5.1 (13.0)
T: 7.8 (19.7)
T: 11.3 (28.7)
BR: 1.4 (3.6)
BE: 1.3 (3.4)
MA: 1.8 (4.5)
WC: 3.0 (7.5)

Barn Owls *(Tytonidae)* and Typical Owls *(Strigidae)*

Emargination is often present in leading three primaries, where it occurs near the tip of the feather. Leading primaries are often curved near the tip. All flight feathers of the wing have a fuzzy appearance. Strong barring or spotting is always present. Tail feathers can be very hard to distinguish from secondaries in some species. Outer tail feathers of larger owls are often strongly curved near the lower end of the feather and have a high degree of camber. (Note: Some hawks and kites also have fuzzy wing feathers, but to a much lesser degree.)

Barn Owl *(Tyto alba)*

F>M

female	
P: 10.6 (26.9)	**S:** 6.9 (17.4)
P: 10.0 (25.5)	**S:** 6.7 (17.1)
P: 7.8 (19.9)	**T:** 5.9 (15.0)
	T: 5.3 (13.5)

continued on page 198

Barn Owl continued

RU: 4.9 (12.4)
RU: 3.9 (9.9)
BE: 2.8 (7.0)
BR: 2.0 (5.1)

Snowy Owl *(Bubo scandiacus)*

F>M

female
P: 11.4 (28.9)
P: 14.2 (36.0)
P: 11.9 (30.1)
P: 10.1 (25.6)
S: 9.2 (23.3)
S: 9.1 (23.2)
TS: 8.7 (22.2)

T: 9.5 (24.2)
T: 10.2 (25.8)
T: 10.0 (25.5)
BR: 4.4 (11.2)
BE: 2.6 (6.6)
RU: 5.3 (13.5)
FL: 6.5 (16.5)

Great Gray Owl *(Strix nebulosa)*

F>M

female
P: 10.2 (26.0)
P: 13.0 (33.0)
P: 14.4 (36.5)
P: 15.4 (39.1)
P: 12.1 (30.8)

continued on page 200

Great Gray Owl continued

S: 11.7 (29.6)
S: 11.2 (28.4)
S: 10.1 (25.6)
TS: 9.1 (23.0)

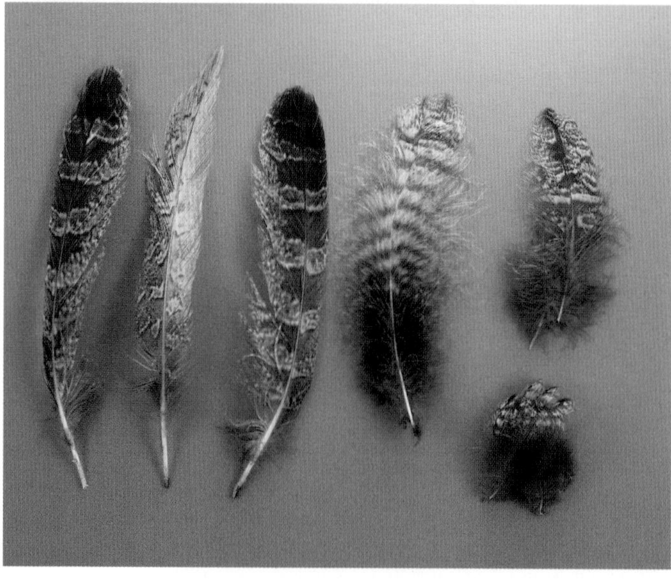

T: 11.5 (29.1)
T: 12.0 (30.6)
T: 11.4 (29.0)
UNTC: 9.3 (23.6)
MA: 6.6 (16.8)
NA: 3.9 (10.0)

Great Horned Owl *(Bubo virginianus)*

F>M

unknown
P: 10.2 (25.9)
P: 12.4 (31.4)
P: 10.3 (26.2)
S: 9.3 (23.7)
S: 8.8 (22.4)

T: 9.1 (23.2)
T: 9.1 (23.0)
T: 8.7 (22.0)
T: 8.7 (22.2)
NA: 1.9 (4.7)
UNWC: 3.3 (8.5)
BE: 4.4 (11.2)

Barred Owl *(Strix varia)*

F>M

male
P: 9.6 (24.5)
P: 11.0 (28.0)
P: 11.1 (28.3)
S: 9.2 (23.4)
TS: 7.4 (18.7)

T: 8.9 (22.5)
T: 8.3 (21.1)
T: 10.5 (26.7)
BR: 2.7 (6.8)
RU: 3.2 (8.2)
WC: 4.6 (11.7)

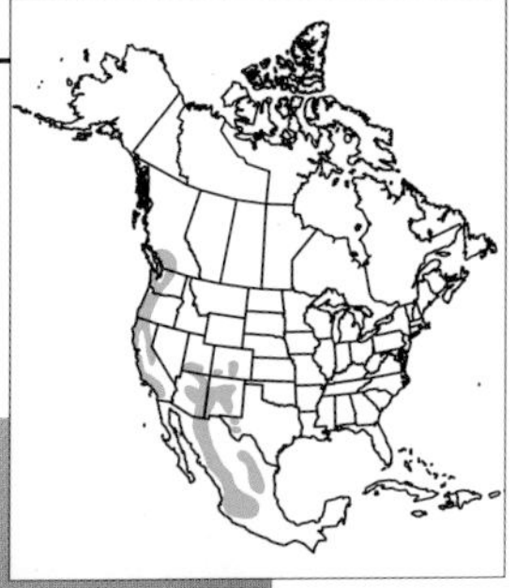

Spotted Owl *(Strix occidentalis)*

■■■■

F>M

male
P: 10.8 (27.4)
P: 9.8 (25.0)
S: 8.0 (20.2)
S: 7.4 (18.9)
S: 6.9 (17.6)
T: 8.5 (21.6)
T: 7.6 (19.3)

Long-eared Owl *(Asio otus)*

■■■■

F>M

unknown
P: 8.5 (21.7)
P: 9.1 (23.0)
S: 6.4 (16.2)
TS: 5.6 (14.2)
T: 6.5 (16.5)
T: 6.5 (16.5)
Ear: 2.3 (5.8)
NA: 2.3 (5.9)
BR: 2.1 (5.4)
FL: 3.3 (8.4)

Short-eared Owl *(Asio flammeus)*

F>M

unknown
P: 8.5 (21.5)
P: 9.7 (24.6)
P: 8.4 (21.4)
P: 6.7 (16.9)
S: 5.6 (14.2)
S: 5.5 (14.0)
T: 5.7 (14.4)
T: 6.0 (15.2)
MA: 3.1 (7.9)
RU: 2.8 (7.0)
CR: 1.9 (4.8)

Burrowing Owl *(Athene cunicularia)*

unknown
P: 4.7 (12.0)
P: 5.5 (14.0)
P: 5.0 (12.8)
S: 4.3 (11.0)
S: 4.3 (10.9)
T: 3.5 (8.8)
T: 3.4 (8.6)
T: 3.3 (8.3)
BR: 1.5 (3.7)
BE: 1.5 (3.7)
MA: 1.7 (4.2)
RU: 1.2 (3.0)

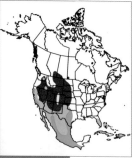

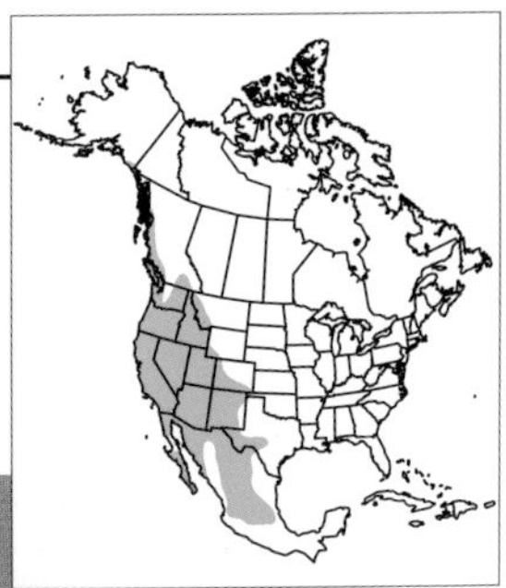

Western Screech-Owl *(Megascops kennicottii)*

(brown Pacific)
F>M

female
P: 5.7 (14.4)
P: 6.2 (15.7)
P: 5.4 (13.8)
S: 5.0 (12.6)
S: 4.2 (10.6)
T: 4.1 (10.3)
T: 3.9 (9.9)
T: 4.0 (10.1)
NA: 1.4 (3.5)
BR: 2.3 (5.9)
BE: 2.9 (7.3)
MA: 3.0 (7.5)

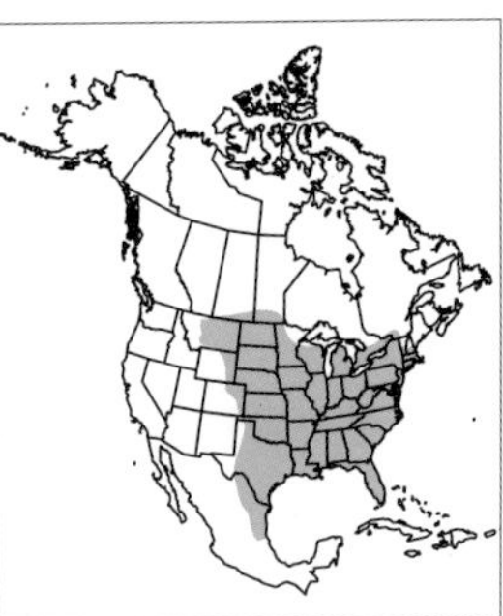

Eastern Screech-Owl *(Megascops asio)*

(gray adult
and red adult)
F>M

unknown,
gray adult
P: 3.3 (8.3)
P: 4.6 (11.8)
P: 4.8 (12.2)
P: 4.1 (10.5)
S: 3.5 (8.8)
S: 3.4 (8.6)
TS: 2.7 (6.9)
T: 3.0 (7.7)
T: 3.0 (7.7)
T: 3.1 (7.9)
NA: 0.6 (1.5)
BR: 1.4 (3.5)
MA: 1.8 (4.5)
RU: 1.1 (2.8)

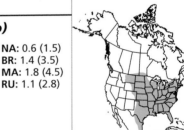

continued on page 206

Eastern Screech-Owl continued

unknown,
red adult
P: 3.3 (8.5)
P: 5.0 (12.6)
P: 5.1 (13.0)
P: 4.8 (12.2)
S: 4.2 (10.6)
S: 3.5 (8.9)
T: 3.3 (8.5)
T: 3.3 (8.5)
T: 3.3 (8.3)
BR: 1.5 (3.8)
BE: 1.6 (4.1)
MA: 2.0 (5.0)

Northern Saw-whet Owl *(Aegolius acadicus)*

F>M

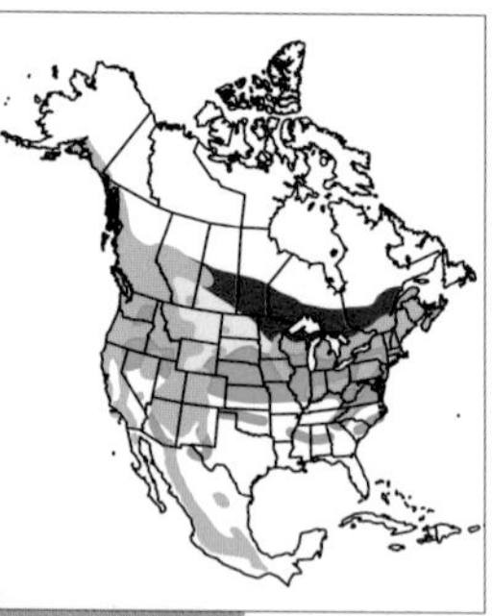

female
P: 3.5 (9.0)
P: 4.4 (11.3)
P: 4.8 (12.1)
P: 4.0 (10.1)
S: 3.7 (9.4)
TS: 3.1 (8.0)
T: 3.0 (7.6)
T: 3.1 (7.8)
CR: 1.1 (2.8)
BE: 1.7 (4.2)
MA: 1.7 (4.2)
RU: 1.3 (3.2)
SC: 1.9 (4.9)

Elf Owl *(Micrathene whitneyi)*

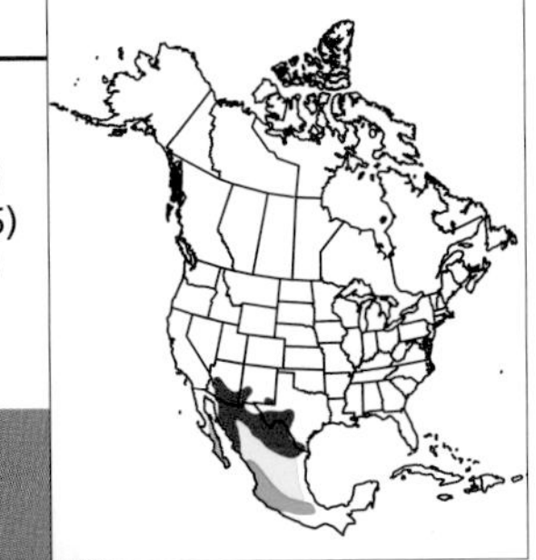

unknown
P: 2.5 (6.4)
P: 3.7 (9.4)
P: 3.6 (9.2)
P: 3.1 (7.8)

S: 3.0 (7.7)
S: 2.9 (7.4)
T: 2.0 (5.0)
T: 2.1 (5.3)

BR: 1.3 (3.2)
MA: 1.4 (3.5)
SC: 1.6 (4.1)
FL: 1.3 (3.4)

Flammulated Owl *(Otus flammeolus)*

unknown
P: 3.3 (8.3)
P: 4.3 (11.0)
P: 4.1 (10.5)

S: 3.3 (8.4)
S: 2.9 (7.4)
T: 2.7 (6.9)
T: 2.6 (6.5)

SC: 1.7 (4.3)
MA: 1.5 (3.7)
BR: 1.4 (3.5)
NA: 0.8 (2.1)

Northern Pygmy-Owl *(Glaucidium gnoma)*

F>M

male
P: 1.7 (4.3)
P: 2.9 (7.3)
P: 2.6 (6.5)
S: 2.4 (6.1)
S: 2.3 (5.9)
T: 2.6 (6.5)
T: 2.7 (6.8)
T: 2.6 (6.6)

MA: 1.1 (2.9)
BR: 1.2 (3.0)
FL: 1.4 (3.5)
RU: 1.2 (3.1)
SC: 1.4 (3.5)

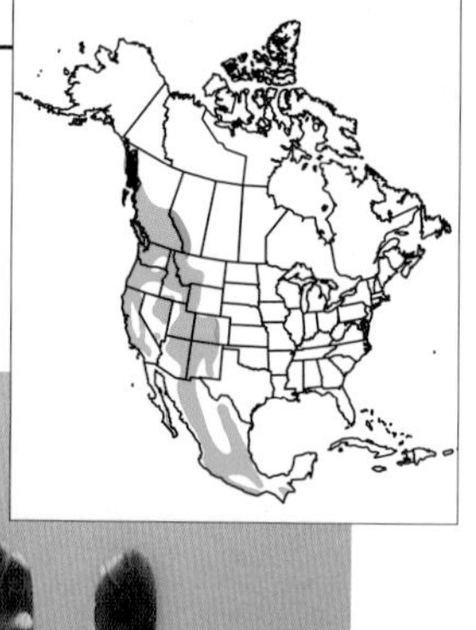

Goatsuckers *(Caprimulgidae)*

Heavily camouflaged flight and body feathers. Outer tail feathers often have white or buff markings.

Lesser Nighthawk *(Chordeiles acutipennis)*

unknown	**S:** 3.1 (8.0)	**MA:** 1.4 (3.5)
P: 6.1 (15.5)	**S:** 3.0 (7.7)	**BR:** 1.0 (2.5)
P: 5.8 (14.8)	**T:** 4.8 (12.2)	**BE:** 1.8 (4.5)
P: 4.8 (12.2)	**T:** 4.4 (11.2)	**RU:** 1.4 (3.5)
P: 4.4 (11.3)	**T:** 4.2 (10.7)	**TH:** 0.7 (1.7)

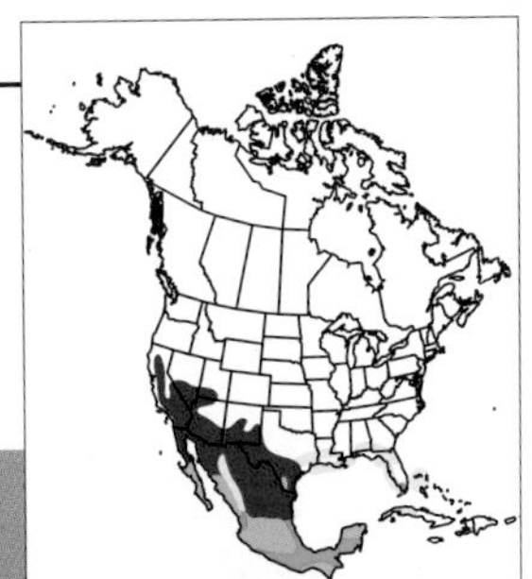

Common Nighthawk *(Chordeiles minor)*

unknown
P: 6.3 (16.1)
P: 5.4 (13.7)
P: 4.3 (10.8)
S: 2.8 (7.0)
T: 4.3 (10.8)
T: 3.5 (8.8)
T: 3.4 (8.7)
T: 4.4 (11.2)
TS: 3.1 (7.9)
BR: 1.0 (2.6)
unknown: 2.3 (5.9)
RU: 1.0 (2.5)

Common Poorwill *(Phalaenoptilus nuttallii)*

(Eastern)

female
P: 4.5 (11.4)
P: 5.0 (12.8)
P: 3.9 (9.8)
S: 3.1 (8.0)

TS: 3.0 (7.5)
T: 3.7 (9.4)
T: 3.7 (9.3)
T: 3.7 (9.3)
NA: 1.2 (3.0)

BR: 1.1 (2.7)
BE: 1.1 (2.8)
RU: 1.0 (2.5)
UPTC: 1.5 (3.9)

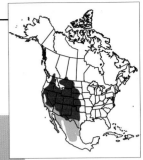

Whip-poor-will *(Caprimulgus vociferus)*

unknown
P: 4.8 (12.3)
P: 5.4 (13.6)
P: 4.9 (12.5)
P: 4.3 (10.8)

S: 3.1 (7.8)
S: 3.0 (7.6)
TS: 2.2 (5.5)
T: 4.8 (12.2)
T: 5.1 (13.0)

bristle: 1.3 (3.3)
RU: 1.2 (3.0)
BE: 1.5 (3.9)
BR: 1.4 (3.5)
NA: 1.1 (2.9)

Swifts *(Apodidae)*

Similar to swallows. Dark primaries and secondaries. Secondaries often notched at top.

Chimney Swift *(Chaetura pelagica)*

unknown
P: 4.1 (10.5)
P: 4.0 (10.1)
P: 3.0 (7.5)
P: 2.0 (5.2)

S: 1.4 (3.6)
TS: 1.4 (3.5)
T: 2.0 (5.0)
T: 2.0 (5.0)
T: 1.9 (4.8)

BR: 0.6 (1.6)
BE: 0.8 (2.1)
RU: 1.1 (2.8)

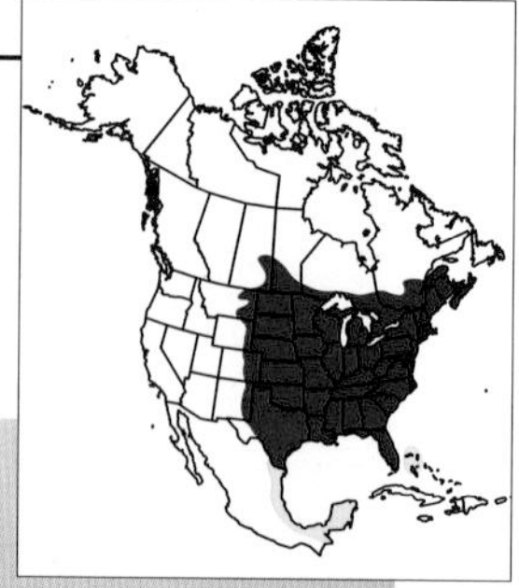

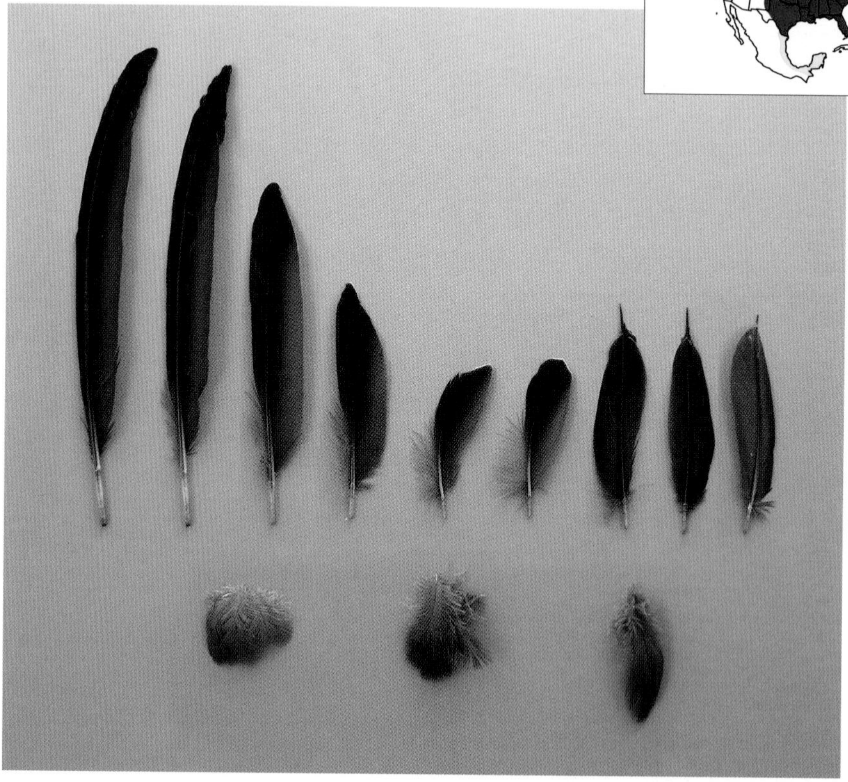

White-throated Swift *(Aeronautes saxatalis)*

unknown	**S:** 1.6 (4.0)	**WC:** 1.0 (2.6)
P: 4.3 (10.9)	**T:** 2.1 (5.3)	**BR:** 0.6 (1.5)
P: 4.1 (10.3)	**T:** 2.4 (6.0)	**RU:** 0.7 (1.8)
P: 2.6 (6.7)	**T:** 2.6 (6.6)	**RU:** 0.8 (2.0)

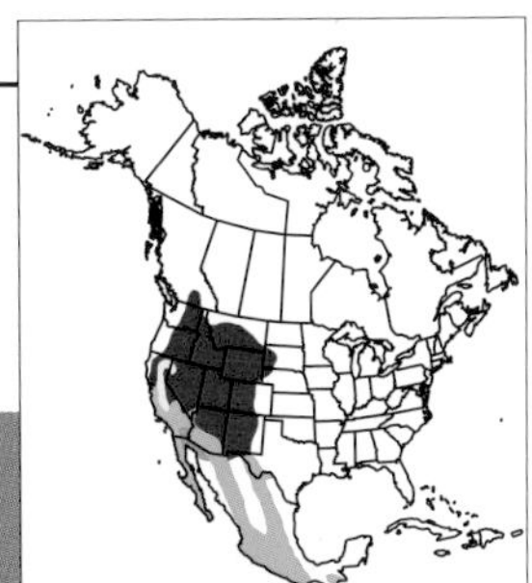

Hummingbirds *(Trochilidae)*

Ruby-throated Hummingbird *(Archilochus colubris)*

F>M

male
P: 1.2 (3.1)
P: 1.2 (3.0)
P: 1.1 (2.7)
P: 0.9 (2.4)
P: 0.7 (1.8)
P: 0.6 (1.4)
S: 0.6 (1.4)
S: 0.5 (1.3)
T: 1.2 (3.0)
T: 1.1 (2.8)
T: 0.8 (2.1)
BE: 0.3 (0.8)
TH: 0.2 (0.6)
RU: 0.4 (0.9)

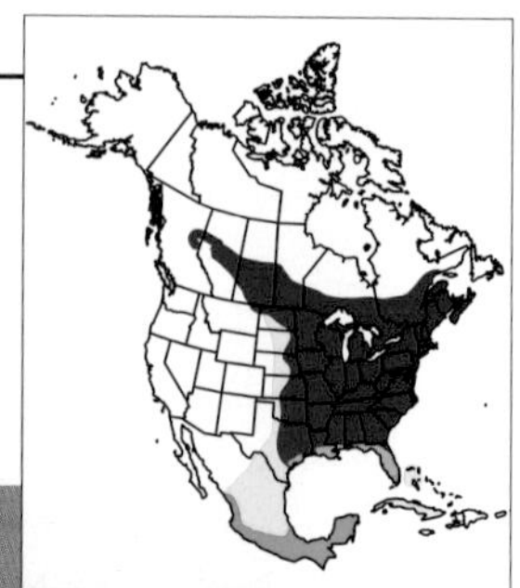

Black-chinned Hummingbird *(Archilochus alexandri)*

F>M

male, throat is iridescent purple; mantle is iridescent green	**P:** 1.4 (3.5)	**T:** 1.1 (2.9)
	P: 1.1 (2.7)	**T:** 1.1 (2.9)
	P: 0.8 (2.0)	**T:** 1.1 (2.9)
	S: 0.6 (1.5)	**TH:** 0.4 (1.1)
	S: 0.6 (1.6)	**MA:** 0.4 (1.1)

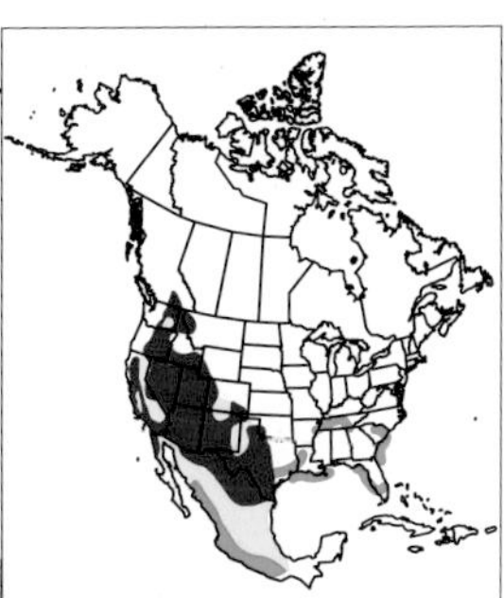

Anna's Hummingbird *(Calypte anna)*

male
P: 1.6 (4.0)
P: 1.5 (3.8)
P: 1.1 (2.8)
S: 0.6 (1.5)
S: 0.6 (1.5)
T: 1.4 (3.6)
T: 1.2 (3.1)
T: 1.1 (2.8)
MA: 0.5 (1.3)
TH: 0.5 (1.3)
UPTC: 1.0 (2.5)

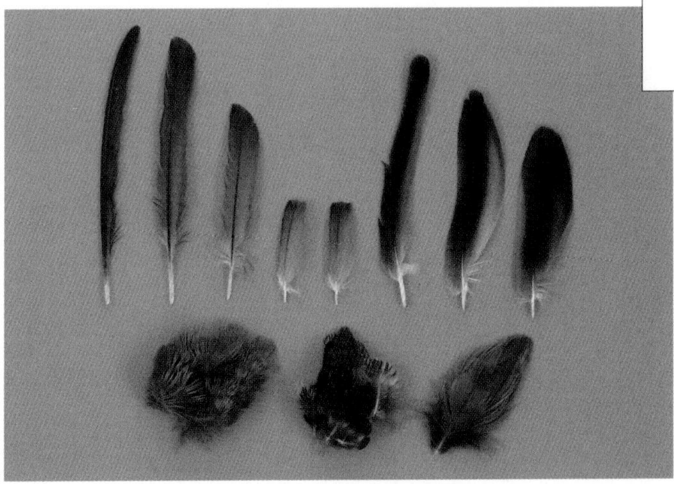

Costa's Hummingbird *(Calypte costae)*

F>M

male	**P:** 0.8 (2.1)	**T:** 1.0 (2.5)
P: 1.3 (3.4)	**S:** 0.6 (1.4)	**T:** 0.9 (2.2)
P: 1.3 (3.3)	**S:** 0.4 (1.1)	**TH:** 0.4 (1.0)
P: 1.1 (2.8)	**T:** 1.0 (2.6)	**MA:** 0.4 (1.0)

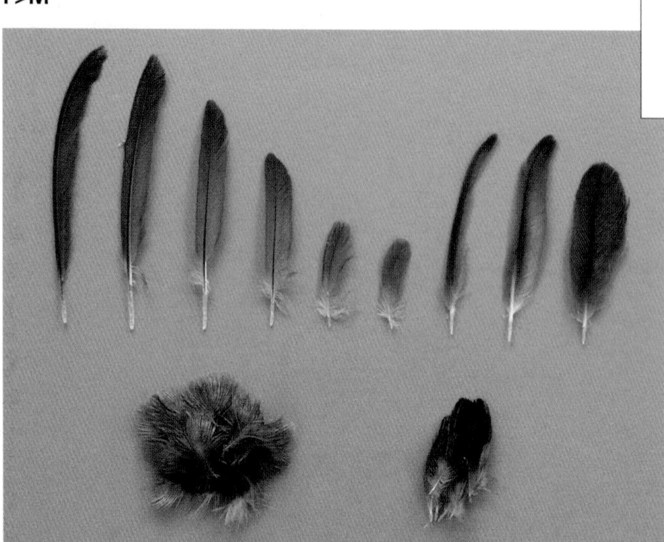

Calliope Hummingbird *(Stellula calliope)*

F>M

male	**S:** 0.5 (1.2)	**T:** 0.8 (2.1)
P: 1.2 (3.0)	**S:** 0.5 (1.3)	**TH:** 0.5 (1.2)
P: 0.9 (2.4)	**T:** 0.9 (2.3)	**MA:** 0.4 (0.9)
P: 0.8 (2.0)	**T:** 0.8 (2.1)	

Rufous Hummingbird *(Selasphorus rufus)*

F>M

male	**P:** 0.6 (1.6)	**T:** 1.1 (2.8)
P: 1.3 (3.3)	**S:** 0.5 (1.3)	**T:** 1.1 (2.9)
P: 1.3 (3.2)	**S:** 0.5 (1.2)	**TH:** 0.4 (0.9)
P: 1.0 (2.6)	**T:** 0.9 (2.3)	**MA:** 0.3 (0.8)

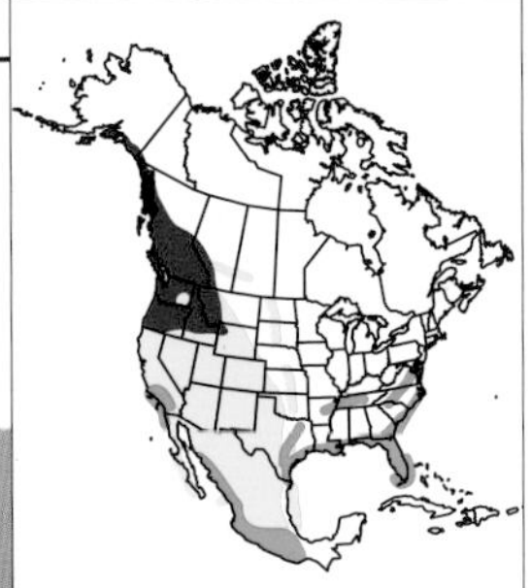

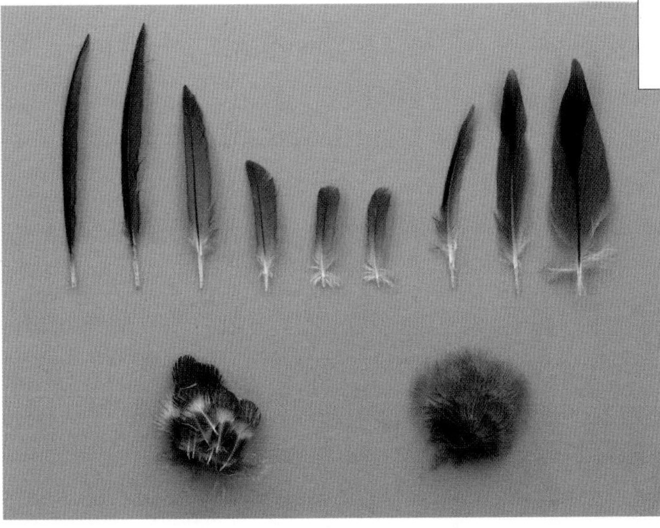

Allen's Hummingbird *(Selasphorus sasin)*

F>M

male	**P:** 0.6 (1.5)	**T:** 1.0 (2.5)
P: 1.2 (3.0)	**S:** 0.5 (1.3)	**T:** 1.1 (2.8)
P: 1.3 (3.2)	**S:** 0.5 (1.2)	**TH:** 0.4 (1.0)
P: 0.9 (2.4)	**T:** 0.9 (2.3)	**NA:** 0.4 (1.1)

Kingfishers *(Alcedinidae)*

Boldly marked. Tail feathers barred. Primaries can easily be mistaken for those of some woodpeckers, some grosbeaks, mockingbirds, shrikes, and Lesser Goldfinch.

Belted Kingfisher *(Megaceryle alcyon)*

male
P: 4.4 (11.1)
P: 5.0 (12.8)
P: 4.8 (12.3)
P: 4.2 (10.7)

S: 3.4 (8.6)
S: 3.5 (8.9)
T: 4.0 (10.1)
T: 3.8 (9.7)
T: 4.0 (10.2)

NA: 1.1 (2.7)
BR: 1.1 (2.8)
BE: 1.3 (3.2)
FL: 1.5 (3.7)
RU: 1.2 (3.1)
CR: 1.4 (3.5)

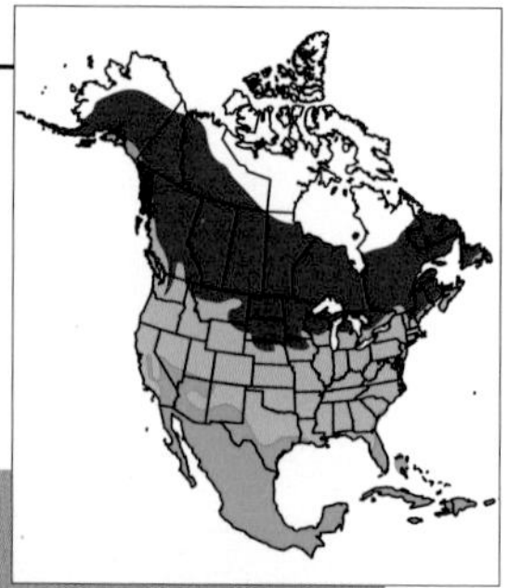

Woodpeckers and Allies *(Picidae)*

Often boldly marked with white and black. White spots often present on flight feathers. Tail feathers stiff and pointed. Primary feathers of some woodpecker species can be mistaken for those of some grosbeaks, mockingbirds, shrikes, kingfishers, and Lesser Goldfinch.

Lewis's Woodpecker *(Melanerpes lewis)*

male	**S:** 4.1 (10.3)	**TH:** 1.1 (2.7)
P: 4.9 (12.5)	**T:** 4.3 (10.8)	**BR:** 1.2 (3.0)
P: 5.9 (14.9)	**T:** 3.9 (9.8)	**BE:** 1.5 (3.8)
S: 4.2 (10.6)	**NA:** 0.7 (1.8)	**MA:** 1.7 (4.3)

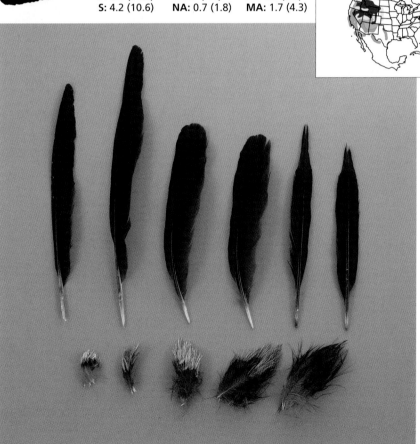

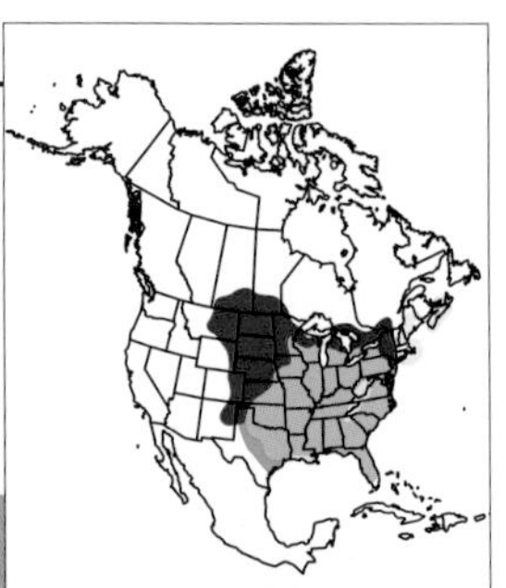

Red-headed Woodpecker *(Melanerpes erythrocephalus)*

male
P: 3.9 (9.8)
S: 3.6 (9.1)
S: 3.5 (8.8)
TS: 2.6 (6.5)

T: 2.9 (7.3)
T: 3.2 (8.1)
T: 3.0 (7.5)
CR: 0.5 (1.3)
TH: 0.8 (2.0)

NA: 0.6 (1.5)
MA: 1.2 (3.1)
RU: 1.0 (2.5)
BR: 1.1 (2.8)

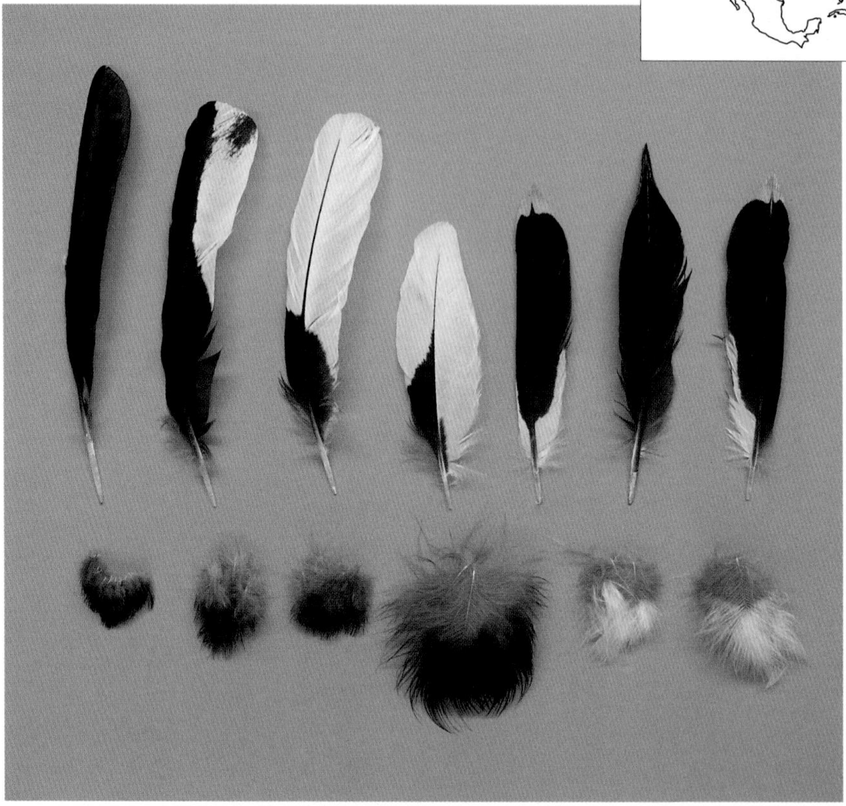

Acorn Woodpecker *(Melanerpes formicivorus)*

unknown
P: 4.7 (11.9)
P: 4.0 (10.2)
S: 3.5 (8.8)
T: 3.6 (9.2)
T: 3.1 (7.8)
T: 2.9 (7.4)
MA: 1.7 (4.3)
UPTC: 2.0 (5.1)
BR: 1.0 (2.5)
FL: 1.7 (4.2)
CR: 0.5 (1.3)

Gila Woodpecker *(Melanerpes uropygialis)*

male
P: 3.7 (9.4)
P: 4.4 (11.3)
P: 4.3 (10.9)
S: 3.6 (9.2)
S: 3.2 (8.1)
T: 3.0 (7.5)
T: 3.5 (8.9)
T: 3.5 (8.8)
BR: 1.1 (2.9)
BE: 1.2 (3.0)
MA: 1.4 (3.6)
CR: 0.5 (1.2)

Red-bellied Woodpecker *(Melanerpes carolinus)*

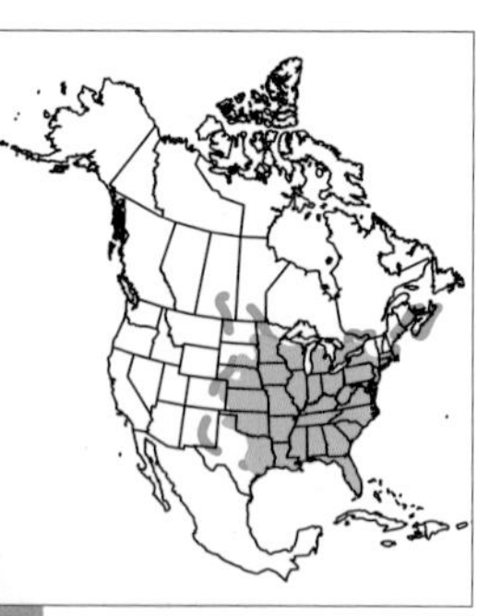

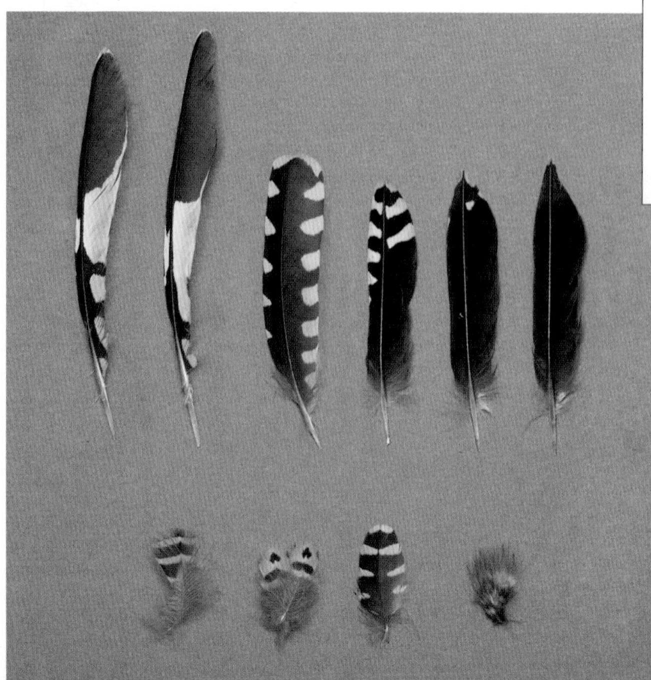

M>F

male
P: 4.2 (10.7)
P: 4.4 (11.3)
S: 3.2 (8.2)
T: 2.9 (7.4)
T: 3.1 (7.9)
T: 3.3 (8.3)
MA: 1.5 (3.7)
RU: 1.1 (2.7)
WC: 1.3 (3.2)
CR: 0.7 (1.8)

Yellow-bellied Sapsucker *(Sphyrapicus varius)*

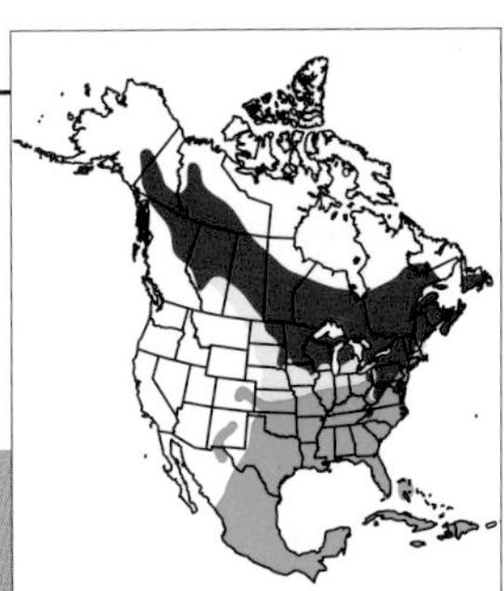

female	**S:** 2.8 (7.0)	**NA:** 0.7 (1.9)
P: 3.5 (9.0)	**T:** 3.3 (8.3)	**BR:** 1.0 (2.5)
P: 4.3 (10.8)	**T:** 3.1 (7.9)	**BE:** 1.1 (2.8)
P: 3.3 (8.3)	**T:** 2.7 (6.8)	**RU:** 1.3 (3.4)
S: 2.9 (7.3)	**T:** 2.5 (6.4)	**WC:** 1.3 (3.4)

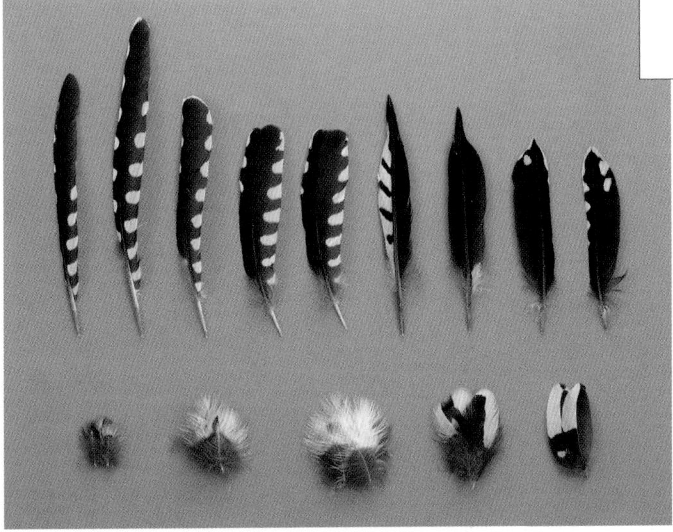

Red-naped Sapsucker *(Sphyrapicus nuchalis)*

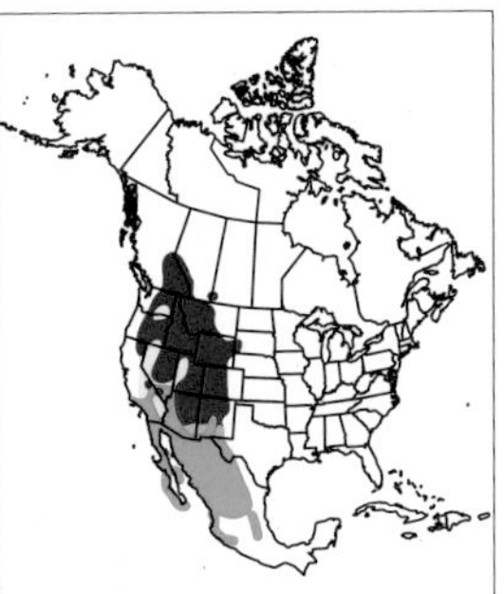

unknown	**S:** 3.1 (8.0)	**T:** 3.2 (8.2)
P: 3.5 (8.8)	**S:** 3.0 (7.5)	**RU:** 1.8 (4.5)
P: 4.3 (10.9)	**T:** 2.5 (6.3)	**BE:** 1.1 (2.8)
P: 3.3 (8.5)	**T:** 2.8 (7.1)	**MA:** 1.2 (3.0)

Red-breasted Sapsucker *(Sphyrapicus ruber)*

unknown
P: 3.5 (9.0)
P: 4.3 (11.0)
S: 3.4 (8.7)
S: 2.9 (7.4)
T: 3.0 (7.7)
T: 3.2 (8.1)
BE: 1.2 (3.1)
MC: 0.7 (1.8)
BR: 0.9 (2.4)
NA: 1.3 (3.2)
CR: 0.5 (1.2)

Ladder-backed Woodpecker *(Picoides scalaris)*

unknown
P: 3.5 (9.0)
P: 3.4 (8.7)
P: 3.2 (8.1)
S: 2.7 (6.9)
S: 2.4 (6.1)
T: 2.5 (6.3)
T: 2.8 (7.0)
BR: 0.8 (2.0)
MA: 1.3 (3.4)

Nuttall's Woodpecker *(Picoides nuttallii)*

male
P: 2.8 (7.2)
P: 3.4 (8.6)
P: 3.5 (8.9)
P: 2.9 (7.3)
S: 2.7 (6.8)
S: 2.5 (6.3)
T: 2.4 (6.2)
T: 2.9 (7.3)
CR: 0.6 (1.4)
BR: 0.7 (1.7)
MA: 1.1 (2.8)

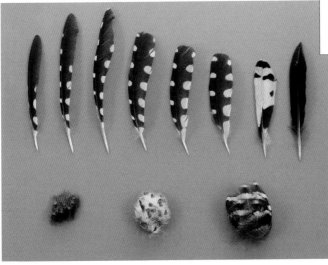

Downy Woodpecker *(Picoides pubescens)*

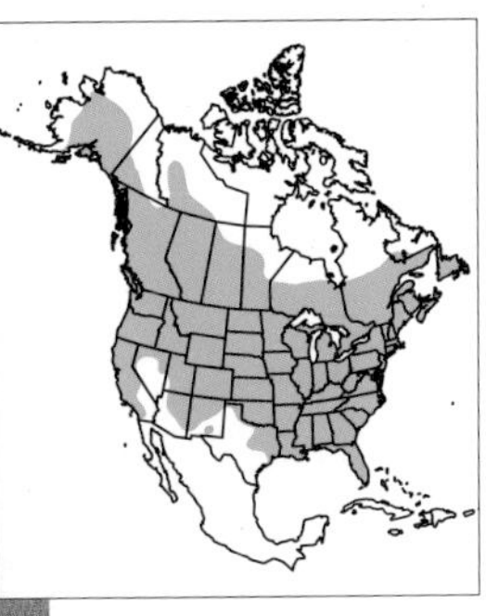

M>F

unknown
P: 3.1 (7.9)
P: 3.2 (8.2)
S: 2.5 (6.3)
T: 2.1 (5.4)
T: 1.9 (4.9)
T: 2.4 (6.0)
WC: 1.0 (2.5)
MA: 1.3 (3.3)
RU: 1.3 (3.3)
BR: 0.6 (1.4)

Hairy Woodpecker *(Picoides villosus)*

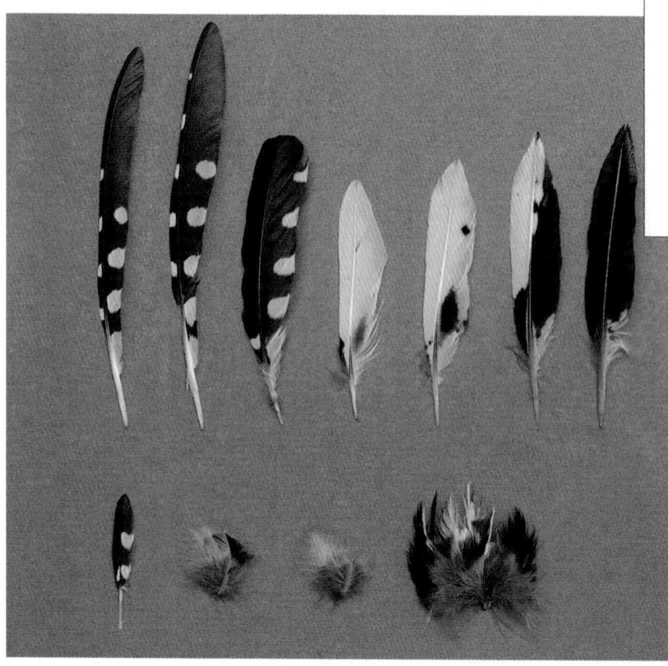

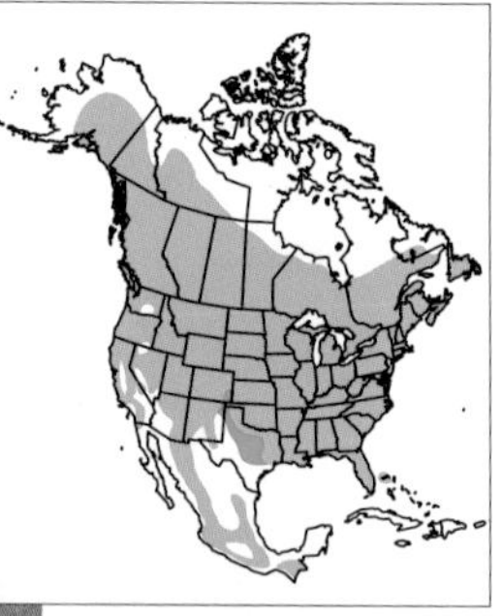

M>F

unknown
P: 4.1 (10.4)
P: 4.4 (11.2)
S: 3.1 (7.9)
T: 2.6 (6.5)
T: 2.9 (7.4)
T: 3.1 (8.0)
T: 3.3 (8.4)
WC: 1.5 (3.9)
BR: 0.9 (2.2)
BE: 0.9 (2.4)
NA: 1.4 (3.5)

White-headed Woodpecker *(Picoides albolarvatus)*

M>F

male
P: 3.9 (10.0)
P: 3.7 (9.4)
S: 3.3 (8.3)
S: 3.0 (7.7)
T: 3.3 (8.3)
T: 3.7 (9.3)
WC: 1.1 (2.8)
UPTC: 1.9 (4.7)
MA: 1.9 (4.7)
CR: 0.7 (1.7)

Northern Flicker (red-shafted)
(Colaptes auratus)

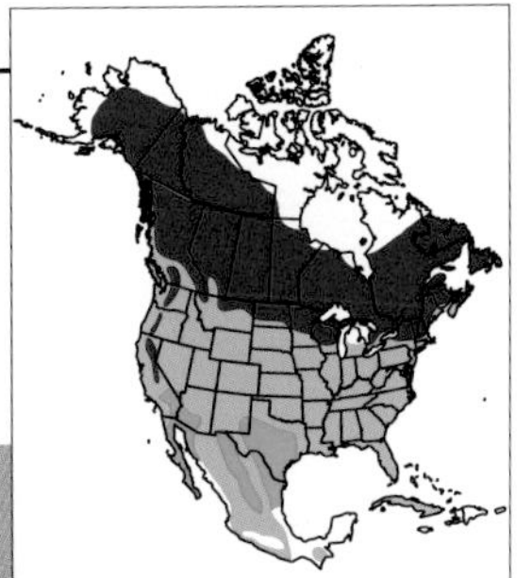

unknown
P: 4.0 (10.1)
P: 4.8 (12.3)
S: 3.9 (10.0)
TS: 3.5 (9.0)
T: 3.3 (8.4)
T: 3.7 (9.3)
T: 4.0 (10.1)
WC: 1.8 (4.5)
BR: 0.8 (2.1)
BE: 0.9 (2.4)
RU: 1.3 (3.3)

Northern Flicker (yellow-shafted)
(Colaptes auratus)

unknown
P: 3.9 (10.0)
P: 4.8 (12.3)
P: 4.8 (12.2)
S: 4.4 (11.2)
S: 4.2 (10.6)
T: 3.4 (8.7)
T: 4.1 (10.4)
T: 3.4 (8.7)
NA: 0.8 (2.1)
BR: 1.0 (2.5)
MA: 1.2 (3.1)
RU: 2.2 (5.5)

Gilded Flicker *(Colaptes chrysoides)*

unknown
P: 4.9 (12.5)
P: 5.2 (13.1)
P: 4.5 (11.4)
P: 4.2 (10.6)
S: 4.0 (10.1)
S: 3.6 (9.2)
T: 3.6 (9.2)
T: 4.2 (10.6)
T: 4.1 (10.5) [underside of feather]
BR: 1.3 (3.4)
MA: 1.3 (3.4)
RU: 0.9 (2.4)
UNTC: 1.4 (3.5)
CR: 0.4 (1.1)

Pileated Woodpecker *(Dryocopus pileatus)*

M>F

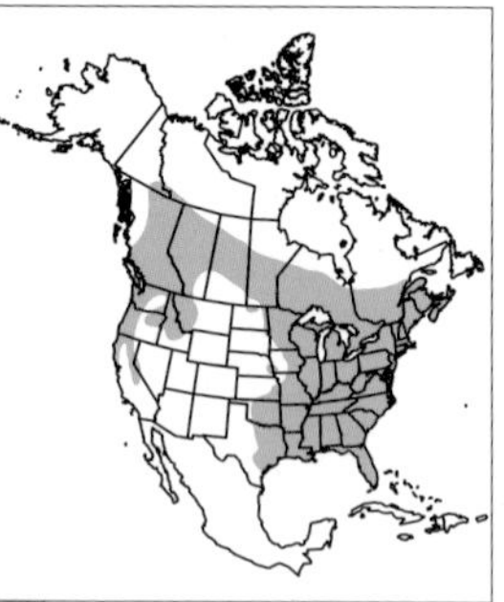

male
P: 2.9 (7.4)
P: 5.8 (14.8)
P: 8.1 (20.6)
P: 7.2 (18.2)
S: 6.7 (17.0)
S: 5.7 (14.5)
T: 5.1 (12.9)
T: 6.5 (16.5)
T: 6.8 (17.2)
NA: 1.0 (2.5)
BR: 1.4 (3.6)
MA: 1.6 (4.1)
CR: 1.1 (2.7)

Tyrant Flycatchers *(Tyrannidae)*

Scissor-tailed Flycatcher and Western Kingbird have sharp, pointed leading primaries.

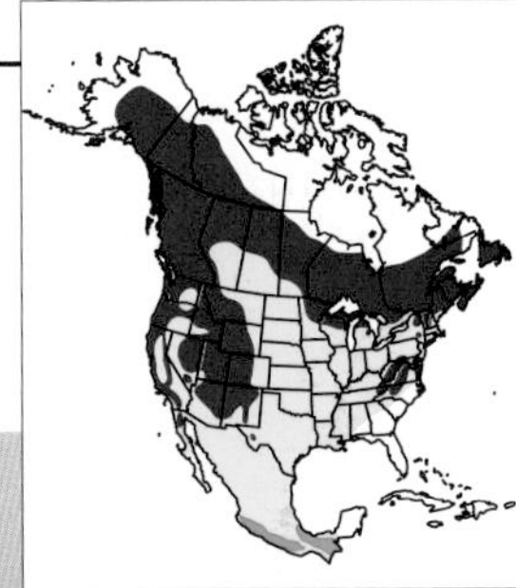

Olive-sided Flycatcher *(Contopus cooperi)*

unknown	**T:** 3.0 (7.7)
P: 3.3 (8.3)	**T:** 3.0 (7.7)
P: 3.5 (8.8)	**FL:** 1.0 (2.5)
P: 2.8 (7.1)	**BE:** 1.2 (3.1)
S: 2.4 (6.1)	**RU:** 1.1 (2.9)

Eastern Wood-Pewee *(Contopus virens)*

unknown	**S:** 2.0 (5.0)	**BE:** 0.9 (2.4)
P: 2.2 (5.5)	**TS:** 1.7 (4.4)	**BR:** 0.7 (1.7)
P: 2.5 (6.4)	**T:** 2.5 (6.3)	**MA:** 0.7 (1.9)
P: 2.0 (5.2)	**T:** 2.5 (6.4)	

Yellow-bellied Flycatcher *(Empidonax flaviventris)*

unknown	**P:** 1.8 (4.6)	**T:** 2.1 (5.4)
P: 1.7 (4.3)	**S:** 1.7 (4.2)	**BR:** 1.0 (2.5)
P: 2.1 (5.4)	**S:** 1.5 (3.8)	**MA:** 1.3 (3.4)
P: 2.2 (5.5)	**T:** 2.2 (5.5)	

Willow Flycatcher or Alder Flycatcher *(Empidonax traillii or Empidonax alnorum)*

unknown
P: 2.0 (5.0)
P: 2.4 (6.0)
P: 2.2 (5.7)
P: 2.1 (5.4)
S: 1.8 (4.6)
S: 1.8 (4.5)
T: 2.4 (6.1)
T: 2.4 (6.2)
BR: 1.3 (3.3)
MA: 0.7 (1.9)
WC: 0.7 (1.9)

species visually indistinguishable

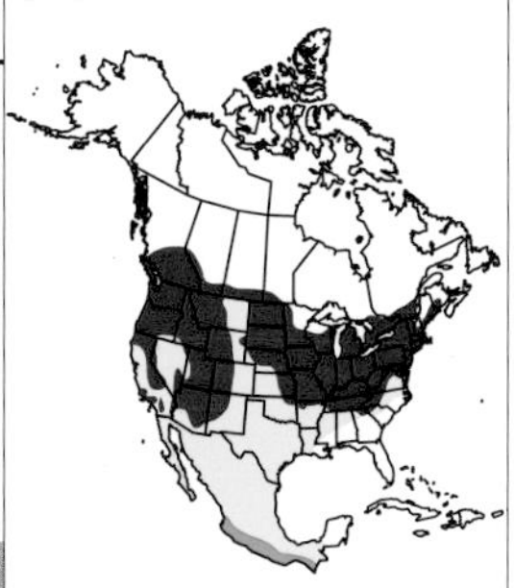

Willow Flycatcher

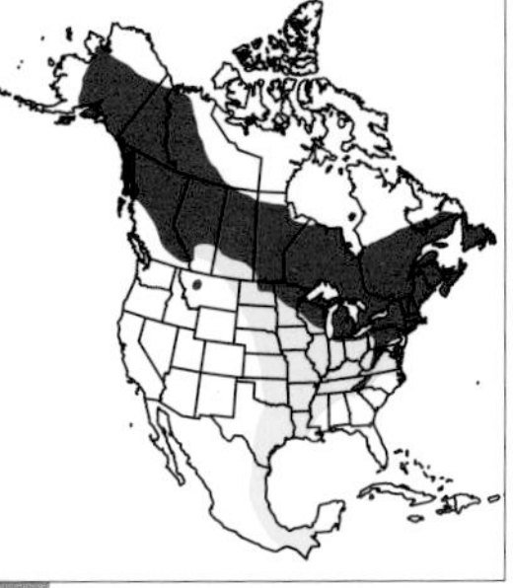

Alder Flycatcher

Least Flycatcher *(Empidonax minimus)*

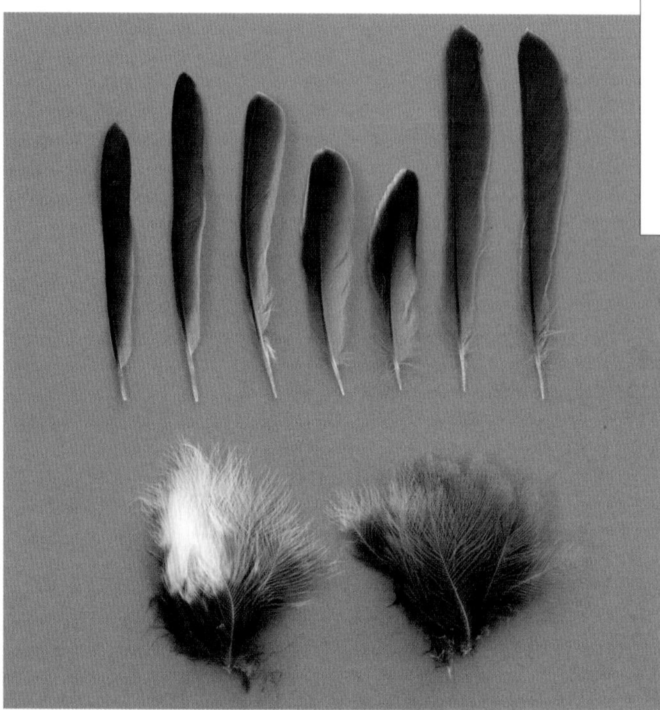

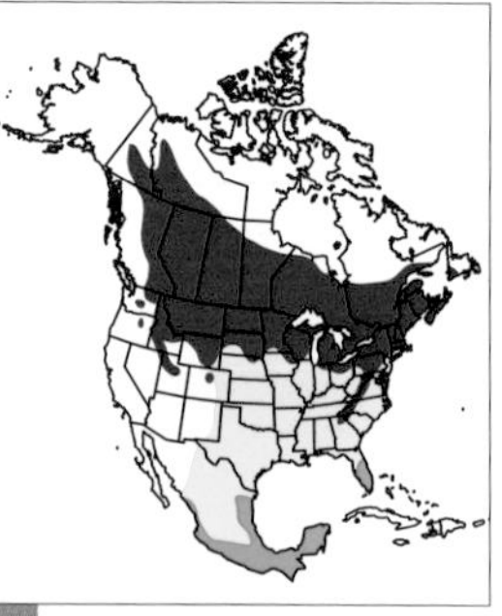

unknown
P: 1.8 (4.5)
P: 2.1 (5.3)
P: 1.9 (4.9)
S: 1.6 (4.0)
TS: 1.4 (3.6)
T: 2.4 (6.0)
T: 2.3 (5.9)
BR: 1.5 (3.7)
RU: 1.1 (2.9)

Hammond's Flycatcher *(Empidonax hammondii)*

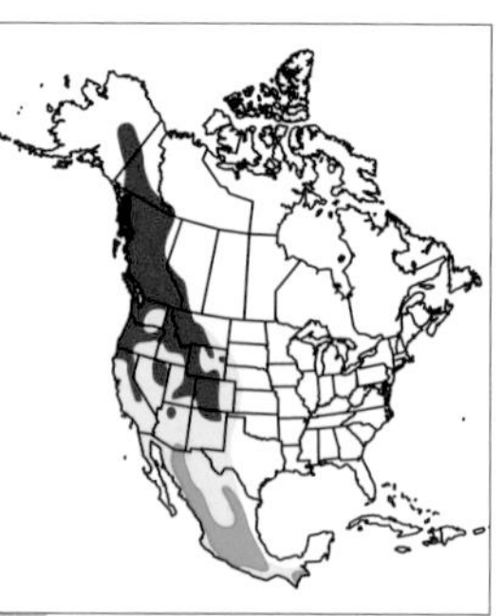

unknown
P: 1.9 (4.9)
P: 2.4 (6.2)
P: 2.0 (5.1)
S: 1.7 (4.3)
T: 2.5 (6.4)
T: 2.4 (6.0)
TS: 1.5 (3.7)
MA: 0.8 (2.0)
BE: 1.0 (2.5)
BR: 0.5 (1.3)

Gray Flycatcher *(Empidonax wrightii)*

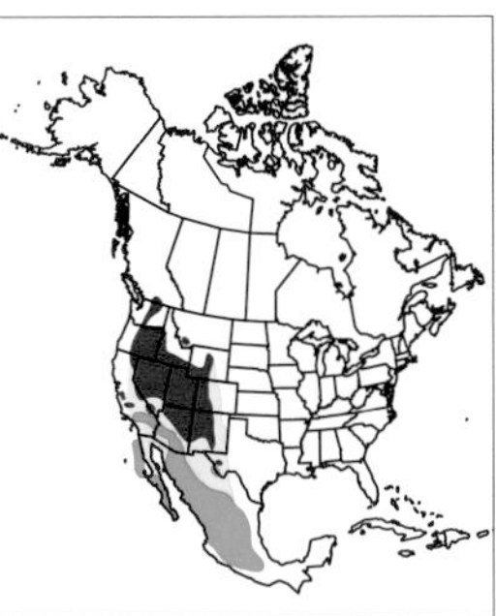

unknown
P: 2.4 (6.1)
P: 2.2 (5.5)
S: 2.1 (5.4)
S: 2.0 (5.1)
T: 2.6 (6.6)
T: 2.6 (6.5)
NA: 0.8 (2.0)
WC: 0.8 (2.1)
BE: 1.1 (2.8)
TH: 0.5 (1.2)
UPTC: 1.3 (3.3)

Pacific-slope Flycatcher *(Empidonax difficilis)*

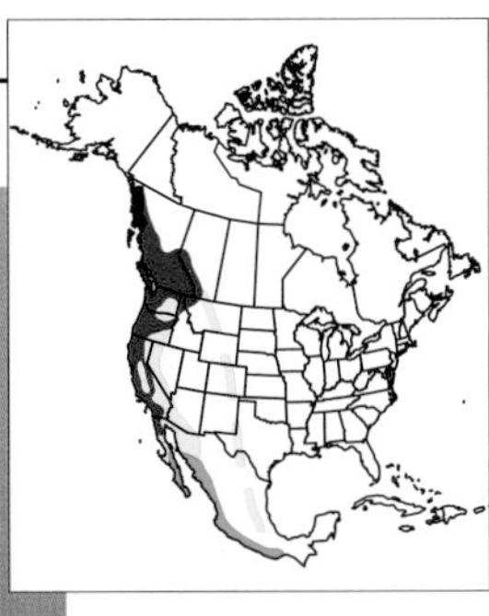

unknown
P: 1.7 (4.4)
P: 2.2 (5.6)
P: 2.0 (5.1)
S: 1.9 (4.8)
TS: 1.6 (4.0)
T: 2.4 (6.2)
T: 2.3 (5.9)
T: 2.4 (6.2)
MA: 0.8 (2.0)
BR: 0.7 (1.8)
BE: 0.8 (2.0)
RU: 0.9 (2.2)

Black Phoebe *(Sayornis nigricans)*

unknown	**S:** 2.5 (6.4)	**T:** 3.2 (8.2)
P: 2.4 (6.1)	**S:** 2.4 (6.1)	**NA:** 1.0 (2.5)
P: 3.0 (7.6)	**T:** 3.2 (8.2)	**BR:** 0.9 (2.4)
P: 2.6 (6.6)	**T:** 3.2 (8.2)	**BE:** 1.2 (3.1)

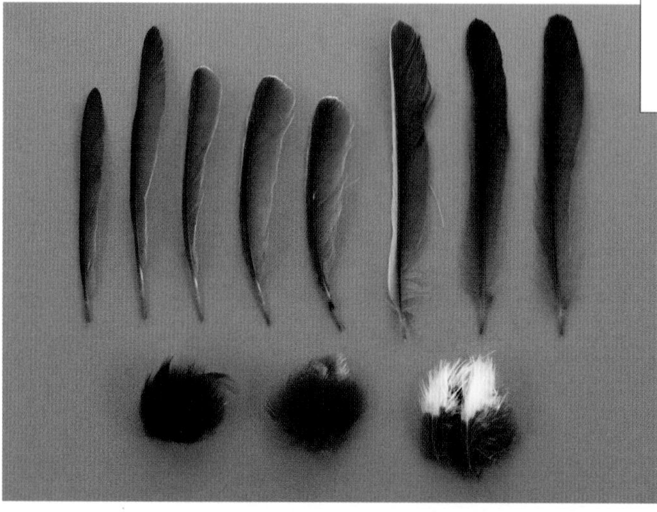

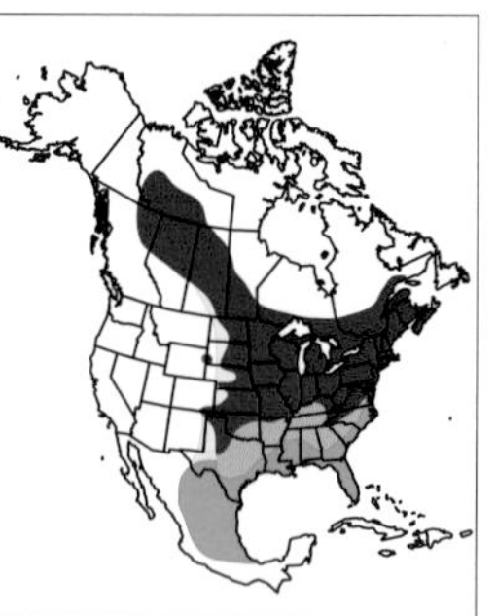

Eastern Phoebe *(Sayornis phoebe)*

unknown	**P:** 2.2 (5.5)	**T:** 2.6 (6.6)
P: 2.3 (5.9)	**S:** 2.0 (5.0)	**BR:** 0.6 (1.6)
P: 2.8 (7.0)	**TS:** 1.8 (4.6)	**BE:** 0.9 (2.2)
P: 2.4 (6.0)	**T:** 2.7 (6.9)	**RU:** 0.9 (2.3)

Say's Phoebe *(Sayornis saya)*

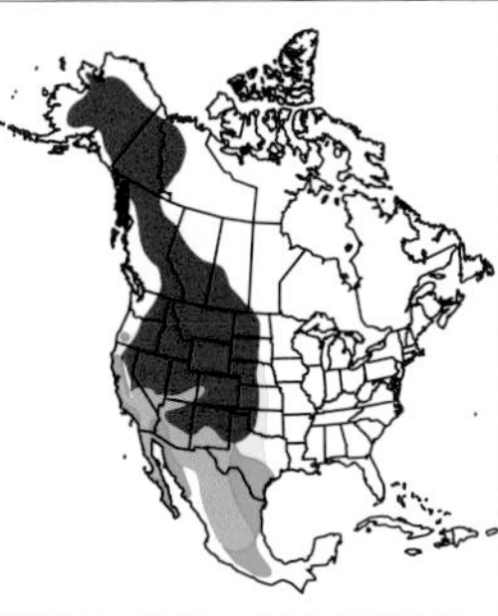

unknown
P: 3.5 (8.9)
P: 3.6 (9.1)
S: 2.4 (6.1)
T: 3.5 (9.0)
T: 3.6 (9.1)
BR: 0.9 (2.4)
BE: 1.6 (4.0)
MA: 1.2 (3.0)

Vermilion Flycatcher *(Pyrocephalus rubinus)*

male
P: 2.8 (7.1)
P: 2.7 (6.8)
S: 2.2 (5.5)
T: 2.6 (6.7)
T: 2.6 (6.6)
BR: 0.8 (2.1)
CR: 0.7 (1.7)
MA: 1.1 (2.9)

Ash-throated Flycatcher *(Myiarchus cinerascens)*

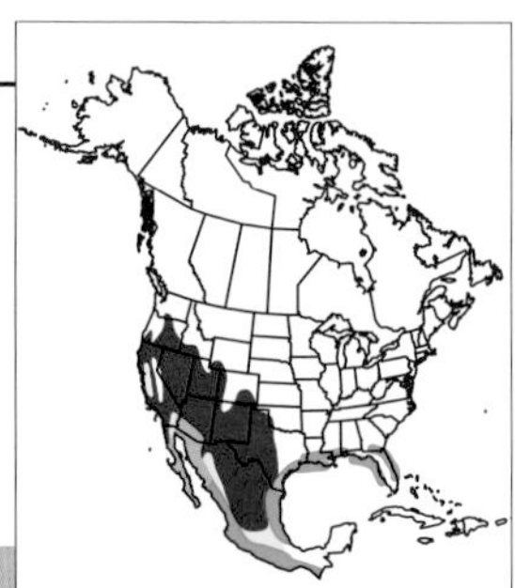

unknown
P: 2.8 (7.0)
P: 3.4 (8.6)
S: 2.8 (7.0)
TS: 2.6 (6.6)
T: 3.9 (9.8)
T: 3.9 (9.8)
TS: 2.4 (6.0)
TH: 0.7 (1.7)
BE: 0.9 (2.2)
UPTC: 1.1 (2.7)
RU: 1.0 (2.5)

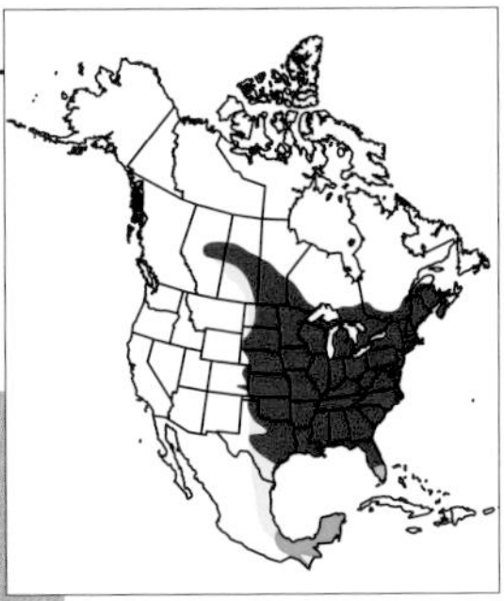

Great Crested Flycatcher *(Myiarchus crinitus)*

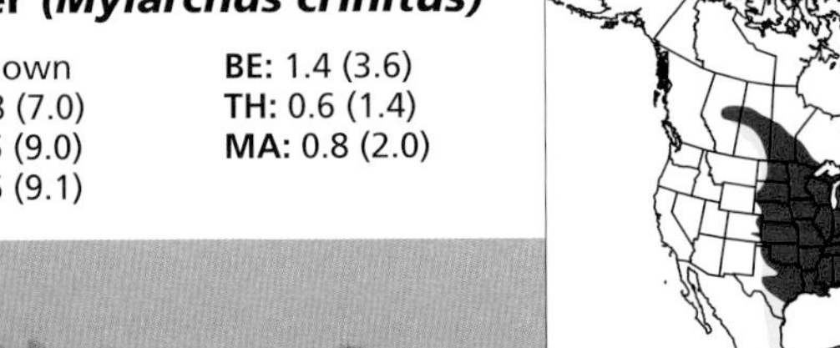

unknown
P: 2.8 (7.0)
T: 3.5 (9.0)
T: 3.6 (9.1)

BE: 1.4 (3.6)
TH: 0.6 (1.4)
MA: 0.8 (2.0)

Western Kingbird *(Tyrannus verticalis)*

unknown
P: 4.3 (10.9)
P: 4.1 (10.3)
P: 3.6 (9.1)
S: 3.1 (8.0)
T: 3.8 (9.7)
T: 3.8 (9.7)
T: 3.8 (9.7)
BR: 0.8 (2.1)
BE: 1.3 (3.3)
MA: 0.7 (1.9)
RU: 0.9 (2.3)
CR: 0.5 (1.2)

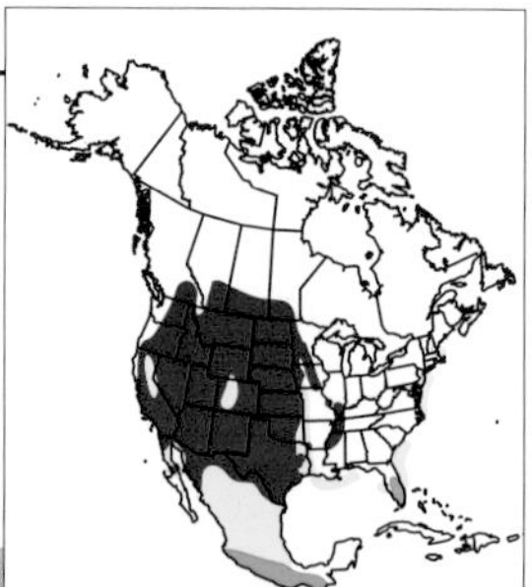

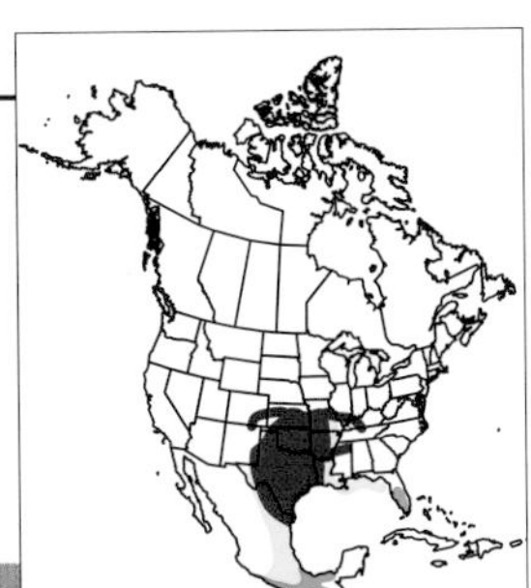

Scissor-tailed Flycatcher *(Tyrannus forficatus)*

male
P: 3.8 (9.7)
P: 4.1 (10.4)
P: 3.9 (9.9)
P: 3.4 (8.7)
S: 2.7 (6.8)
T: 5.8 (14.7)
T: 8.0 (20.2)
T: 8.6 (21.8)
MC: 0.5 (1.3)
BR: 0.8 (2.0)
BE: 1.4 (3.5)
MA: 1.0 (2.5)

Shrikes *(Laniidae)*

Boldly marked black and white tail. Primaries can be mistaken for some woodpeckers, some grosbeaks, mockingbirds, kingfishers, and Lesser Goldfinch.

Loggerhead Shrike *(Lanius ludovicianus)*

unknown
P: 3.4 (8.7)
P: 3.2 (8.2)
S: 2.8 (7.2)
S: 2.5 (6.3)
T: 3.3 (8.5)
T: 3.9 (9.8)
T: 3.9 (9.9)
T: 4.0 (10.2)
WC: 1.2 (3.0)
MA: 1.4 (3.6)
BR: 1.4 (3.5)

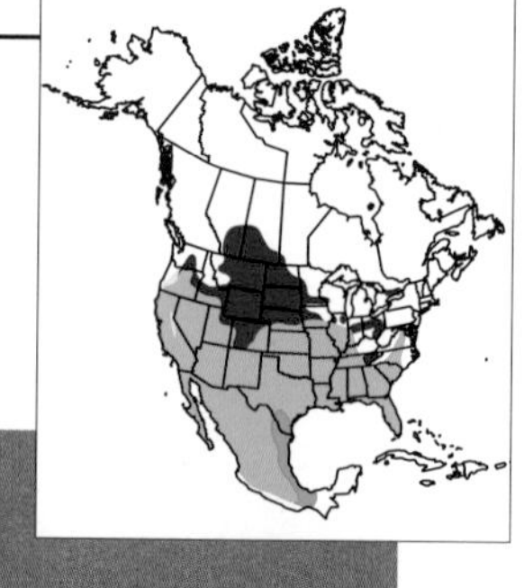

Vireos (*Vireonidae*)

White-eyed Vireo (*Vireo griseus*)

unknown
P: 1.6 (4.1)
P: 2.0 (5.1)
S: 1.5 (3.9)
S: 1.4 (3.6)
T: 2.0 (5.1)
T: 2.0 (5.1)
FL: 1.1 (2.8)
BE: 0.8 (2.1)

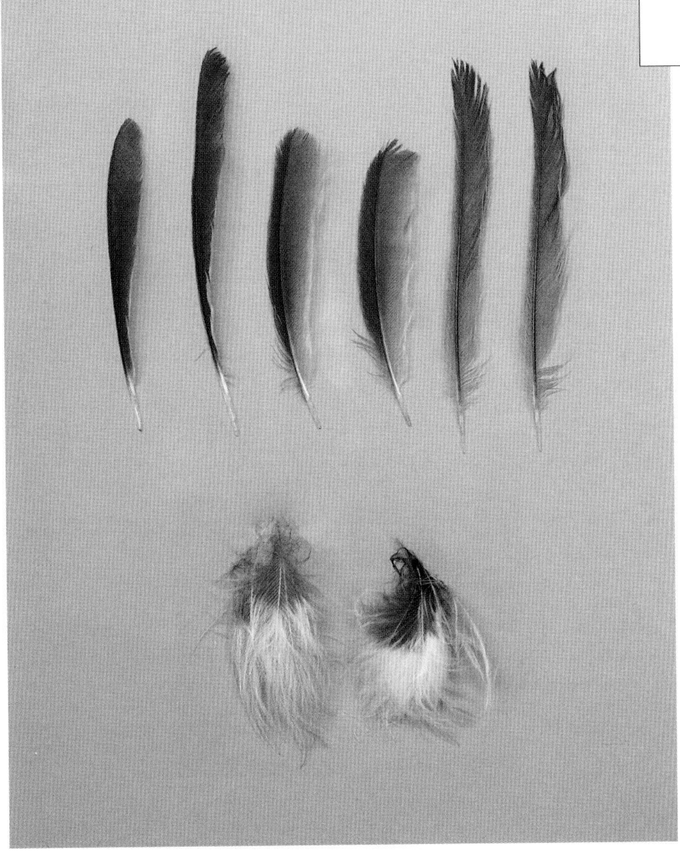

Bell's Vireo *(Vireo bellii)*

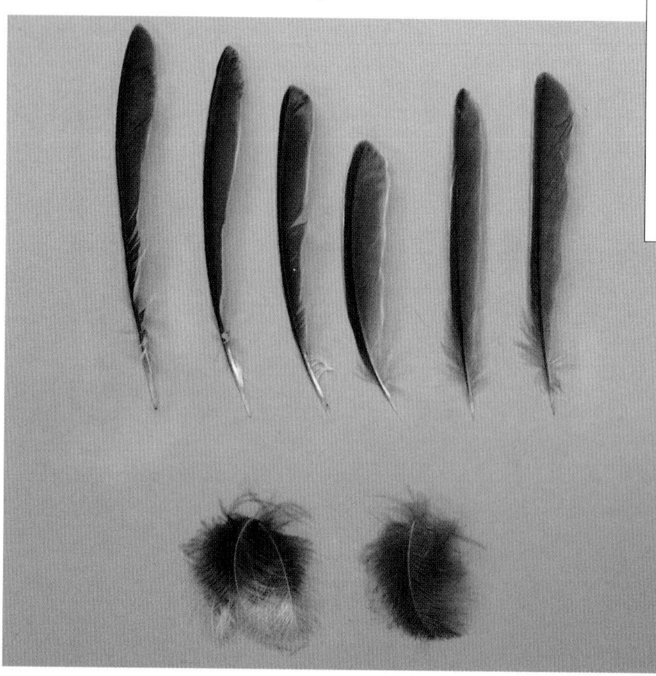

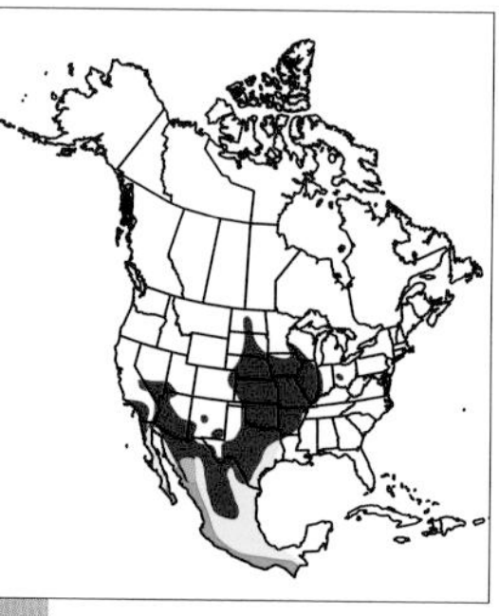

unknown
P: 2.5 (6.3)
P: 2.4 (6.1)
P: 2.2 (5.5)
S: 1.8 (4.6)
T: 2.0 (5.2)
T: 2.2 (5.6)
BE: 0.9 (2.3)
RU: 0.9 (2.2)

Yellow-throated Vireo *(Vireo flavifrons)*

unknown
P: 2.6 (6.7)
P: 2.4 (6.2)
P: 2.0 (5.1)
S: 1.9 (4.7)
TS: 1.7 (4.3)
T: 2.0 (5.0)
T: 1.9 (4.9)
NA: 0.5 (1.3)
BR: 0.5 (1.3)
BE: 0.9 (2.3)
RU: 0.8 (2.1)

Cassin's Vireo *(Vireo cassinii)*

unknown
P: 2.6 (6.5)
P: 2.1 (5.4)
S: 1.8 (4.6)
TS: 1.7 (4.4)
T: 2.2 (5.5)
T: 2.2 (5.5)
TS: 1.4 (3.6)
BE: 0.8 (2.1)
FL: 1.0 (2.6)

Blue-headed Vireo *(Vireo solitarius)*

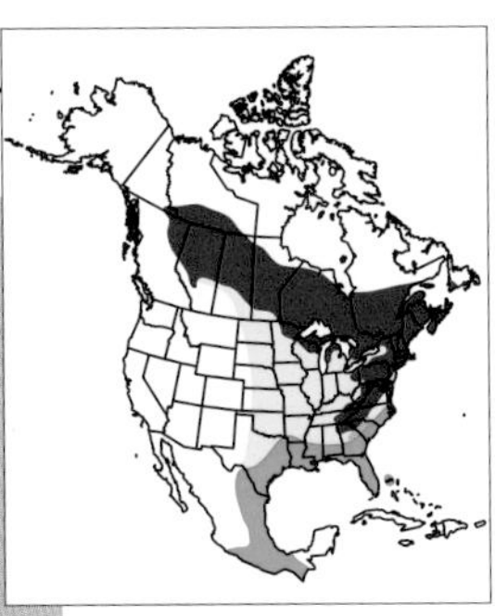

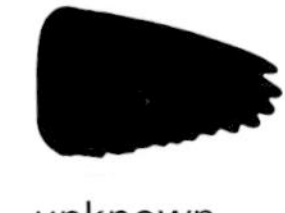

unknown
P: 2.8 (7.0)
P: 2.4 (6.1)
S: 2.1 (5.4)
TS: 1.8 (4.6)
T: 2.4 (6.0)
T: 2.5 (6.4)
TS: 1.7 (4.3)
WC: 0.5 (1.2)
NA: 0.4 (1.1)
BE: 0.9 (2.2)
RU: 1.1 (2.9)

Warbling Vireo *(Vireo gilvus)*

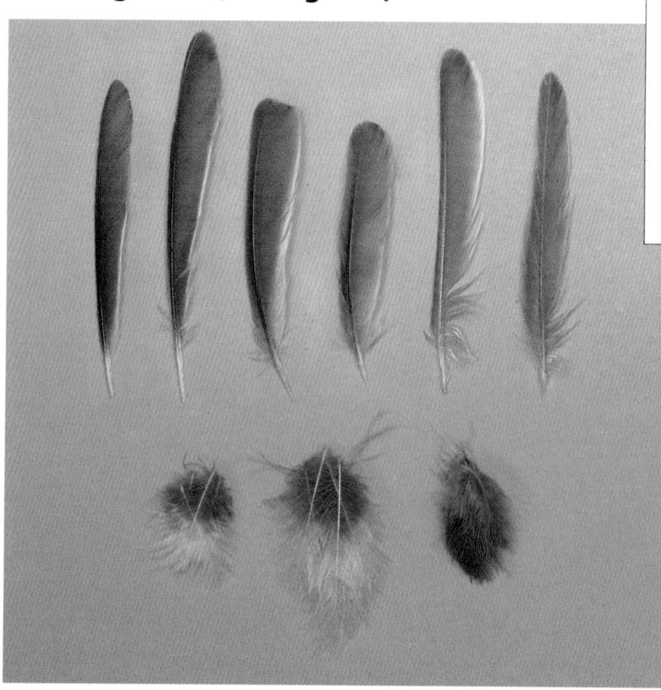

unknown
P: 2.0 (5.1)
P: 2.3 (5.9)
S: 1.9 (4.8)
S: 1.7 (4.3)
T: 2.2 (5.6)
T: 2.1 (5.4)
BE: 0.7 (1.7)
FL: 1.3 (3.4)
RU: 0.8 (2.0)

Philadelphia Vireo *(Vireo philadelphicus)*

unknown
P: 2.0 (5.0)
P: 2.3 (5.8)
P: 2.1 (5.3)
S: 1.8 (4.5)
TS: 1.6 (4.0)
T: 2.1 (5.3)
T: 2.0 (5.1)
BR: 1.2 (3.0)
MA: 1.1 (2.9)

Red-eyed Vireo *(Vireo olivaceus)*

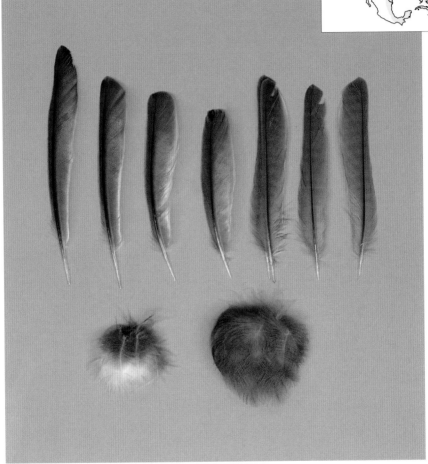

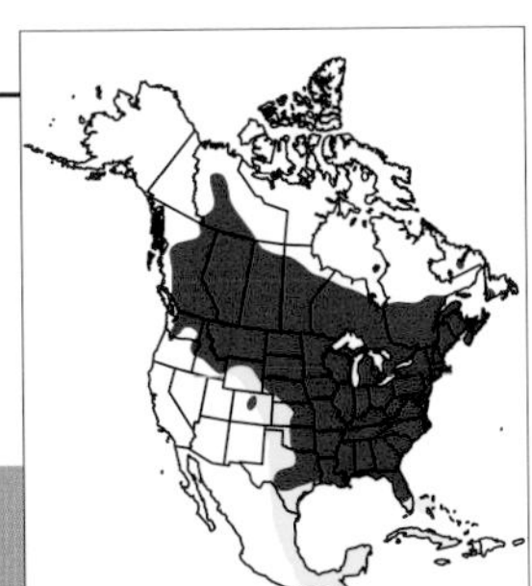

unknown
P: 2.7 (6.8)
P: 2.4 (6.0)
S: 2.1 (5.4)
S: 1.9 (4.7)

T: 2.2 (5.7)
T: 2.2 (5.6)
T: 2.2 (5.7)
BR: 0.7 (1.8)
MA: 0.9 (2.3)

Jays, Crows, and Allies *(Corvidae)*

Crows, ravens, and magpies have iridescence in their flight and body feathers.

Steller's Jay *(Cyanocitta stelleri)*

unknown
P: 4.4 (11.3)
P: 4.9 (12.4)
S: 3.7 (9.5)
TS: 2.7 (6.8)
T: 5.2 (13.1)

T: 5.2 (13.2)
CR: 1.7 (4.3)
BE: 1.9 (4.7)
RU: 1.6 (4.0)
MA: 2.0 (5.0)

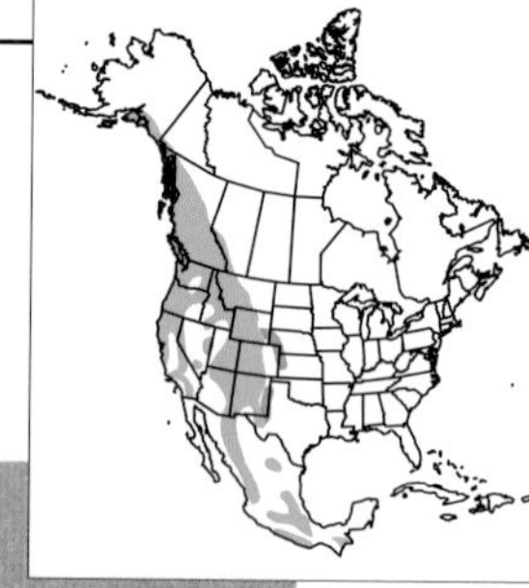

Blue Jay *(Cyanocitta cristata)*

unknown
P: 4.5 (11.5)
P: 4.7 (11.9)
S: 3.5 (8.9)
T: 4.9 (12.4)
T: 5.1 (12.9)
T: 5.2 (13.2)
CR: 1.3 (3.4)
MA: 1.9 (4.9)
WC: 1.7 (4.3)
UPTC: 1.8 (4.6)

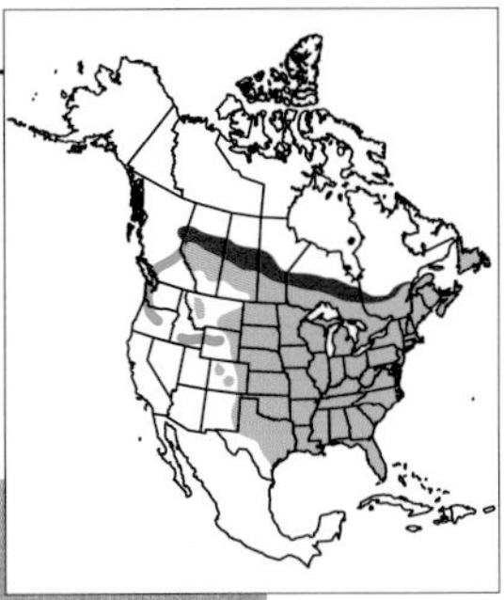

Western Scrub-Jay *(Aphelocoma californica)*

unknown
P: 4.4 (11.1)
P: 4.4 (11.1)
S: 3.9 (9.8)
S: 3.9 (9.8)
T: 4.8 (12.1)
T: 5.5 (13.9)
CR: 0.5 (1.2)
WC: 1.4 (3.6)
TH: 0.7 (1.7)
NA: 0.7 (1.9)
MA: 1.9 (4.9)
BE: 1.6 (4.0)

Mexican Jay *(Aphelocoma ultramarina)*

unknown
P: 4.8 (12.1)
P: 5.9 (14.9)
S: 5.0 (12.8)
T: 6.1 (15.6)
T: 6.3 (15.9)
WC: 1.7 (4.3)
MA: 1.6 (4.0)
BR: 1.3 (3.3)

Black-billed Magpie *(Pica hudsonia)*

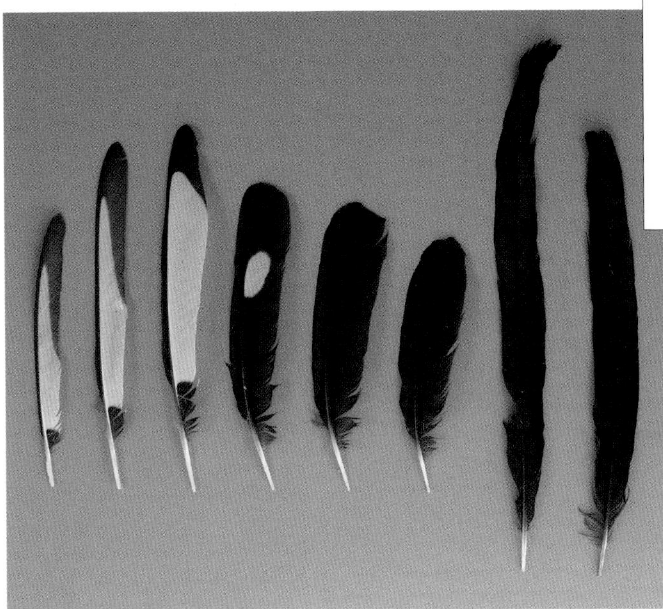

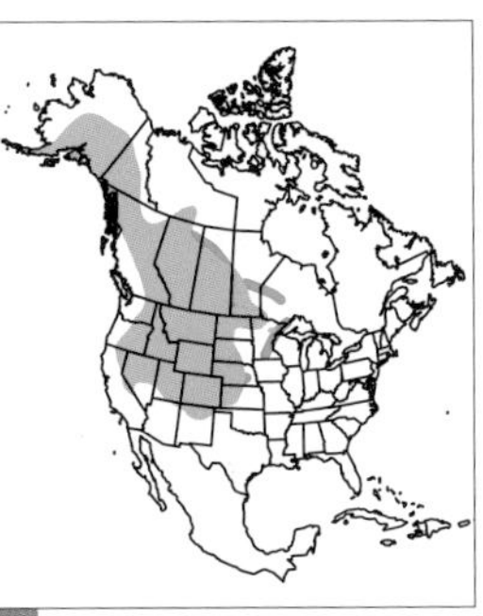

M>F

unknown
P: 5.3 (13.4)
P: 6.8 (17.2)
P: 7.1 (18.1)
P: 6.0 (15.2)
S: 5.6 (14.3)
TS: 5.0 (12.6)
T: 10.4 (26.3)
T: 8.7 (22.0)

American Crow *(Corvus brachyrhynchos)*

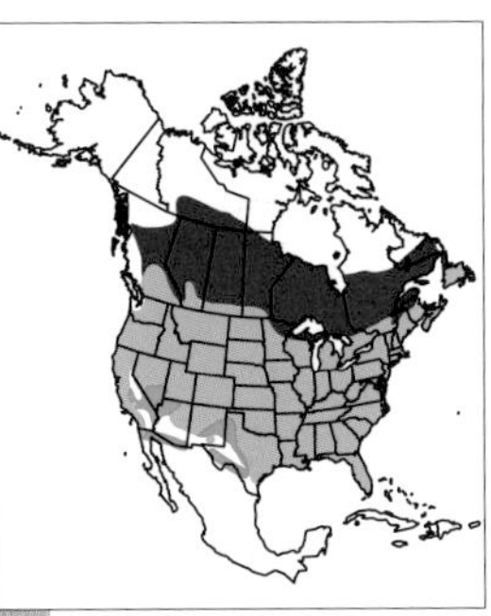

unknown
P: 8.2 (20.8)
P: 10.3 (26.2)
P: 10.3 (26.1)
P: 9.0 (22.8)
S: 7.3 (18.5)
T: 6.6 (16.7)
T: 7.6 (19.4)
BR: 2.2 (5.6)
RU: 2.4 (6.1)
MA: 2.5 (6.3)

Common Raven *(Corvus corax)*

M>F

unknown	**P:** 10.3 (26.2)
P: 8.0 (20.3)	**S:** 8.9 (22.6)
P: 11.7 (29.7)	**TS:** 8.4 (21.3)
P: 13.4 (34.0)	**T:** 9.1 (23.0)
P: 12.4 (31.5)	**T:** 9.4 (23.8)

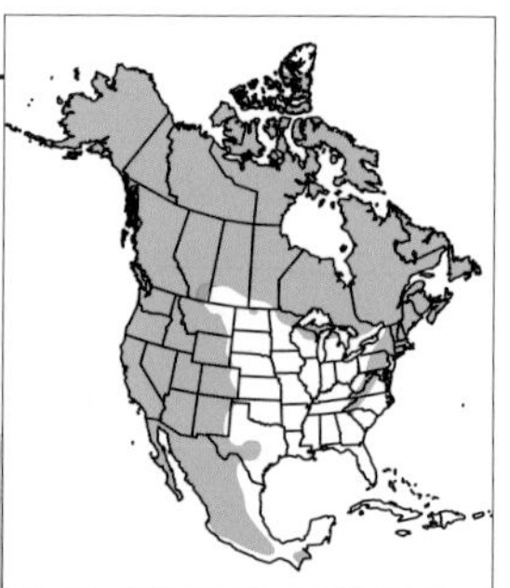

Larks *(Alaudidae)*

Horned Lark *(Eremophila alpestris)*

unknown
P: 3.3 (8.5)
P: 2.9 (7.3)
P: 2.6 (6.5)
S: 2.7 (6.8)
T: 3.0 (7.6)
T: 2.9 (7.3)
TS: 2.2 (5.7)
BE: 1.1 (2.9)
WC: 1.2 (3.0)
WC: 1.2 (3.0)

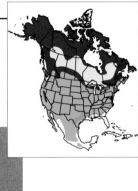

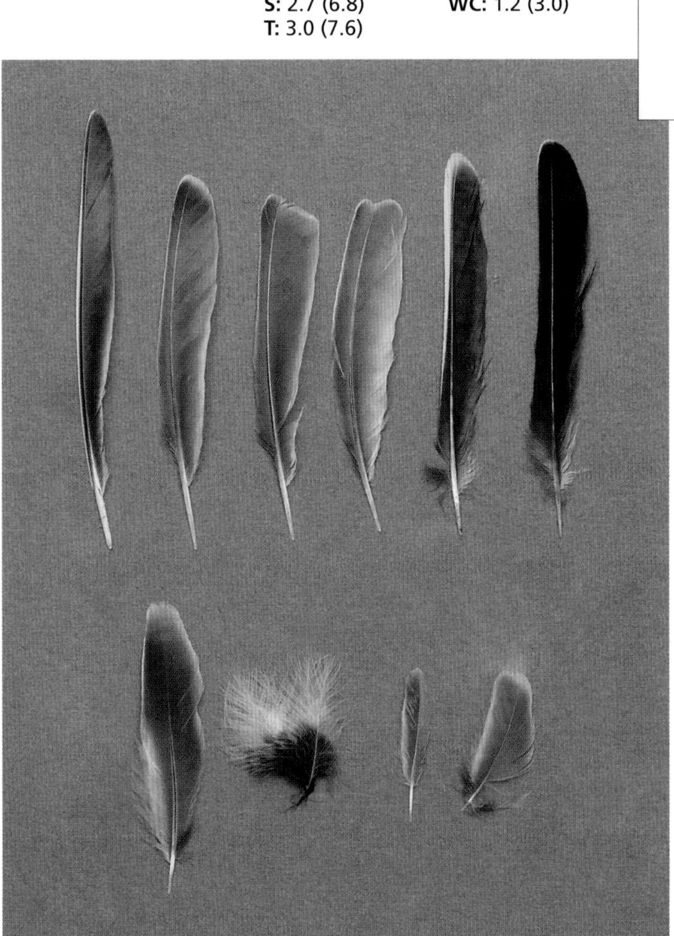

Swallows *(Hirundinidae)*

Similar to swifts. Dark, unmarked primaries and secondaries. Secondaries often notched at top.

Purple Martin *(Progne subis)*

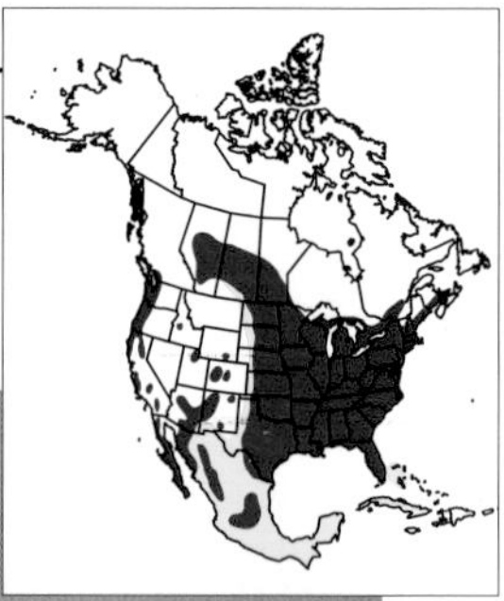

female	**P:** 2.7 (6.8)	**T:** 2.3 (5.8)
P: 4.6 (11.8)	**S:** 2.3 (5.8)	**BR:** 0.6 (1.5)
P: 4.4 (11.3)	**TS:** 2.2 (5.5)	**BE:** 0.9 (2.2)
P: 3.8 (9.7)	**T:** 2.8 (7.2)	**MA:** 0.9 (2.2)

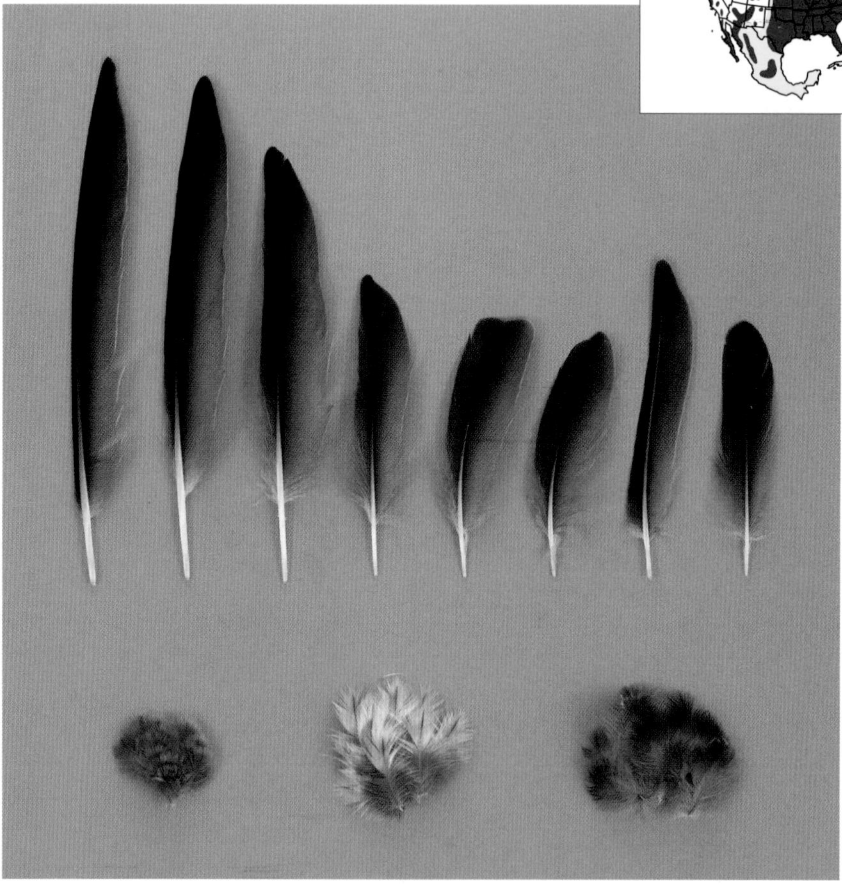

Tree Swallow *(Tachycineta bicolor)*

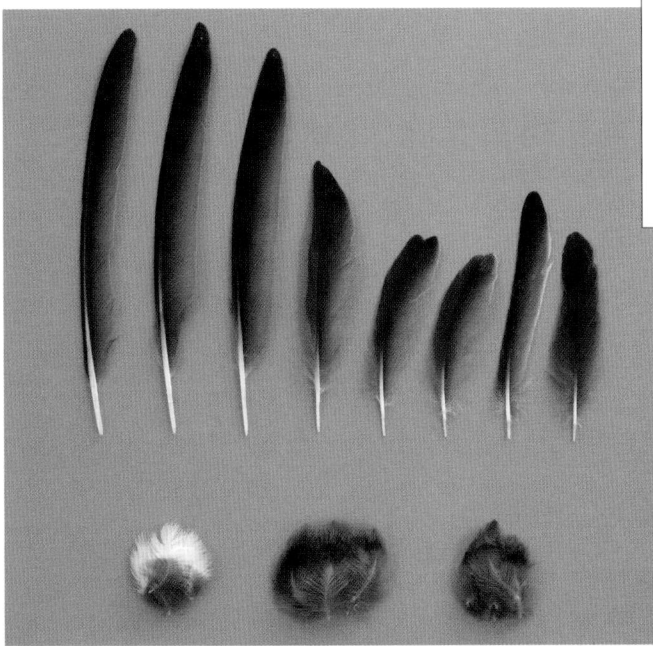

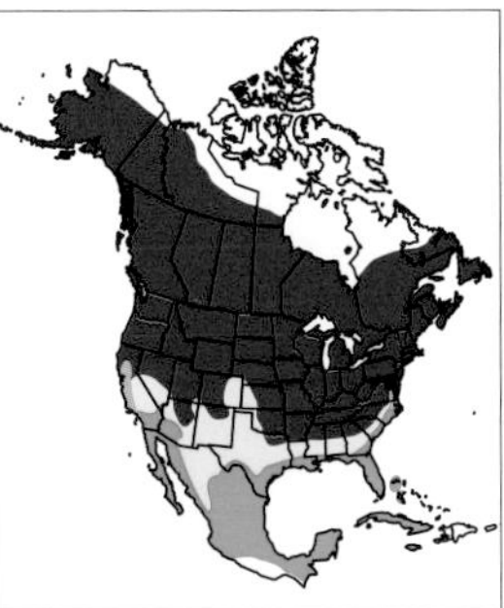

male
P: 3.8 (9.7)
P: 3.9 (9.8)
P: 3.7 (9.3)
P: 2.6 (6.5)
S: 1.9 (4.9)
TS: 1.8 (4.6)
T: 2.4 (6.0)
T: 2.0 (5.2)
BR: 0.7 (1.9)
RU: 0.9 (2.4)
MA: 0.9 (2.3)

Violet-green Swallow *(Tachycineta thalassina)*

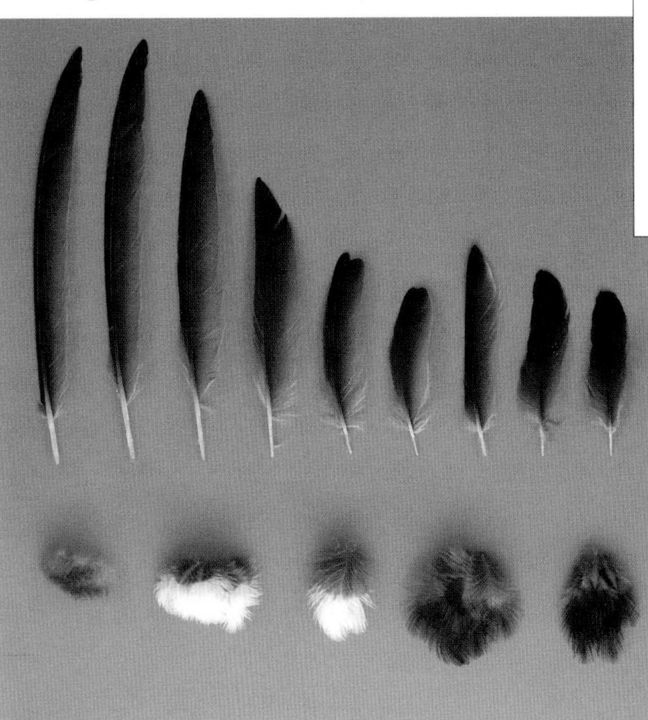

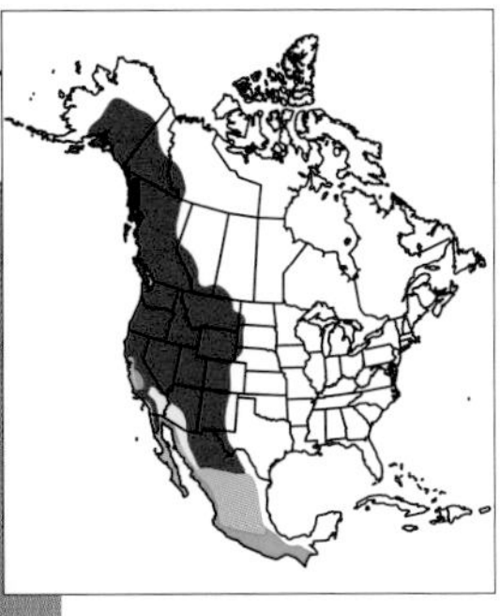

male
P: 4.1 (10.3)
P: 4.1 (10.3)
P: 3.5 (9.0)
P: 2.7 (6.9)
S: 1.9 (4.9)
TS: 1.6 (4.1)
T: 2.0 (5.2)
T: 1.6 (4.0)
T: 1.8 (4.5)
NA: 0.4 (1.0)
BR: 0.6 (1.6)
BE: 0.9 (2.2)
RU: 1.1 (2.7)
UPTC: 0.8 (2.0)

Northern Rough-winged Swallow *(Stelgidopteryx serripennis)*

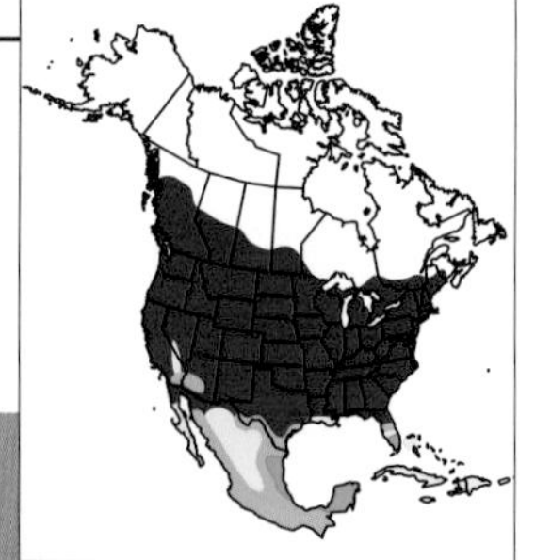

unknown	**S:** 1.9 (4.9)	**MA:** 1.0 (2.6)
P: 3.7 (9.5)	**S:** 1.9 (4.7)	**BR:** 0.9 (2.3)
P: 3.3 (8.4)	**T:** 2.2 (5.7)	**RU:** 0.9 (2.2)
P: 2.5 (6.3)	**T:** 2.1 (5.3)	

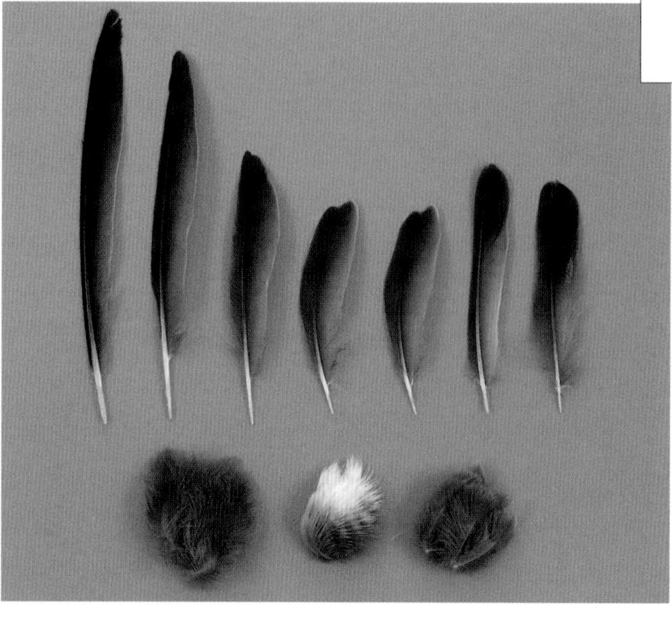

Cliff Swallow *(Petrochelidon pyrrhonota)*

unknown	**S:** 1.9 (4.9)	**CR:** 0.4 (0.9)
P: 3.7 (9.4)	**S:** 1.9 (4.8)	**MA:** 0.9 (2.2)
P: 3.5 (9.0)	**T:** 2.2 (5.6)	**BR:** 0.7 (1.9)
P: 2.8 (7.1)	**T:** 2.1 (5.4)	**BE:** 1.2 (3.1)
P: 2.3 (5.9)	**T:** 2.1 (5.3)	**RU:** 0.9 (2.3)
		TH: 0.3 (0.7)

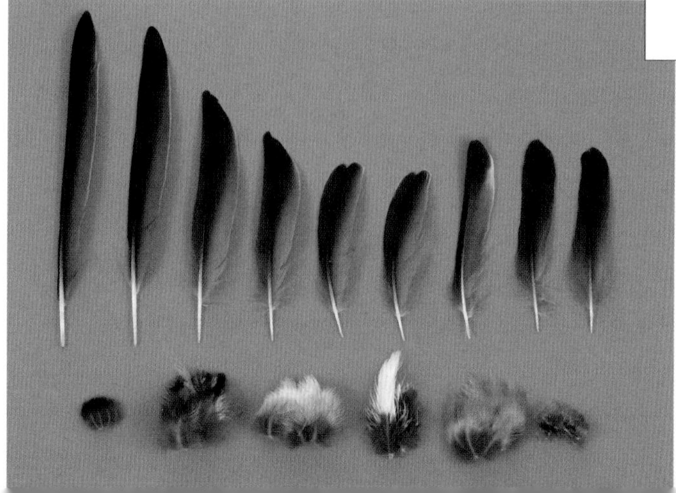

Barn Swallow *(Hirundo rustica)*

male
P: 4.0 (10.2)
P: 3.8 (9.7)
P: 3.1 (7.8)
S: 1.9 (4.8)
S: 1.9 (4.8)
T: 3.9 (10.0)
T: 2.6 (6.6)
T: 2.3 (5.8)
TH: 0.3 (0.7)
MA: 0.9 (2.2)
NA: 0.6 (1.6)
BR: 0.7 (1.8)

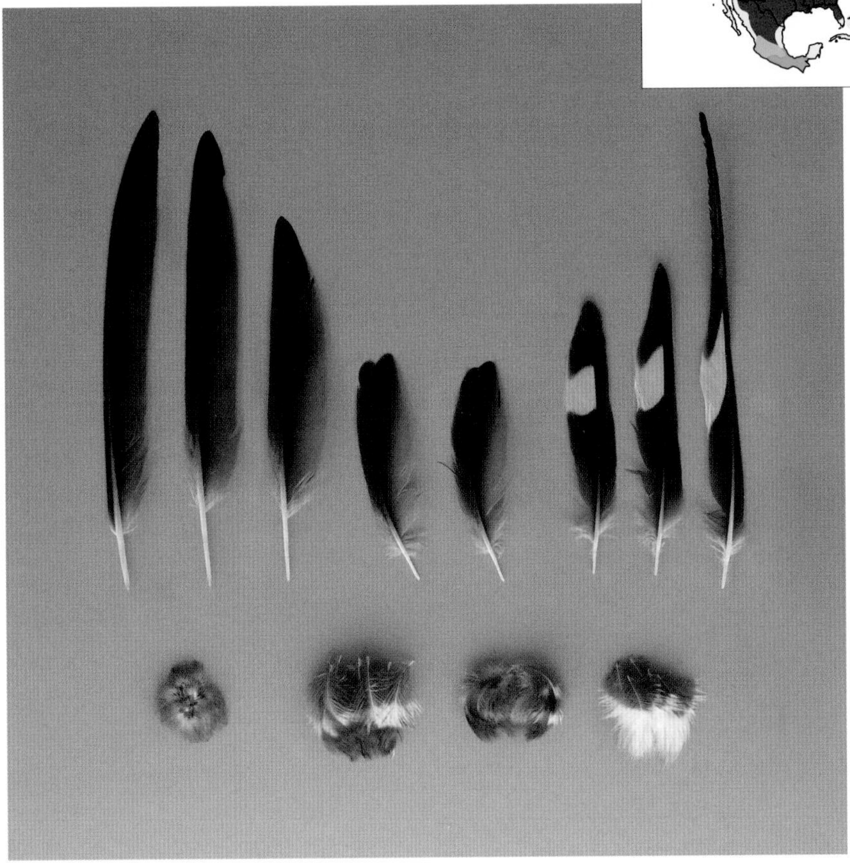

Chickadees and Titmice *(Paridae)*

Black-capped Chickadee *(Poecile atricapillus)*

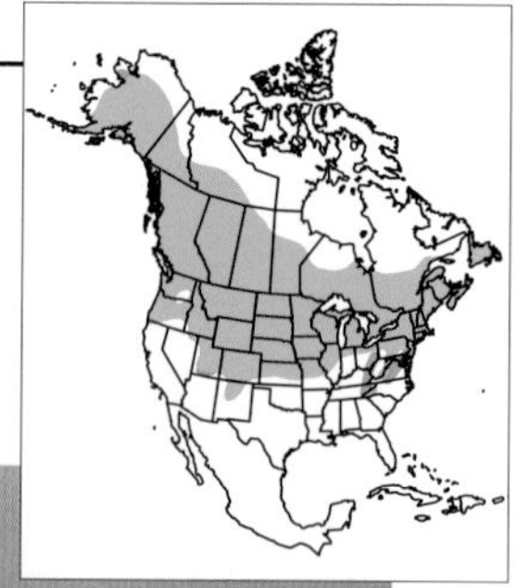

unknown
P: 1.7 (4.2)
P: 2.0 (5.2)
P: 2.0 (5.1)
S: 1.5 (3.9)
S: 1.5 (3.7)
T: 2.2 (5.5)
T: 2.1 (5.4)
T: 2.2 (5.5)
CR: 0.3 (0.8)
BE: 0.7 (1.7)
MA: 0.7 (1.7)

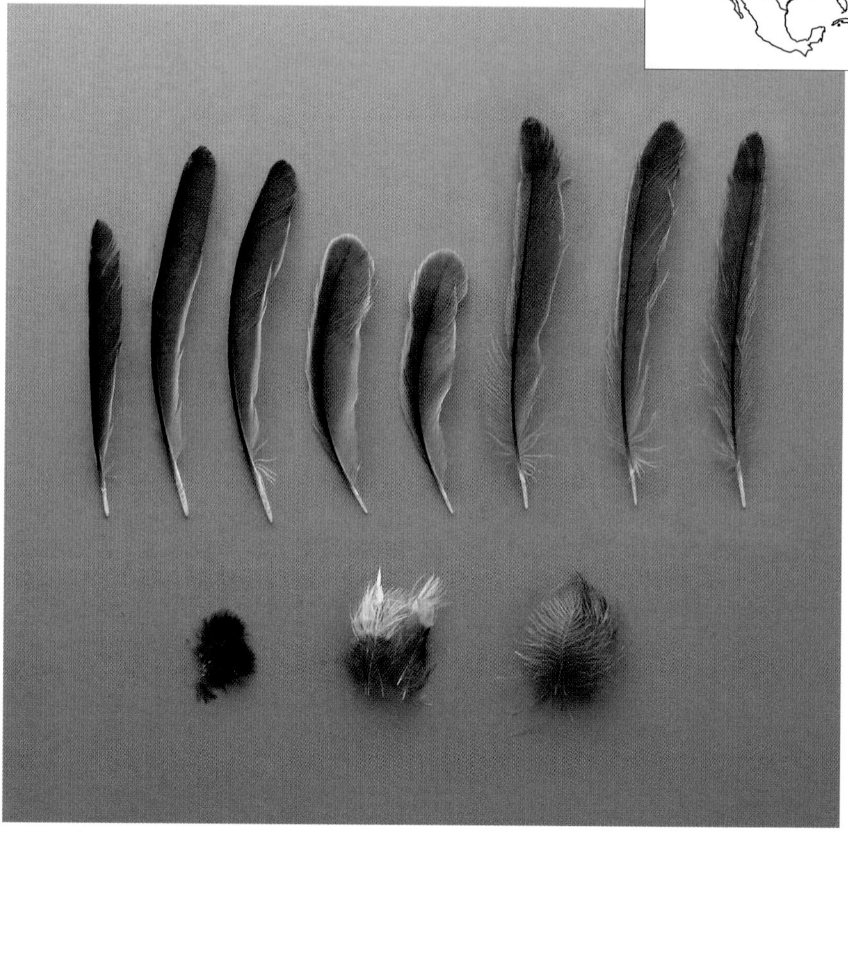

Chestnut-backed Chickadee *(Poecile rufescens)*

unknown
P: 2.2 (5.5)
P: 2.0 (5.1)
TS: 1.4 (3.6)
T: 2.2 (5.6)
T: 2.2 (5.5)
MA: 0.7 (1.9)
BE: 0.9 (2.3)
CR: 0.4 (1.1)

Oak Titmouse *(Baeolophus inornatus)*

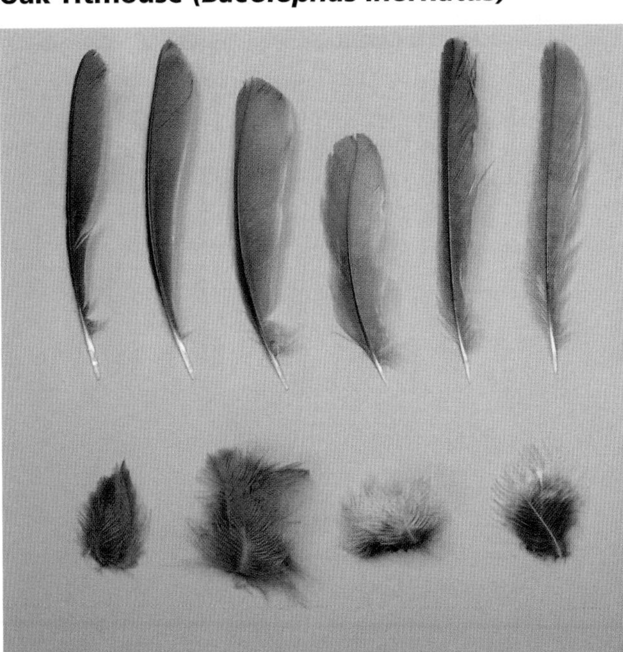

unknown
P: 2.3 (5.8)
P: 2.4 (6.0)
S: 2.2 (5.5)
TS: 1.8 (4.5)
T: 2.4 (6.0)
T: 2.3 (5.9)
CR: 0.7 (1.8)
MA: 0.8 (2.0)
TH: 0.6 (1.4)
BE: 0.8 (2.0)

Tufted Titmouse *(Baeolophus bicolor)*

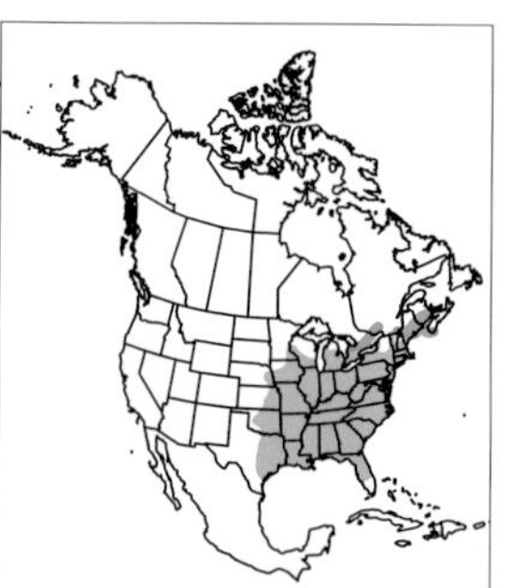

unknown
P: 2.8 (7.0)
P: 2.9 (7.3)
S: 2.5 (6.3)
TS: 2.1 (5.4)
T: 2.9 (7.3)
T: 3.1 (7.8)
CR: 0.9 (2.2)
NA: 1.0 (2.6)
TH: 0.5 (1.3)
FL: 1.1 (2.9)
UPTC: 1.3 (3.3)

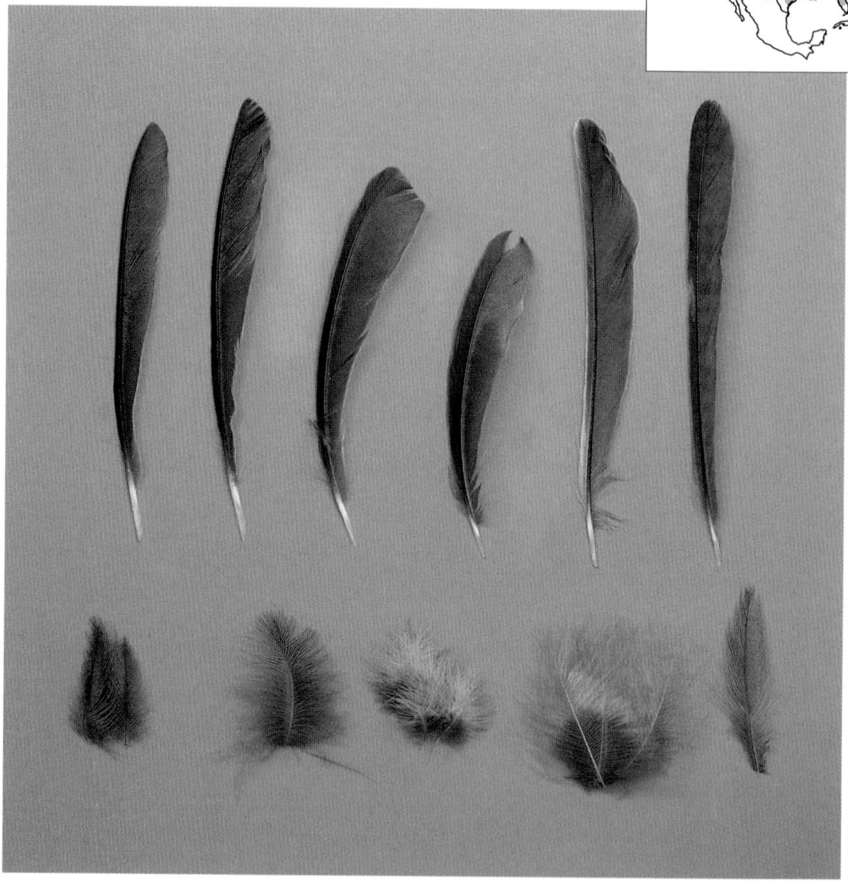

Verdin *(Remizidae)*

Verdin *(Auriparus flaviceps)*

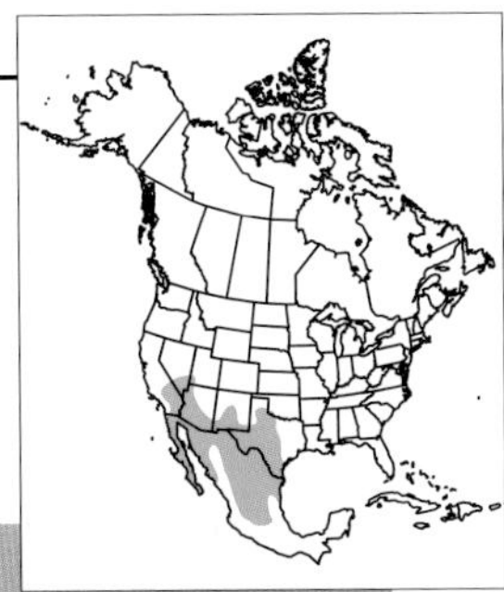

male
P: 1.7 (4.3)
P: 1.8 (4.6)
S: 1.7 (4.2)
T: 1.9 (4.8)
T: 2.0 (5.0)
T: 1.9 (4.8)

BR: 0.4 (1.1)
BE: 0.6 (1.5)
MA: 0.6 (1.6)
MC: 0.2 (0.6)
TH: 0.2 (0.6)
CR: 0.2 (0.6)

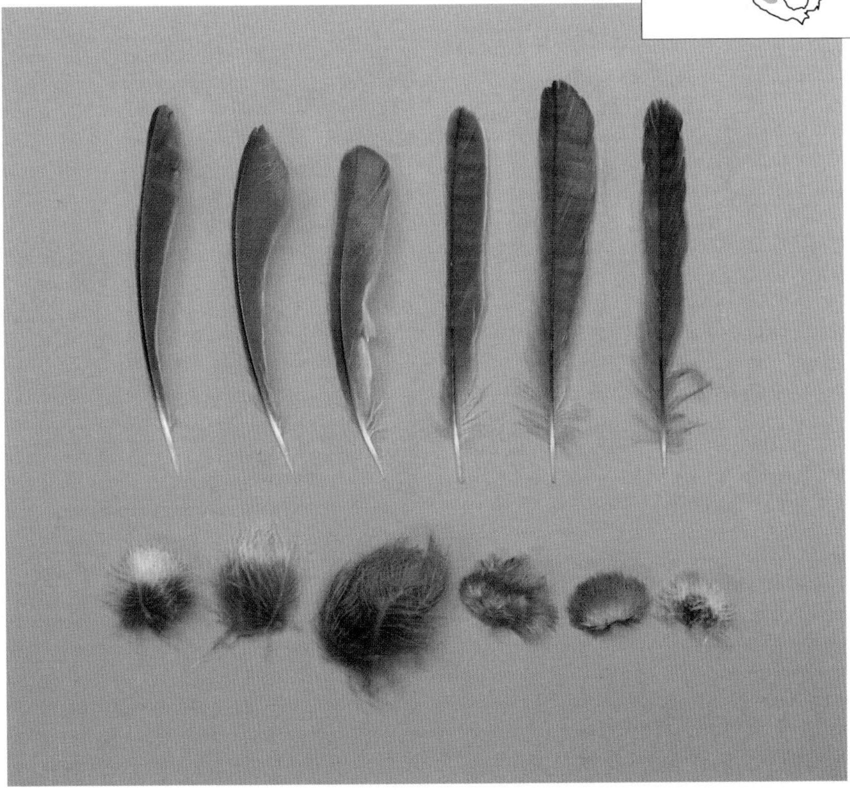

Bushtits *(Aegithalidae)*

Bushtit *(Psaltriparus minimus)*

unknown
P: 1.4 (3.5)
P: 1.6 (4.1)
P: 1.3 (3.3)
S: 1.2 (3.0)

TS: 1.0 (2.5)
T: 1.7 (4.3)
T: 1.8 (4.6)
MA: 0.8 (2.1)
BR: 0.6 (1.5)

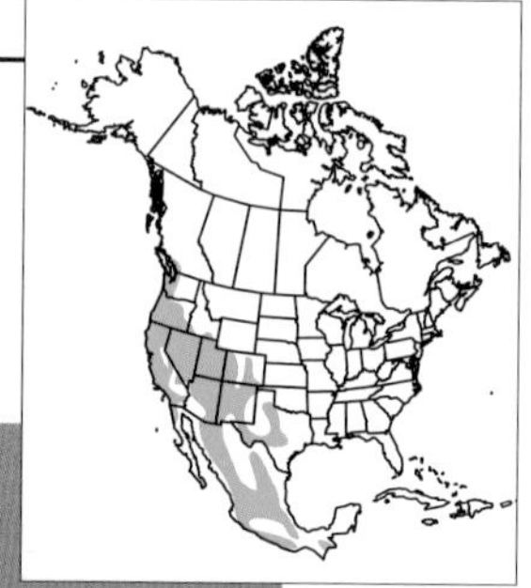

Nuthatches *(Sittidae)*

White stripe and dark tip on outer tail feathers. Similar to warbler tails.

Red-breasted Nuthatch *(Sitta canadensis)*

male
P: 2.4 (6.2)
P: 2.0 (5.1)
S: 1.8 (4.5)
T: 1.7 (4.2)
T: 1.7 (4.2)
T: 1.7 (4.2)
T: 1.7 (4.2)
MA: 0.8 (2.1)
TH: 0.4 (1.0)
CR: 0.4 (1.1)
BR: 0.5 (1.3)
UNTC: 1.3 (3.2)
WC: 0.9 (2.2)

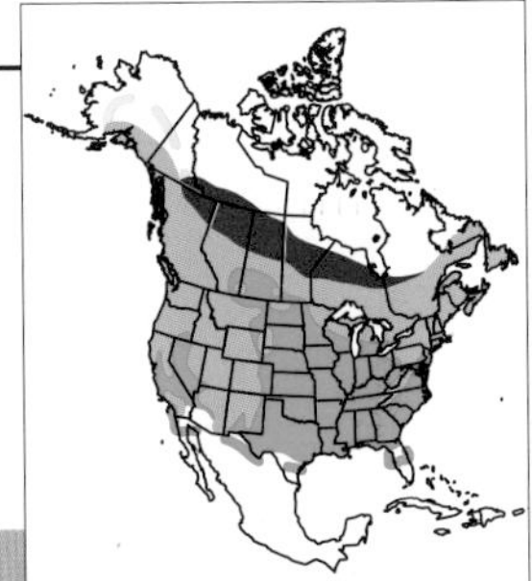

White-breasted Nuthatch *(Sitta carolinensis)*

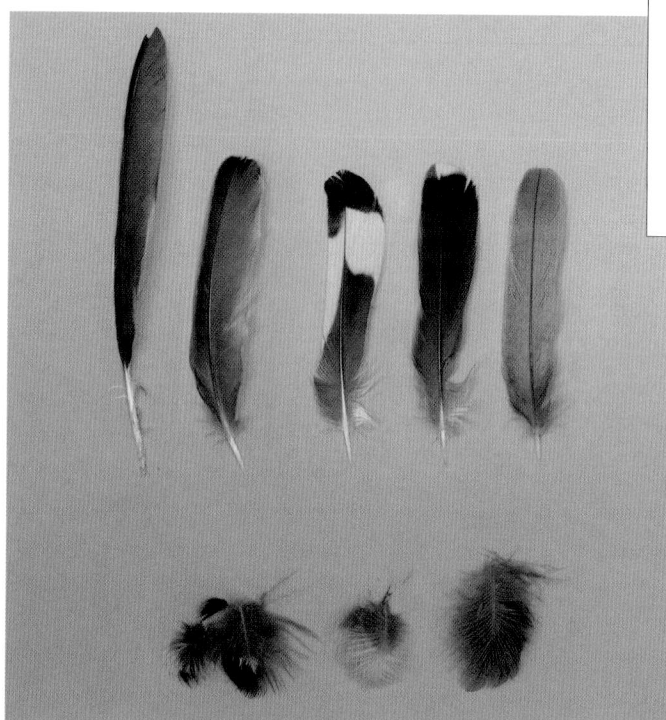

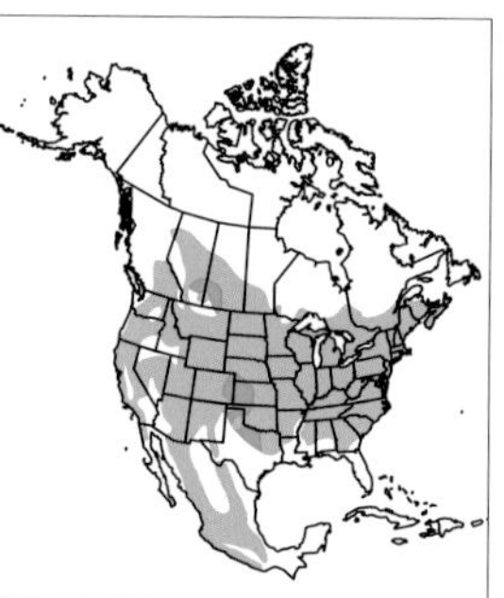

unknown
P: 3.1 (7.8)
S: 2.2 (5.5)
T: 2.0 (5.2)
T: 2.1 (5.3)
T: 2.1 (5.3)
NA: 0.6 (1.6)
BR: 0.5 (1.2)
RU: 0.7 (1.8)

Pygmy Nuthatch *(Sitta pygmaea)*

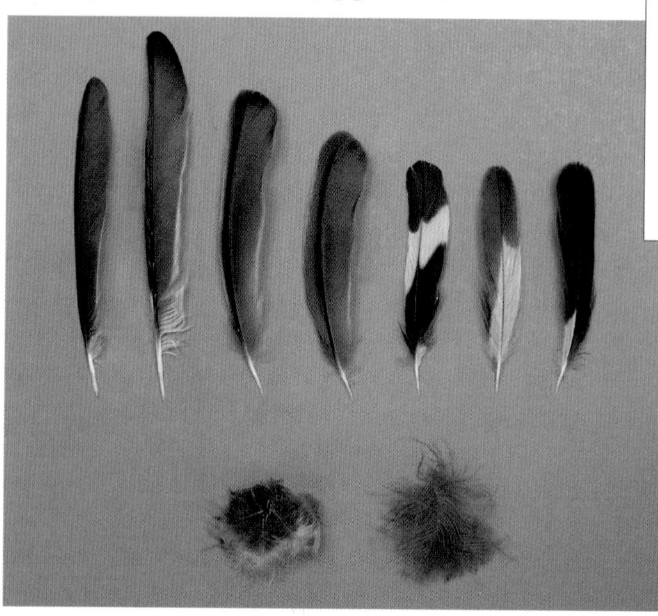

unknown
P: 2.0 (5.1)
P: 2.4 (6.0)
S: 2.0 (5.1)
S: 1.7 (4.4)
T: 1.5 (3.7)
T: 1.5 (3.7)
T: 1.5 (3.7)
BR: 0.6 (1.4)
RU: 0.8 (2.1)

Creepers *(Certhiidae)*

Pointed tail feathers. Primaries and secondaries smaller but similar in pattern to those of some thrushes.

Brown Creeper *(Certhia americana)*

unknown
P: 1.6 (4.1)
P: 2.2 (5.5)
S: 1.9 (4.9)
S: 1.9 (4.7)

T: 1.9 (4.9)
T: 2.2 (5.5)
T: 2.2 (5.6)
BR: 0.7 (1.8)
NA: 0.8 (2.0)
RU: 0.9 (2.4)

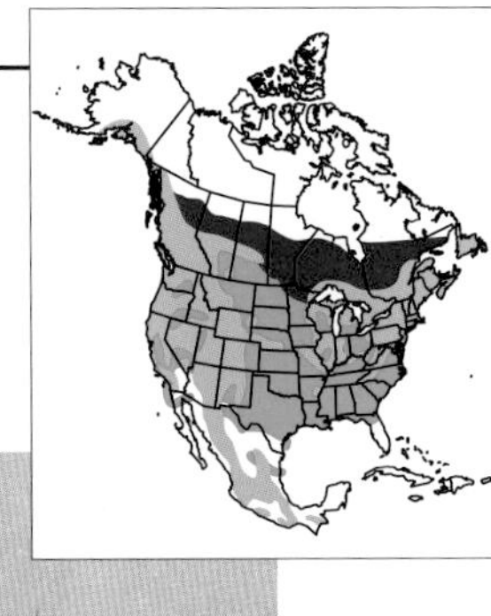

Wrens (*Troglodytidae*)

Wrens have distinctly spotted or barred flight feathers.

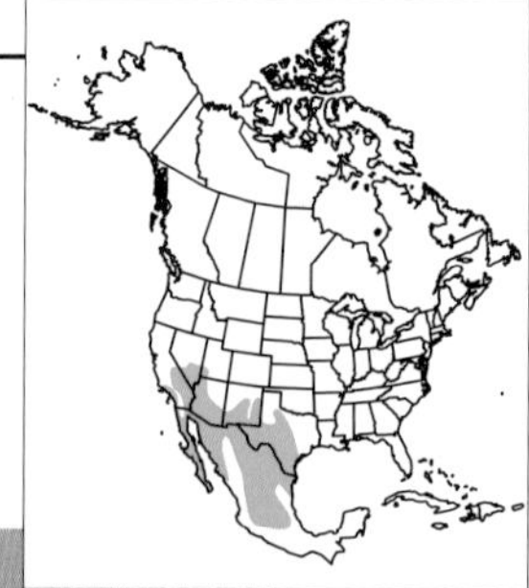

Cactus Wren (*Campylorhynchus brunneicapillus*)

unknown
P: 2.5 (6.3)
P: 2.9 (7.3)
P: 2.9 (7.4)
S: 2.7 (6.9)
TS: 2.3 (5.9)
T: 3.1 (7.9)
T: 3.4 (8.6)
NA: 0.7 (1.7)
BR: 0.7 (1.9)
MA: 1.0 (2.6)
UNTC: 1.3 (3.3)

Carolina Wren *(Thryothorus ludovicianus)*

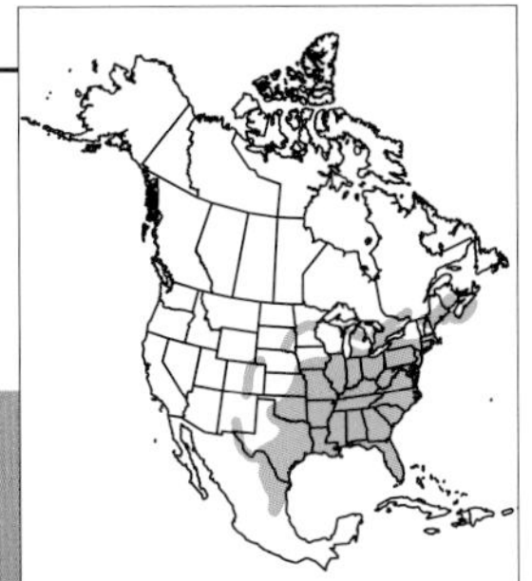

unknown	**S:** 1.7 (4.4)	**BR:** 0.6 (1.5)
P: 1.7 (4.2)	**TS:** 1.5 (3.7)	**BE:** 0.8 (2.0)
P: 2.0 (5.0)	**T:** 2.0 (5.0)	**MA:** 0.9 (2.4)
P: 2.2 (5.6)	**T:** 2.1 (5.4)	**RU:** 1.1 (2.8)

Bewick's Wren *(Thryomanes bewickii)*

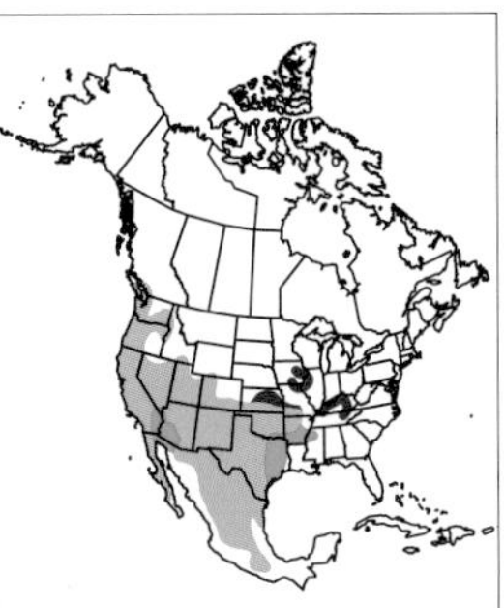

unknown	**S:** 1.8 (4.5)	**T:** 1.7 (4.2)
P: 1.4 (3.5)	**TS:** 1.4 (3.5)	**BR:** 0.7 (1.7)
P: 1.7 (4.3)	**T:** 2.0 (5.2)	**MA:** 0.7 (1.7)
P: 1.8 (4.5)	**T:** 1.9 (4.8)	

House Wren *(Troglodytes aedon)*

unknown
P: 0.8 (2.0)
P: 1.7 (4.2)
P: 1.7 (4.3)
P: 1.7 (4.3)
S: 1.5 (3.9)
S: 1.5 (3.8)
T: 1.8 (4.5)
T: 1.8 (4.5)
BR: 1.1 (2.9)
MA: 0.8 (2.0)

Winter Wren *(Troglodytes troglodytes)*

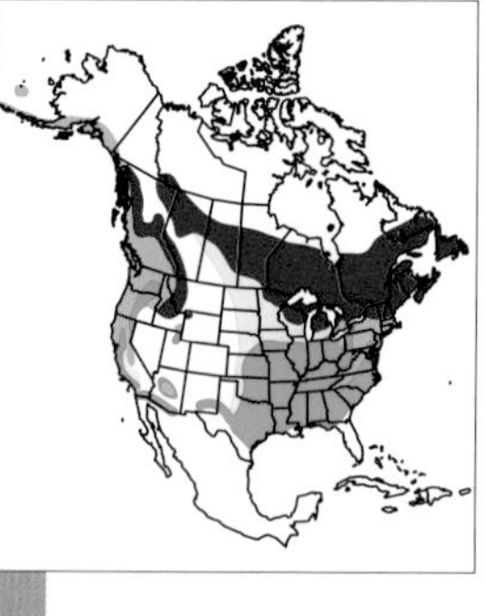

unknown
P: 1.2 (3.2)
P: 1.6 (4.1)
P: 1.5 (3.8)
S: 1.3 (3.4)
S: 1.3 (3.2)
T: 1.3 (3.2)
T: 1.3 (3.3)
BR: 0.7 (1.8)
BE: 0.6 (1.4)
RU: 0.8 (2.0)

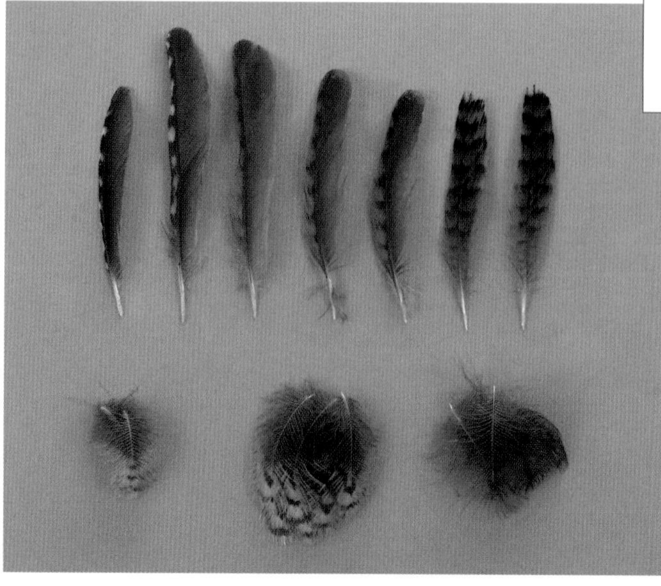

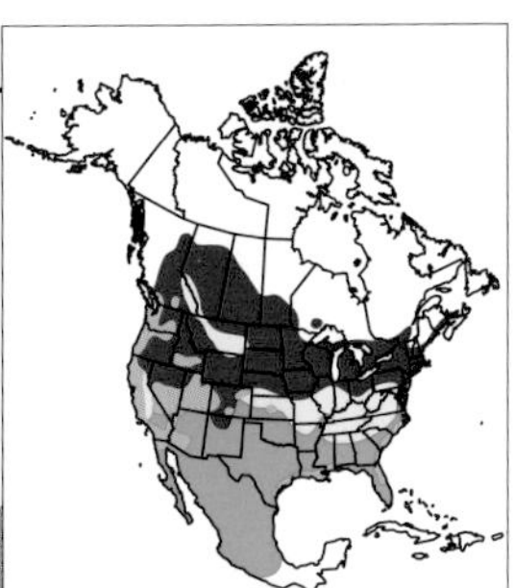

Marsh Wren *(Cistothorus palustris)*

unknown
P: 1.4 (3.5)
P: 1.7 (4.3)
P: 1.6 (4.1)
S: 1.5 (3.9)
TS: 1.3 (3.2)

T: 1.5 (3.8)
T: 1.8 (4.5)
T: 1.8 (4.5)
BR: 0.8 (2.1)
MA: 0.7 (1.9)
CR: 0.7 (1.7)

Kinglets *(Regulidae)*

Wing feathers similar to those of warblers.

Golden-crowned Kinglet *(Regulus satrapa)*

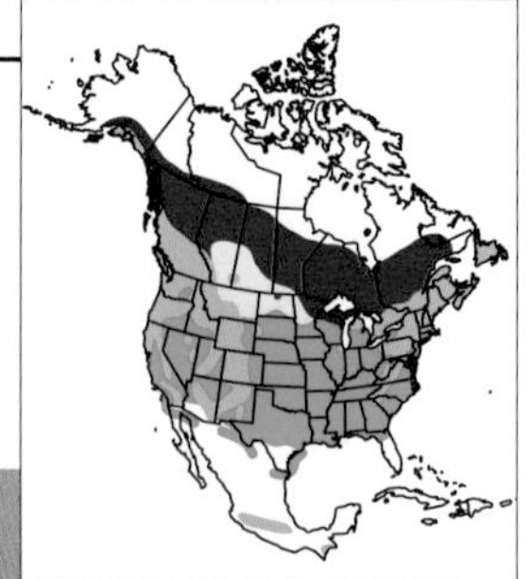

male
P: 1.7 (4.2)
P: 2.0 (5.1)
P: 1.8 (4.6)
S: 1.6 (4.0)
TS: 1.4 (3.5)
T: 1.7 (4.4)
T: 1.9 (4.7)
BE: 1.0 (2.6)
MA: 0.7 (1.8)
NA: 0.5 (1.3)
CR: 0.4 (0.9)

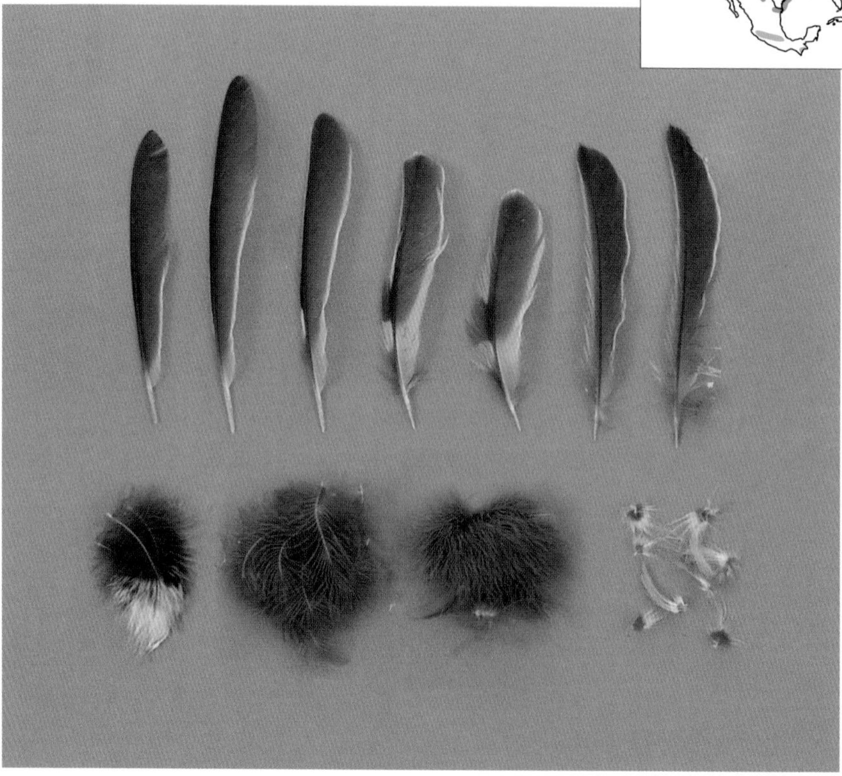

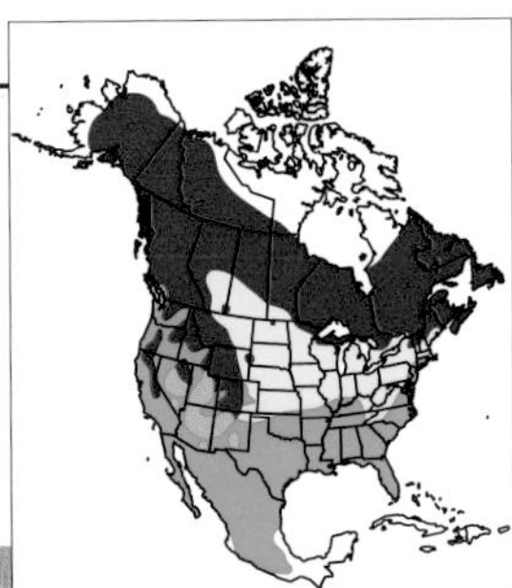

Ruby-crowned Kinglet *(Regulus calendula)*

male
P: 1.7 (4.4)
P: 2.0 (5.2)
P: 2.0 (5.2)
P: 1.9 (4.8)
S: 1.7 (4.3)
S: 1.1 (2.9)

T: 2.0 (5.0)
T: 2.0 (5.0)
CR: 0.5 (1.3)
MA: 0.7 (1.9)
BE: 1.0 (2.5)
BR: 0.7 (1.8)

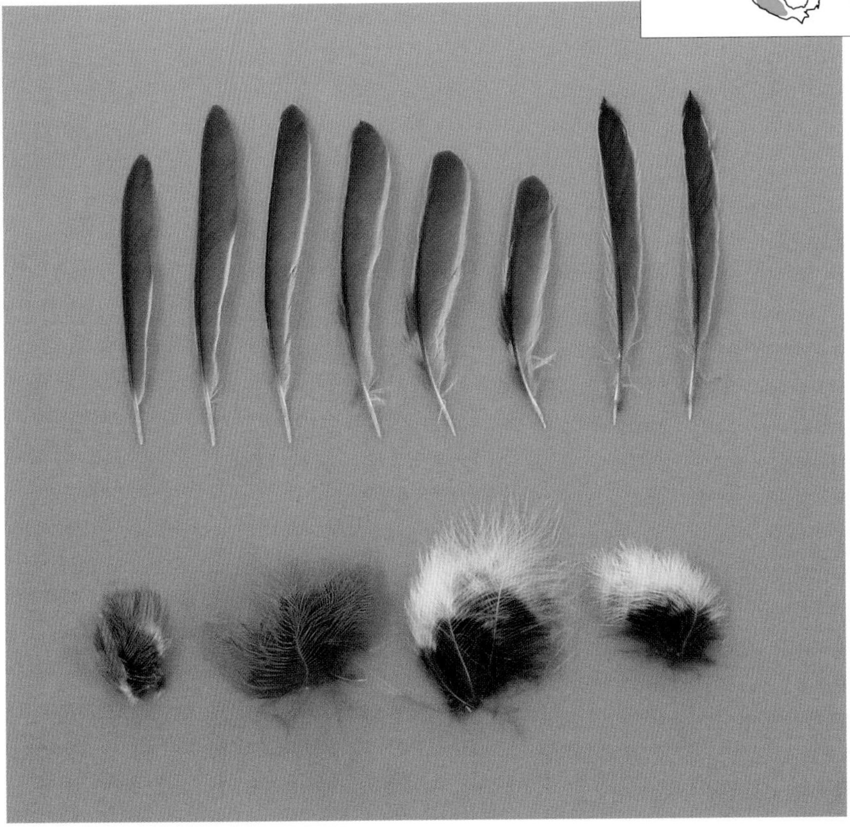

Gnatcatchers *(Sylviidae)*

Blue-gray Gnatcatcher *(Polioptila caerulea)*

unknown
P: 1.9 (4.7)
P: 1.8 (4.6)
S: 1.6 (4.1)
S: 1.6 (4.1)
T: 2.0 (5.1)

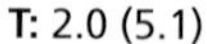

T: 2.0 (5.1)
T: 2.0 (5.1)
MA: 0.9 (2.2)
BE: 0.8 (2.0)
WC: 0.6 (1.5)

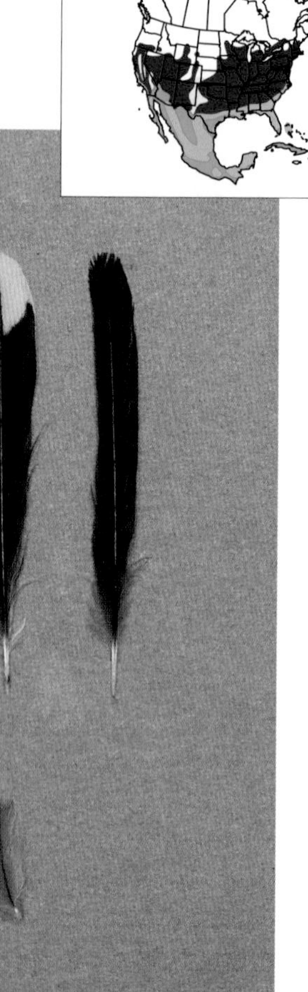

Thrushes *(Turdidae)*

Eastern Bluebird *(Sialia sialis)*

male
P: 3.3 (8.4)
S: 2.6 (6.5)
TS: 2.1 (5.4)
T: 2.9 (7.4)
T: 2.9 (7.3)
T: 2.6 (6.6)
NA: 1.0 (2.6)
RU: 1.3 (3.3)
CR: 0.8 (2.0)
BR: 0.7 (1.9)

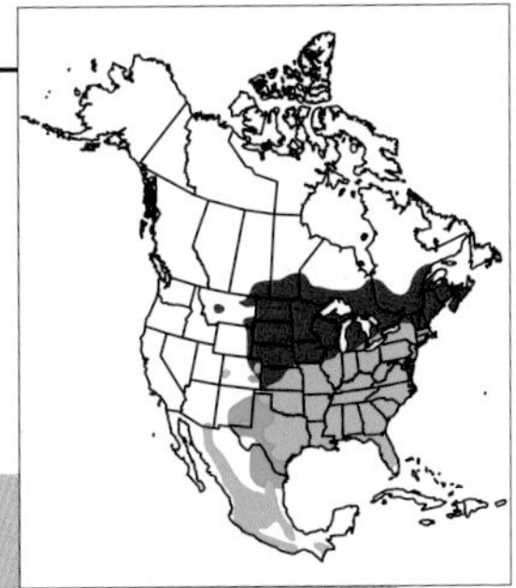

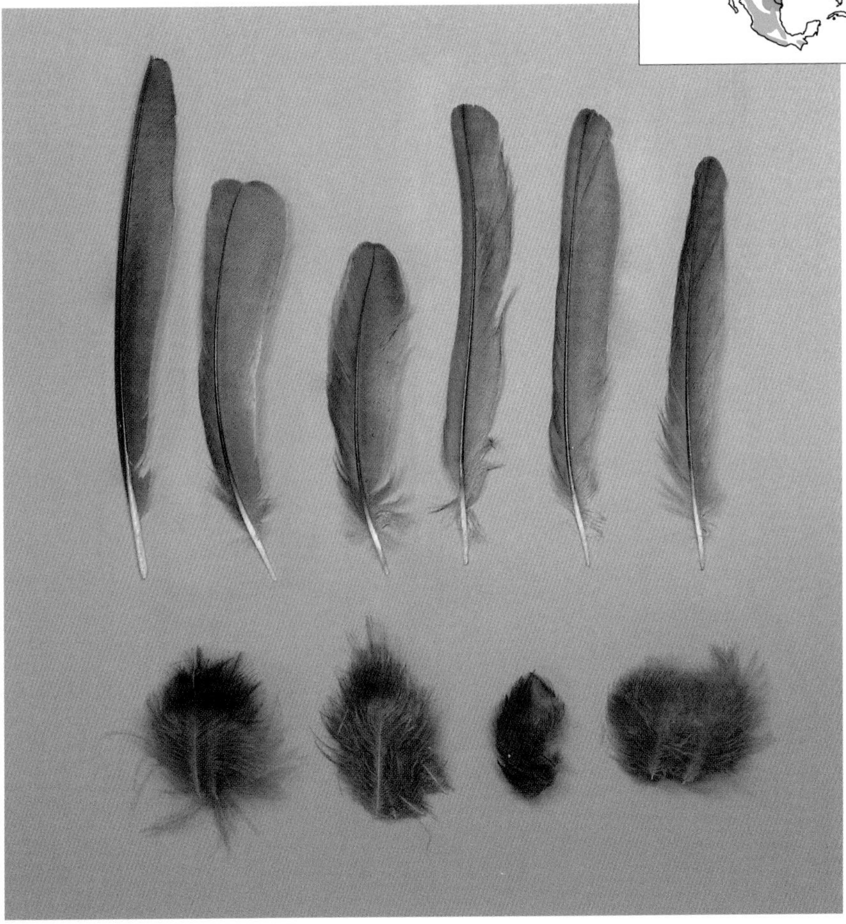

Western Bluebird *(Sialia mexicana)*

male
P: 3.5 (8.8)
S: 2.4 (6.0)
TS: 2.1 (5.3)
T: 2.9 (7.3)
T: 2.9 (7.3)
T: 2.9 (7.3)
BR: 0.6 (1.5)
NA: 0.4 (1.1)
RU: 1.0 (2.5)

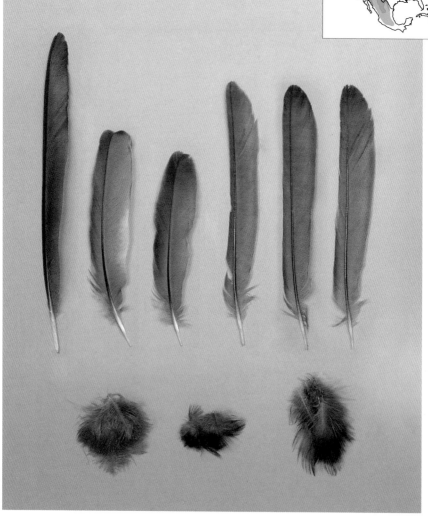

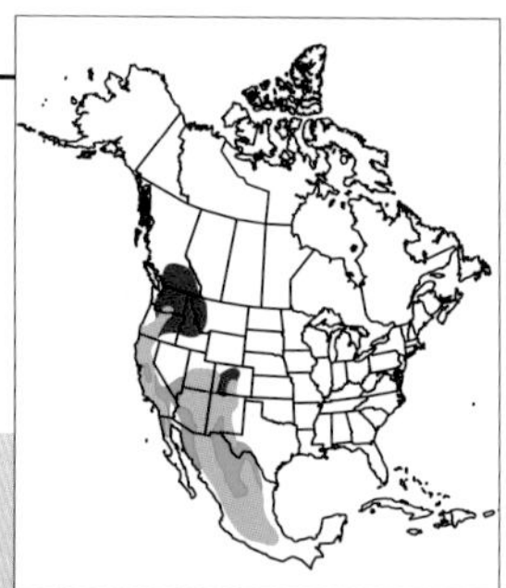

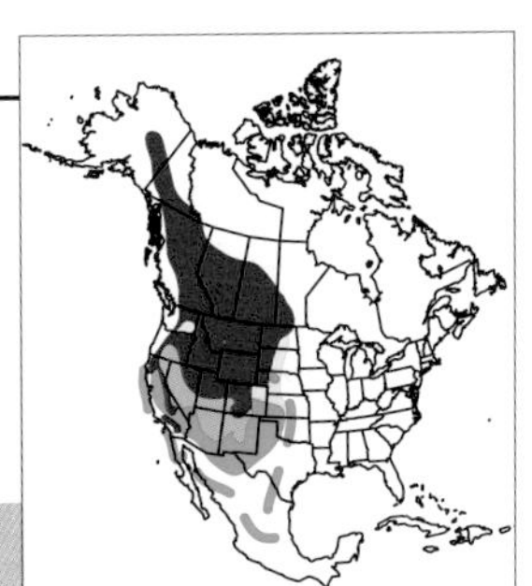

Mountain Bluebird *(Sialia currucoides)*

female on left, juvenile male on right (feathers still sheathed)

P: 3.6 (9.2)
P: 3.7 (9.3)
TS: 2.2 (5.7)
T: 2.9 (7.3)
P: 3.3 (8.3)
T: 3.2 (8.1)

T: 2.1 (5.3)
NA: 0.8 (2.0)
WC: 1.0 (2.6)
TH: 0.6 (1.4)
NA: 0.8 (2.0)
WC: 1.1 (2.7)

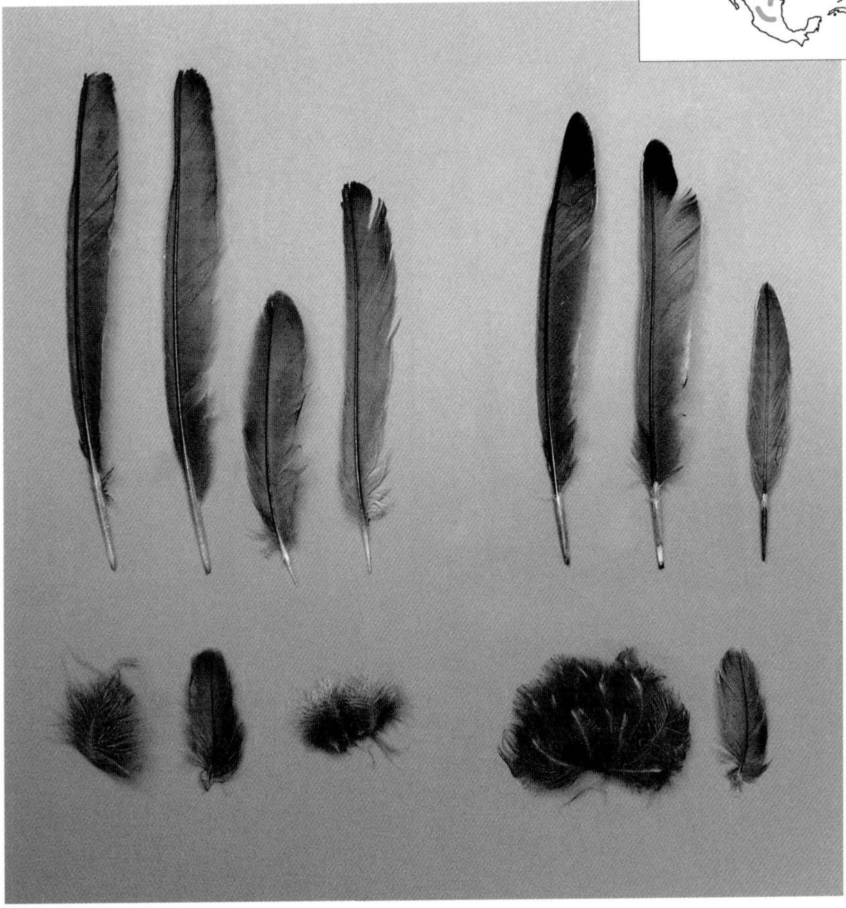

Veery *(Catharus fuscescens)*

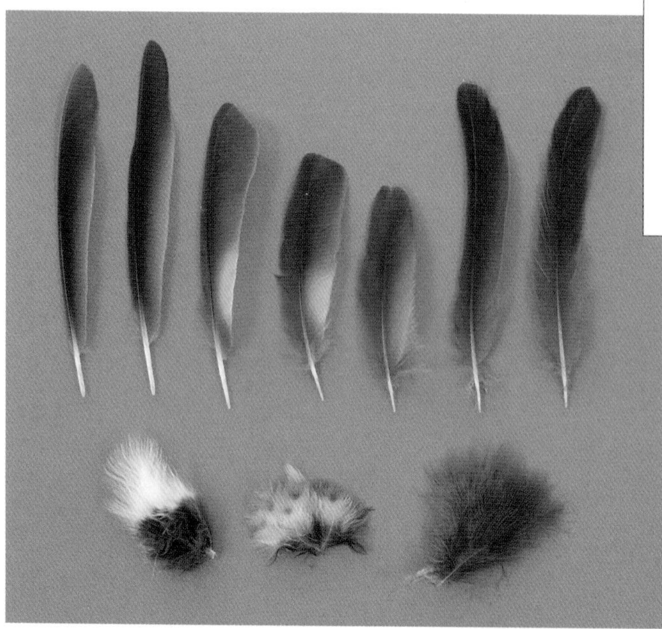

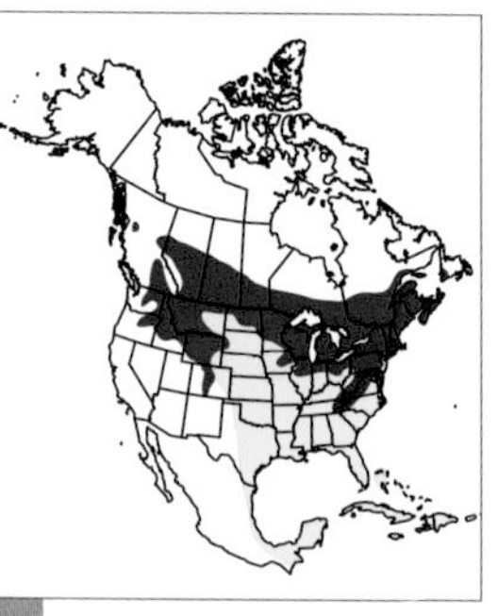

unknown
P: 3.1 (8.0)
P: 3.3 (8.5)
P: 2.9 (7.3)
S: 2.3 (5.9)
TS: 2.1 (5.4)
T: 3.1 (7.9)
T: 3.1 (7.8)
BE: 1.1 (2.8)
CR: 0.6 (1.6)
MA: 1.1 (2.9)

Gray-cheeked Thrush *(Catharus minimus)*

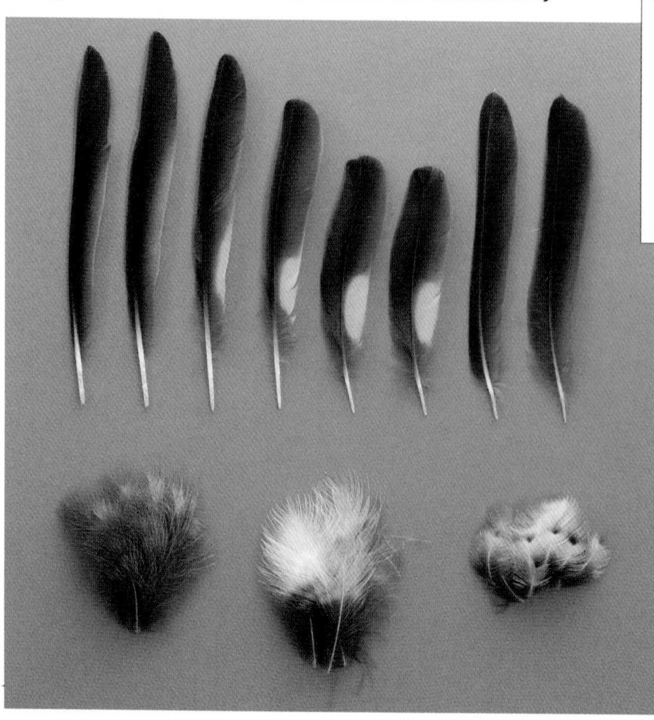

unknown
P: 3.4 (8.6)
P: 3.5 (9.0)
P: 3.3 (8.5)
P: 2.9 (7.4)
S: 2.4 (6.2)
S: 2.4 (6.0)
T: 3.1 (7.8)
T: 3.1 (7.8)
MA: 1.6 (4.1)
BE: 1.7 (4.2)
BR: 0.9 (2.2)

Swainson's Thrush *(Catharus ustulatus)*

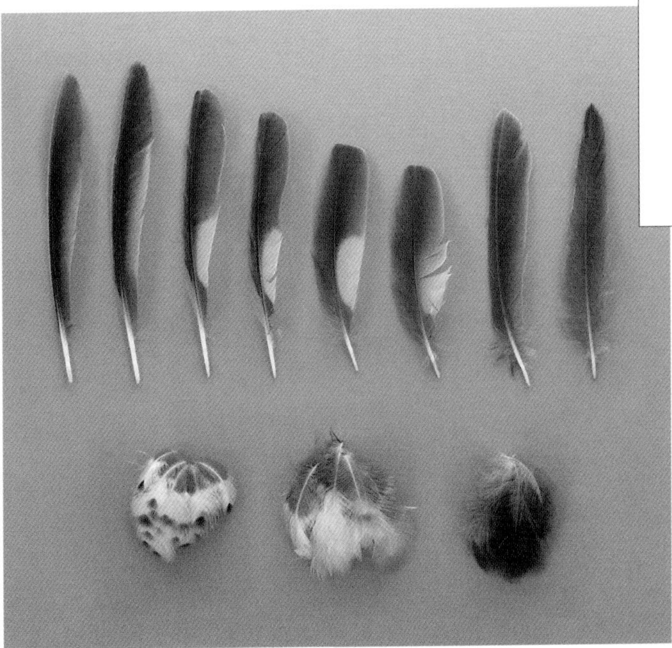

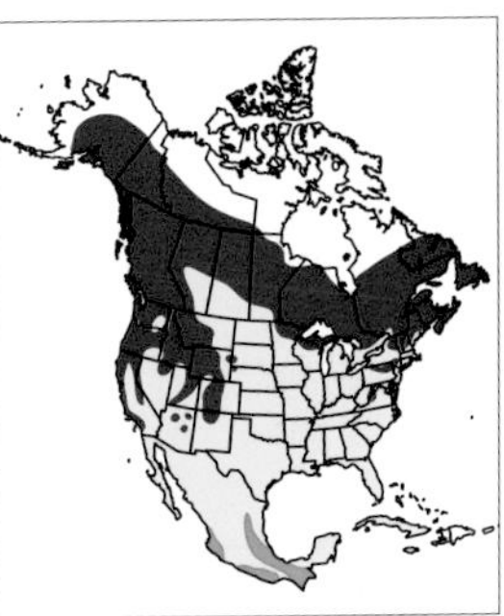

unknown
P: 3.1 (7.9)
P: 3.2 (8.1)
P: 2.9 (7.3)
P: 2.6 (6.6)
S: 2.2 (5.7)
S: 2.1 (5.4)
T: 2.8 (7.1)
T: 2.7 (6.9)
BR: 0.8 (2.0)
BE: 1.0 (2.5)
RU: 1.1 (2.8)

Hermit Thrush *(Catharus guttatus)*

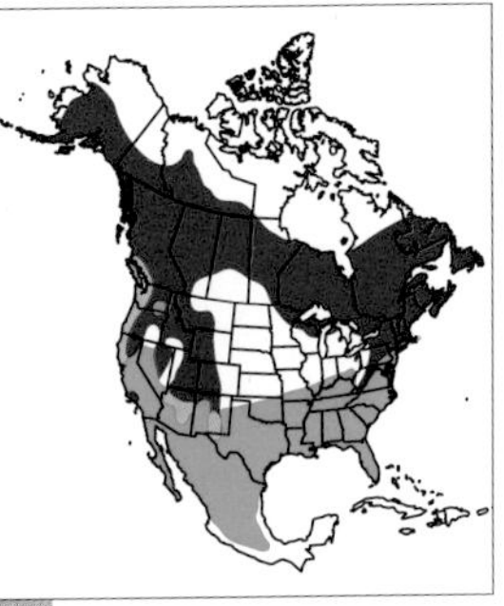

unknown
P: 3.1 (8.0)
S: 2.6 (6.6)
S: 2.4 (6.1)
TS: 2.2 (5.5)
T: 3.0 (7.5)
T: 3.0 (7.5)
UNTC: 1.9 (4.9)
BR: 0.7 (1.7)
MA: 1.1 (2.8)
UPTC: 1.2 (3.1)

Wood Thrush *(Hylocichla mustelina)*

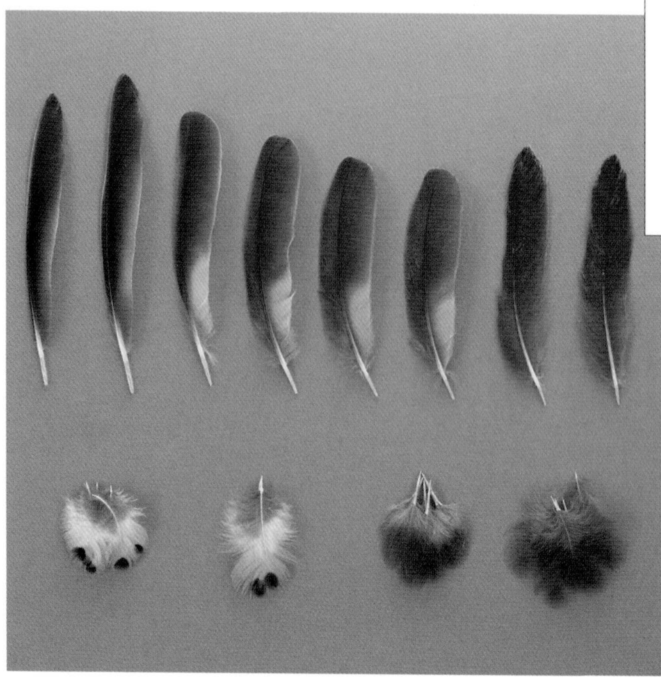

unknown
P: 3.3 (8.4)
P: 3.6 (9.1)
P: 3.1 (8.0)
P: 3.0 (7.5)
S: 2.8 (7.0)
S: 2.7 (6.8)
T: 3.0 (7.6)
T: 2.9 (7.3)
BR: 1.0 (2.6)
BE: 1.3 (3.4)
RU: 1.1 (2.9)
MA: 1.1 (2.8)

American Robin *(Turdus migratorius)*

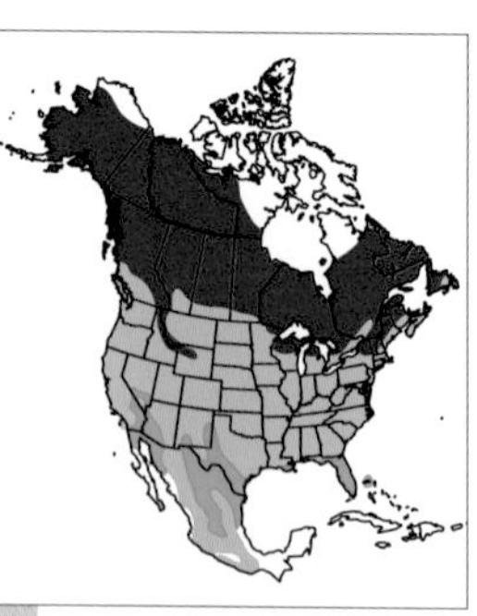

male
P: 4.6 (11.7)
P: 4.8 (12.2)
P: 3.9 (9.8)
S: 3.6 (9.2)
T: 4.5 (11.4)
T: 4.5 (11.5)
BR: 0.7 (1.9)
BE: 1.8 (4.5)
MA: 1.6 (4.0)
tibial: 1.6 (4.0)
WC: 1.4 (3.6)

Varied Thrush *(Ixoreus naevius)*

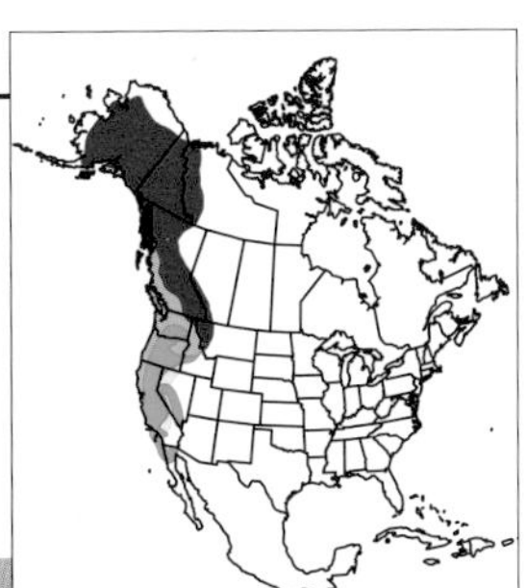

male
P: 4.2 (10.6)
P: 3.8 (9.6)
P: 3.3 (8.3)
S: 2.9 (7.4)
S: 2.8 (7.0)
T: 3.7 (9.4)
T: 3.7 (9.4)
T: 3.7 (9.4)
NA: 1.4 (3.6)
BR: 1.2 (3.0)
BE: 1.5 (3.8)
WC: 1.2 (3.0)

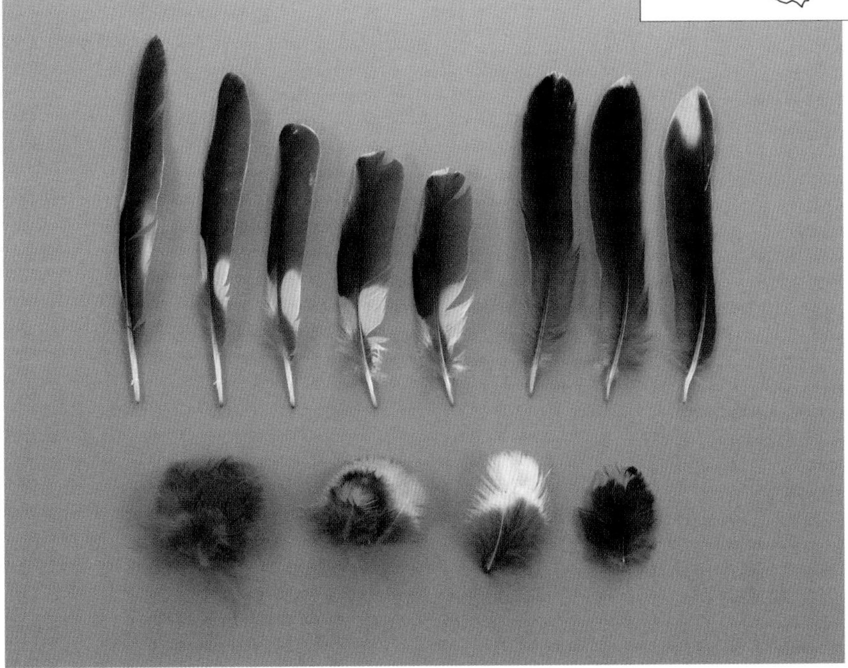

Mockingbirds and Thrashers *(Mimidae)*

Long tail feathers. Flight feathers similar to those of cuckoos and towhees. Some mockingbird primaries similar to some grosbeaks, kingfishers, woodpeckers, and Lesser Goldfinch.

Gray Catbird *(Dumetella carolinensis)*

unknown
P: 1.3 (3.4)
P: 3.0 (7.5)
P: 2.8 (7.2)
S: 2.7 (6.9)
S: 2.5 (6.3)
T: 3.1 (7.9)
T: 3.7 (9.5)
T: 3.4 (8.7)
CR: 0.6 (1.4)
NA: 1.1 (2.8)
RU: 1.2 (3.0)
BE: 1.1 (2.8)
UNTC: 1.8 (4.5)

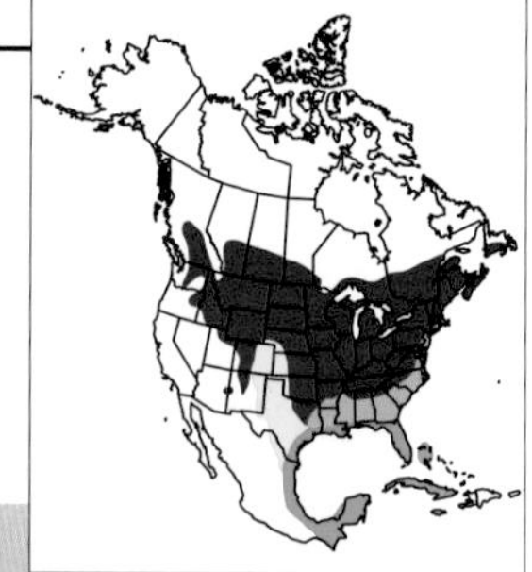

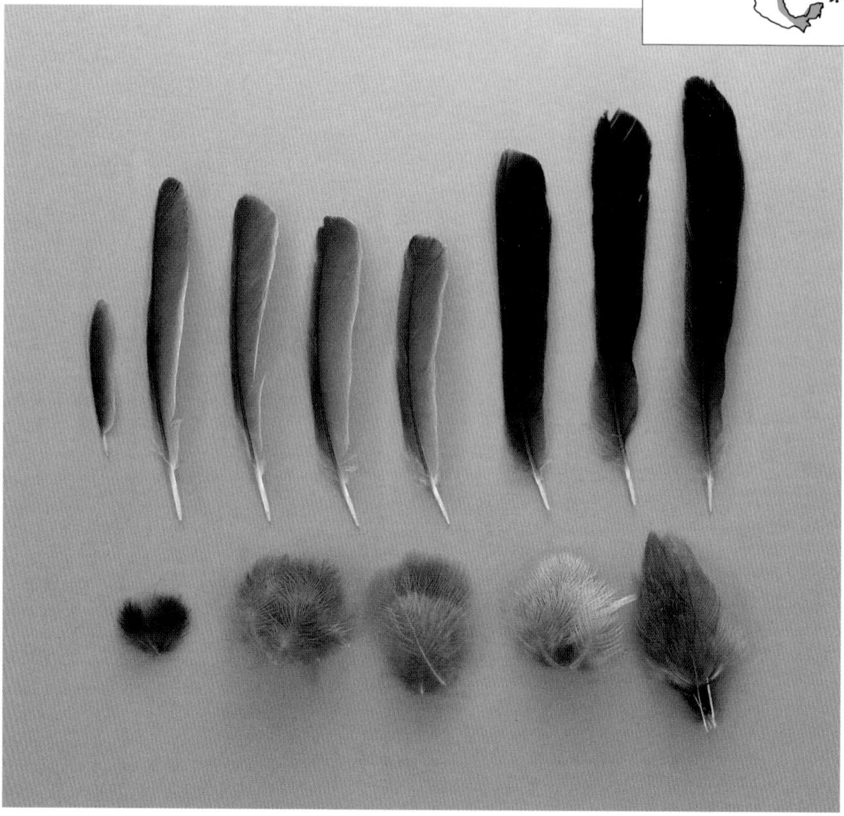

Northern Mockingbird *(Mimus polyglottos)*

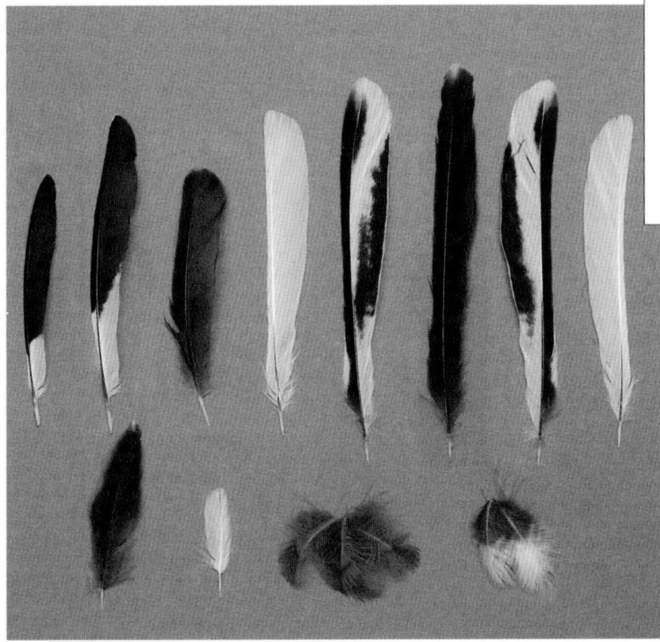

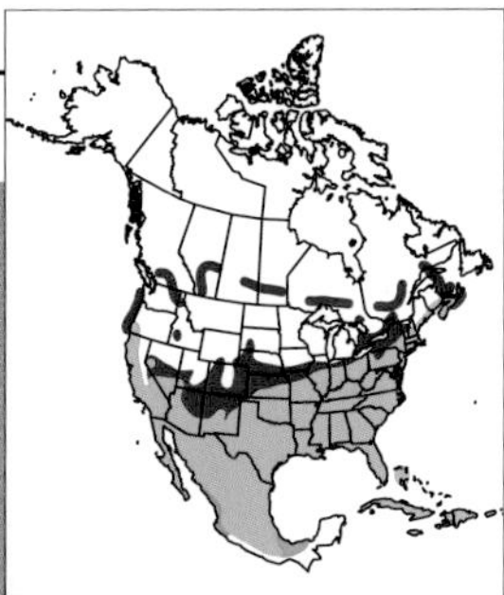

unknown
P: 3.2 (8.2)
P: 4.1 (10.4)
S: 3.2 (8.2)
T: 4.1 (10.4)
T: 4.9 (12.4)
T: 5.0 (12.8)
T: 4.9 (12.5)
T: 4.3 (10.8)
TS: 2.4 (6.2)
WC: 1.5 (3.7)
MA: 1.1 (2.8)
BR: 1.2 (3.0)

Sage Thrasher *(Oreoscoptes montanus)*

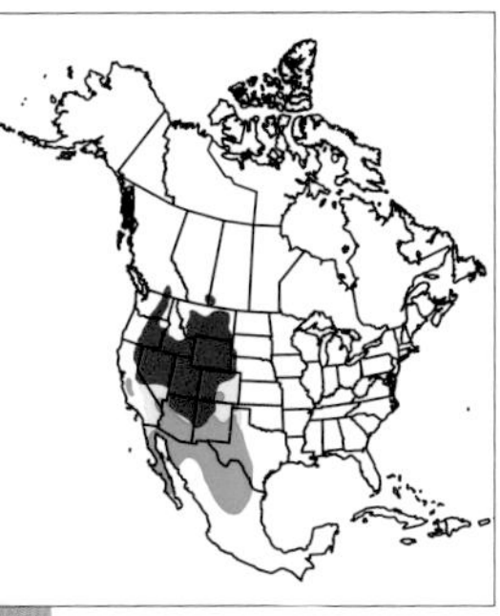

unknown
P: 2.9 (7.4)
P: 3.3 (8.4)
S: 2.8 (7.0)
T: 3.7 (9.5)
T: 3.9 (9.8)
WC: 1.2 (3.0)
TH: 0.7 (1.9)
UPTC: 1.1 (2.8)

Brown Thrasher *(Toxostoma rufum)*

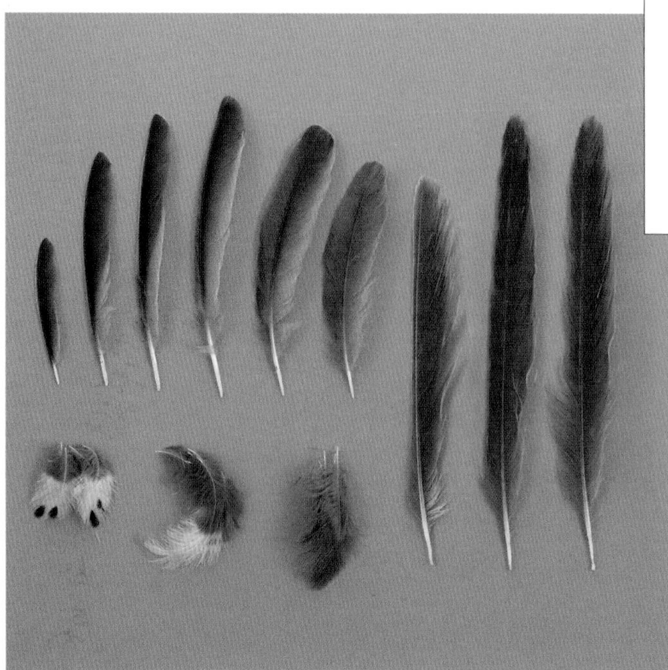

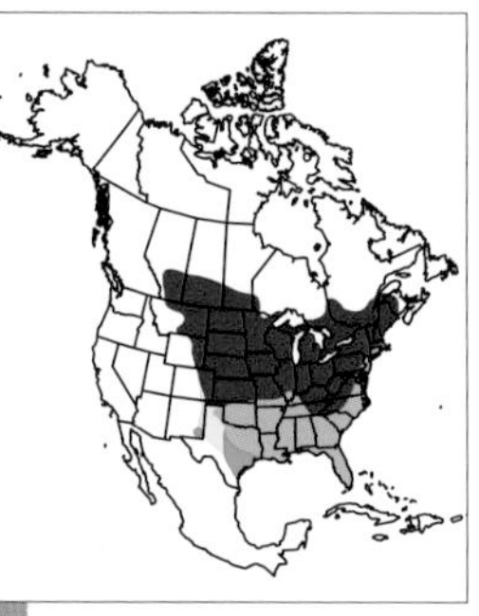

unknown
P: 1.7 (4.4)
P: 2.7 (6.9)
P: 3.3 (8.4)
P: 3.5 (9.0)
S: 3.2 (8.1)
TS: 2.8 (7.1)
T: 4.5 (11.5)
T: 5.3 (13.5)
T: 5.3 (13.5)
BR: 0.9 (2.3)
BE: 1.4 (3.6)
MA: 1.6 (4.0)

Curve-billed Thrasher *(Toxostoma curvirostre)*

unknown
P: 3.1 (7.8)
P: 3.6 (9.2)
P: 3.8 (9.6)
S: 3.2 (8.2)
TS: 2.7 (6.9)
T: 4.0 (10.1)
T: 4.6 (11.6)
BR: 1.5 (3.9)
BE: 1.6 (4.1)
FL: 2.3 (5.9)

California Thrasher *(Toxostoma redivivum)*

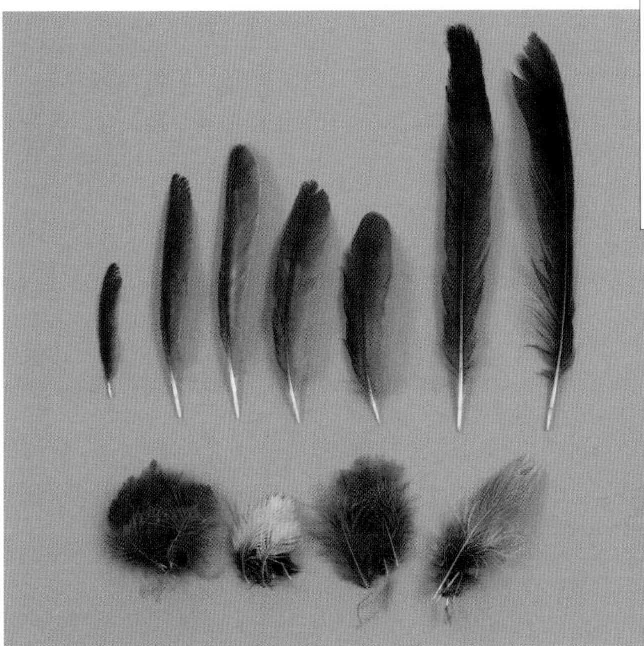

unknown
P: 1.8 (4.6)
P: 3.2 (8.2)
P: 3.5 (9.0)
S: 3.1 (7.9)
TS: 2.8 (7.1)
T: 5.3 (13.5)
T: 5.2 (13.2)
MA: 1.3 (3.3)
BR: 1.0 (2.6)
RU: 1.7 (4.3)
UNTC: 2.0 (5.2)

Crissal Thrasher *(Toxostoma crissale)*

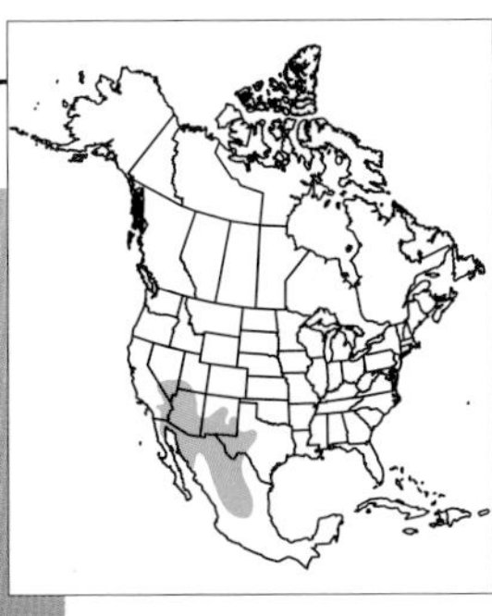

unknown
P: 2.7 (6.8)
P: 3.2 (8.2)
P: 3.4 (8.6)
P: 3.3 (8.5)
S: 3.2 (8.2)
TS: 2.8 (7.1)
T: 4.7 (12.0)
T: 5.7 (14.4)
BR: 1.6 (4.0)
BE: 1.5 (3.7)
RU: 1.6 (4.1)
UNTC: 1.7 (4.3)

Starlings *(Sturnidae)*

Iridescence in body feathers of starlings.

European Starling *(Sturnus vulgaris)*

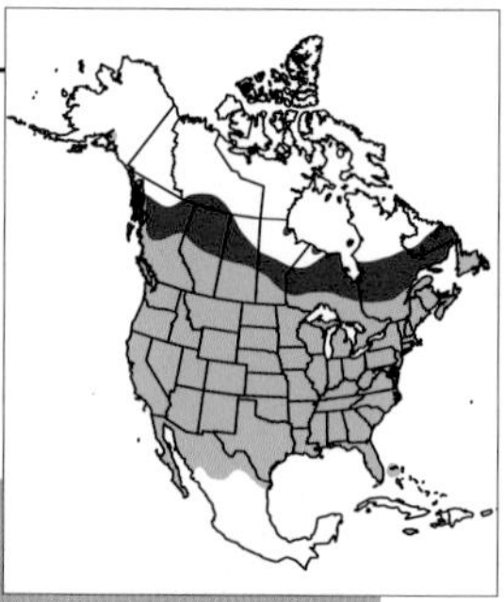

M>F

unknown, non-breeding
P: 4.1 (10.4)
P: 4.1 (10.4)
P: 3.7 (9.4)
P: 3.3 (8.3)

S: 2.7 (6.9)
TS: 2.4 (6.2)
T: 2.8 (7.0)
T: 2.6 (6.5)
NA: 0.6 (1.6)
BR: 0.9 (2.4)

BE: 1.2 (3.0)
MA: 1.3 (3.3)
RU: 1.2 (3.1)
WC: 1.3 (3.2)

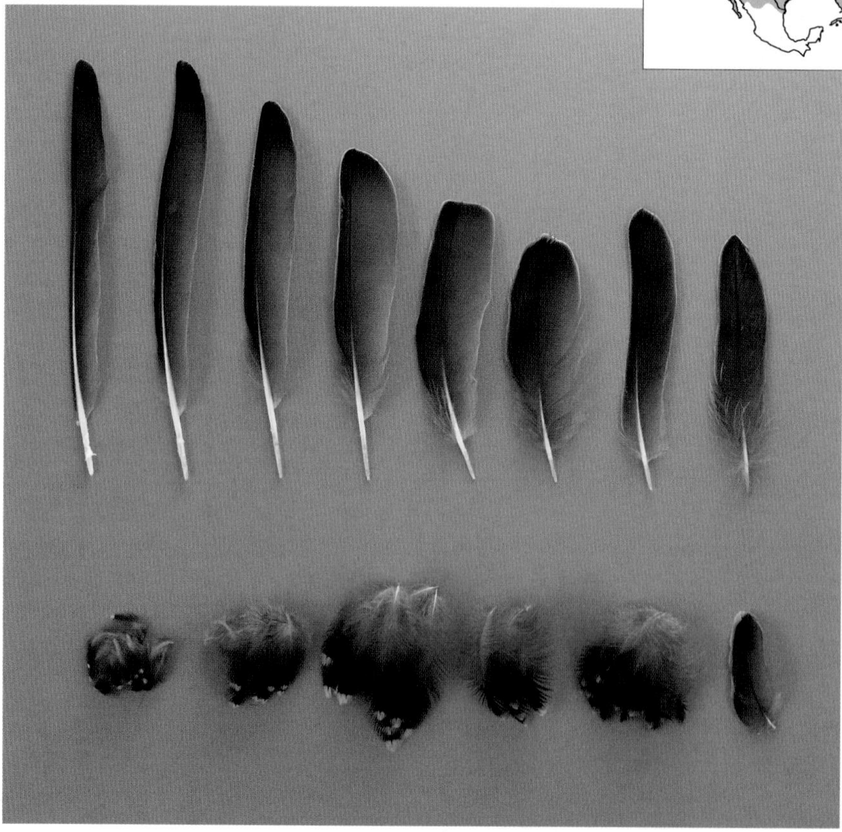

Waxwings *(Bombycillidae)*

Yellow-tipped tails. Red "wax" on secondaries of adult birds.

Cedar Waxwing *(Bombycilla cedrorum)*

unknown
P: 2.9 (7.4)
P: 3.0 (7.6)
P: 2.4 (6.2)
S: 2.3 (5.8)
S: 2.2 (5.7)
TS: 1.5 (3.7)
T: 2.4 (6.2)
T: 2.4 (6.2)
T: 2.4 (6.1)
MA: 1.0 (2.6)
BR: 0.7 (1.9)
BE: 1.1 (2.9)
RU: 0.9 (2.4)

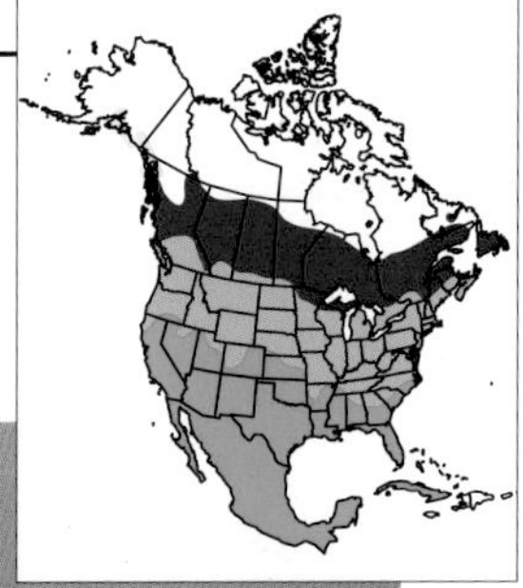

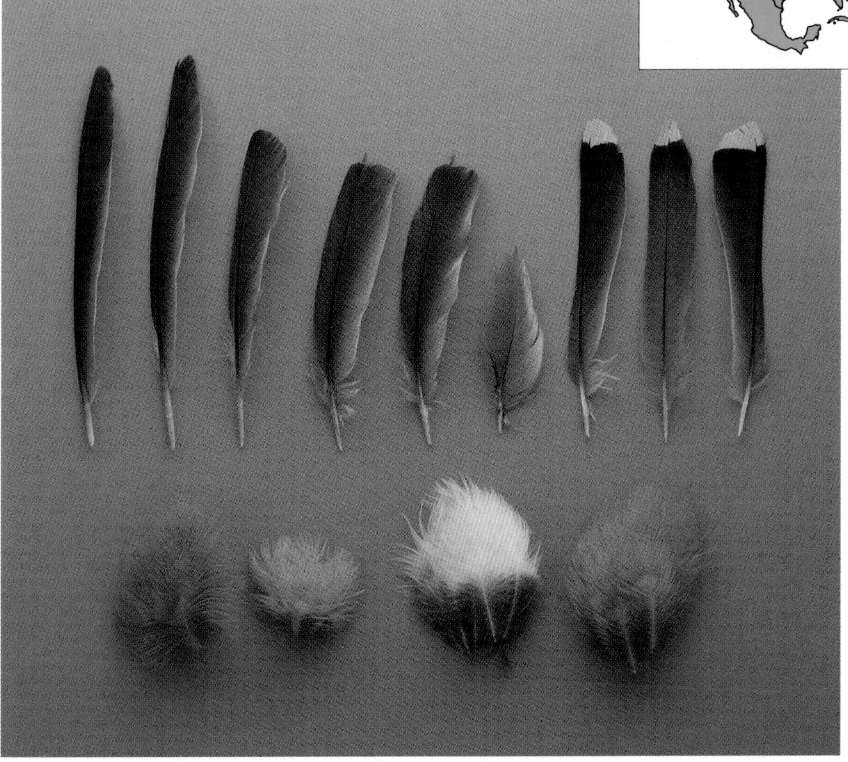

Silky-flycatchers *(Ptilogonatiade)*

Black and white primaries; similar in appearance to those of magpies, only much smaller.

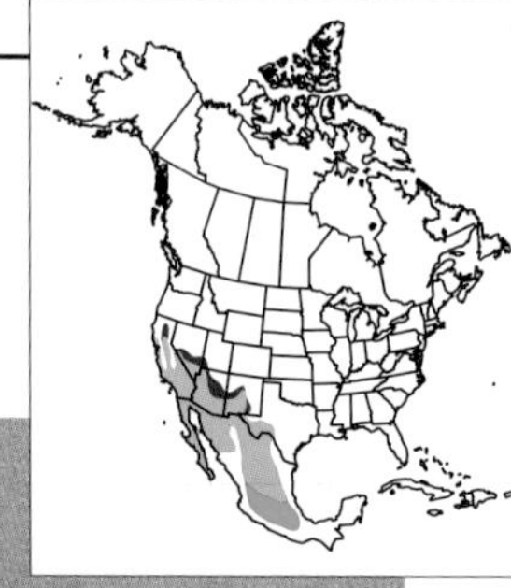

Phainopepla *(Phainopepla nitens)*

male
P: 3.3 (8.3)
P: 3.3 (8.3)
S: 2.9 (7.4)
S: 2.6 (6.7)
T: 3.6 (9.1)
T: 3.8 (9.7)
MA: 0.7 (1.8)
BE: 0.7 (1.8)

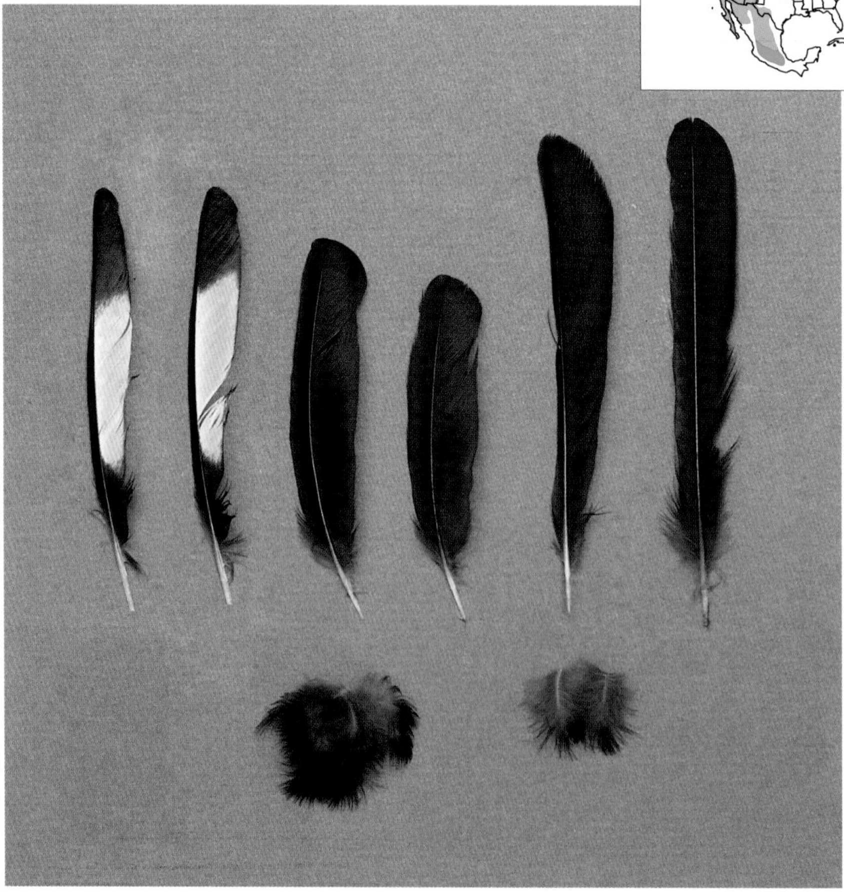

Wood-Warblers *(Parulidae)*

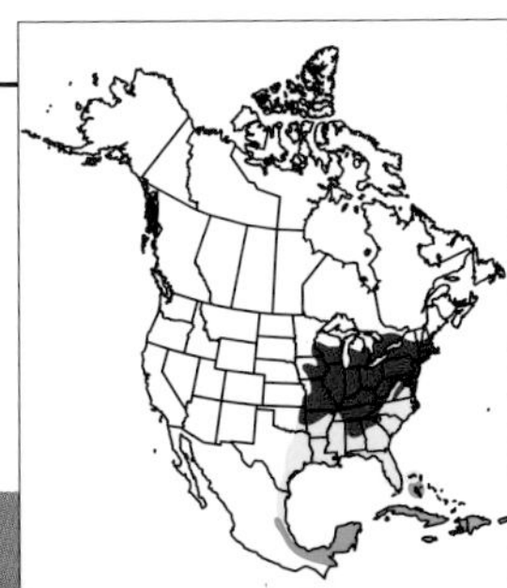

Blue-winged Warbler *(Vermivora pinus)*

unknown
P: 2.0 (5.1)
P: 2.1 (5.4)
P: 1.9 (4.7)
S: 1.7 (4.4)
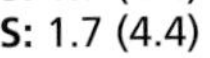
S: 1.7 (4.4)

T: 2.1 (5.3)
T: 2.2 (5.5)
T: 2.2 (5.5)
BR: 1.0 (2.5)
MA: 0.9 (2.2)

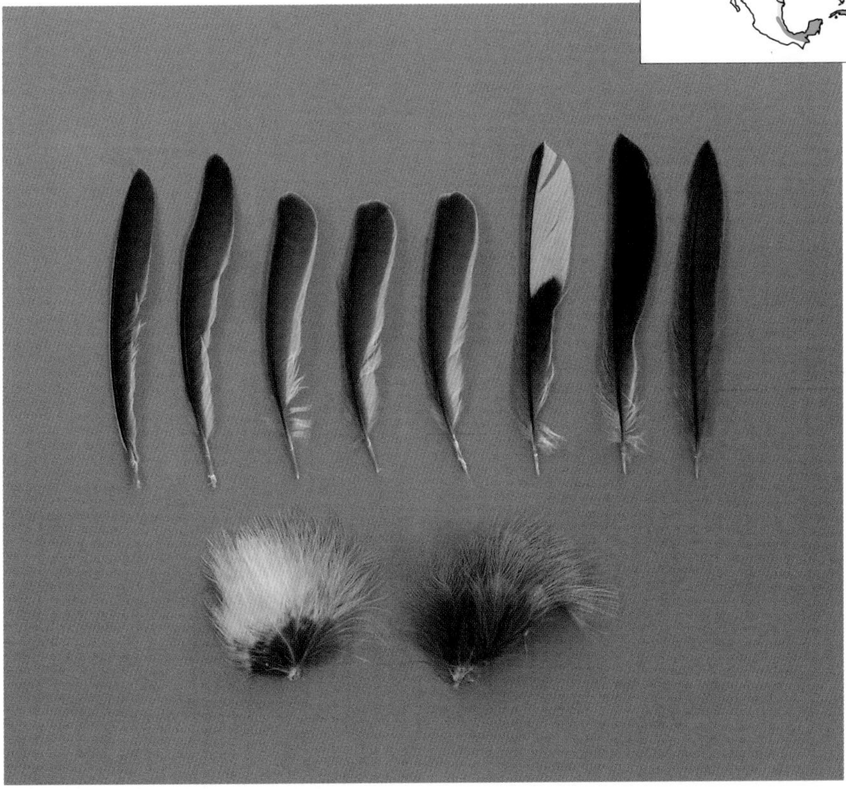

Tennessee Warbler *(Vermivora peregrina)*

male, breeding
P: 2.2 (5.6)
P: 2.3 (5.8)
P: 1.9 (4.9)
P: 1.8 (4.5)
S: 1.7 (4.3)
S: 1.6 (4.1)
T: 1.9 (4.7)
T: 1.8 (4.5)
BR: 1.1 (2.7)
BE: 1.2 (3.1)
MA: 1.0 (2.5)
RU: 0.8 (2.1)

Orange-crowned Warbler *(Vermivora celata)*

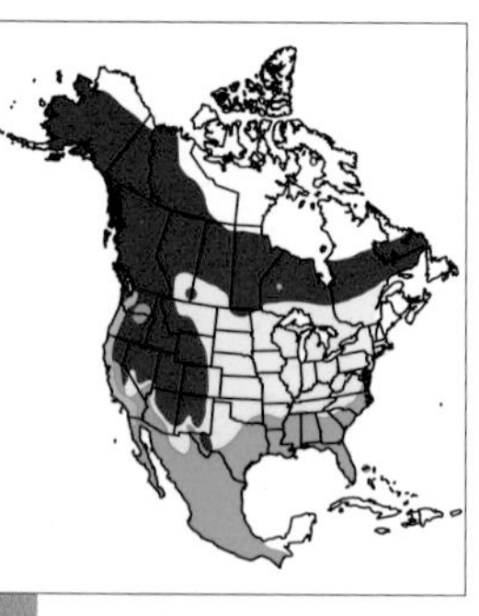

male
P: 1.9 (4.8)
P: 2.0 (5.1)
S: 1.5 (3.9)
T: 1.9 (4.7)
T: 1.9 (4.8)
T: 1.9 (4.9)
UPTC: 1.2 (3.1)
MA: 0.9 (2.4)
BR: 0.5 (1.3)
BE: 0.7 (1.9)
CR: 0.4 (1.0)

Nashville Warbler *(Vermivora ruficapilla)*

unknown
P: 2.0 (5.0)
P: 2.0 (5.1)
P: 1.8 (4.6)
P: 1.7 (4.4)
S: 1.6 (4.1)
S: 1.6 (4.0)
T: 1.9 (4.7)
T: 1.9 (4.7)
BR: 0.7 (1.7)
MA: 0.9 (2.3)
RU: 0.9 (2.2)

Northern Parula *(Parula americana)*

unknown
P: 1.9 (4.9)
P: 2.0 (5.1)
P: 1.7 (4.2)
S: 1.4 (3.6)
TS: 1.3 (3.2)
T: 1.7 (4.2)
T: 1.7 (4.2)
BR: 0.6 (1.6)
RU: 0.9 (2.2)
WC: 0.6 (1.4)

Yellow Warbler *(Dendroica petechia)*

male
P: 2.2 (5.5)
P: 2.2 (5.7)
P: 2.1 (5.3)
P: 1.9 (4.9)
S: 1.8 (4.6)
TS: 1.6 (4.0)
T: 1.9 (4.9)
T: 1.9 (4.7)
BR: 1.0 (2.5)
MA: 0.6 (1.5)
RU: 0.7 (1.7)

Chestnut-sided Warbler *(Dendroica pensylvanica)*

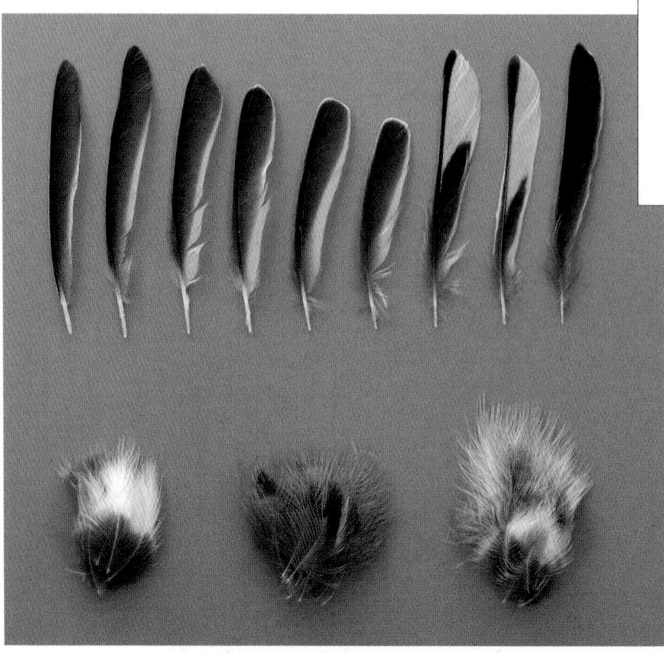

unknown
P: 1.9 (4.9)
P: 2.1 (5.3)
P: 1.9 (4.8)
P: 1.8 (4.5)
S: 1.6 (4.1)
S: 1.5 (3.8)
T: 2.0 (5.0)
T: 1.9 (4.9)
T: 2.0 (5.0)
BR: 0.9 (2.4)
MA: 0.9 (2.2)
FL: 1.1 (2.7)

Magnolia Warbler *(Dendroica magnolia)*

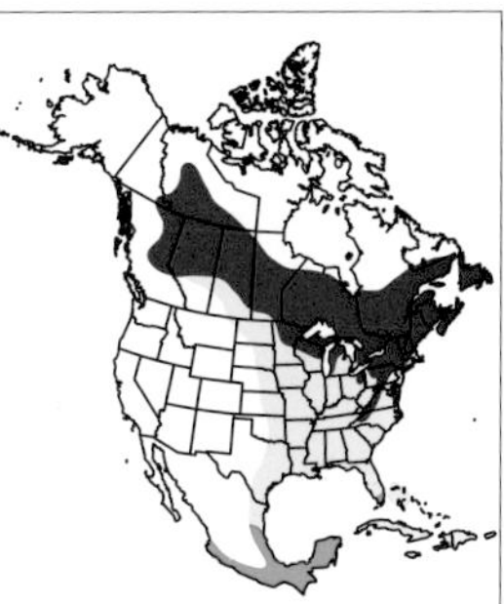

unknown
P: 1.9 (4.7)
P: 2.0 (5.1)
S: 1.5 (3.7)
S: 1.5 (3.8)
T: 2.0 (5.1)
T: 2.0 (5.0)
T: 2.0 (5.0)
BR: 0.7 (1.9)
BE: 1.0 (2.6)
RU: 0.9 (2.2)

Cape May Warbler *(Dendroica tigrina)*

male
P: 2.3 (5.9)
P: 2.3 (5.9)
P: 2.1 (5.3)
S: 1.7 (4.3)
TS: 1.5 (3.9)
T: 2.0 (5.1)
T: 1.9 (4.8)
BR: 0.9 (2.4)
MA: 0.9 (2.4)
RU: 0.9 (2.3)

Black-throated Blue Warbler *(Dendroica caerulescens)*

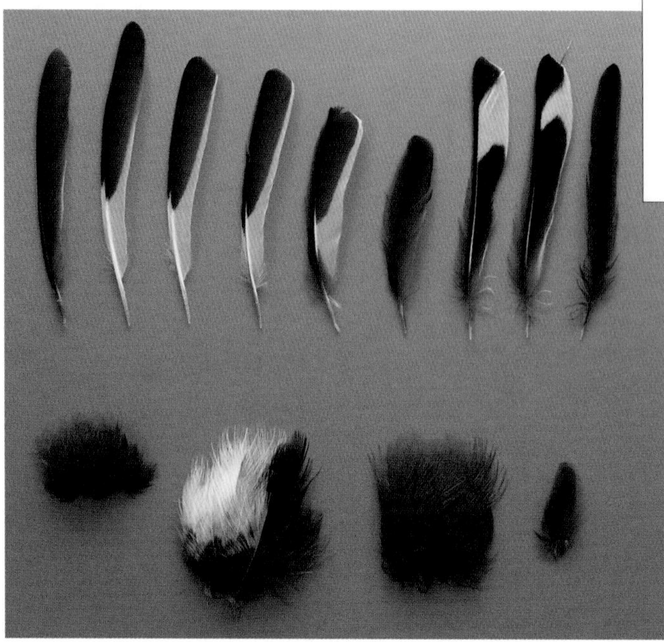

male
P: 2.1 (5.3)
P: 2.3 (5.9)
P: 2.1 (5.3)
P: 1.9 (4.9)
S: 1.7 (4.3)
TS: 1.5 (3.9)
T: 2.1 (5.4)
T: 2.1 (5.4)
T: 2.1 (5.3)
NA: 0.4 (1.1)
BR: 1.3 (3.3)
MA: 1.0 (2.6)
WC: 0.7 (1.9)

Yellow-rumped Warbler *(Dendroica coronata)*

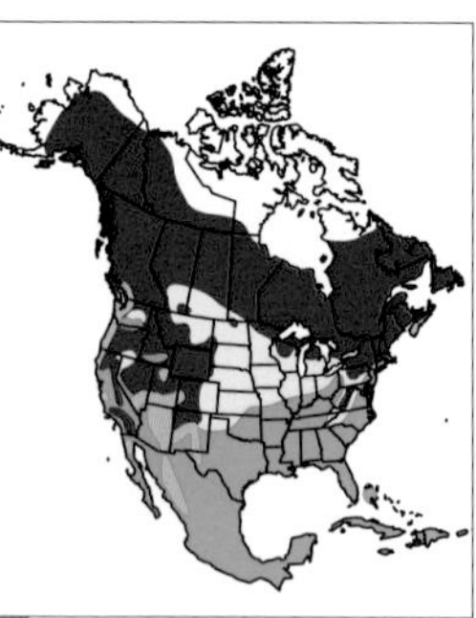

female, breeding
P: 2.5 (6.4)
P: 2.7 (6.8)
P: 2.3 (5.9)
S: 2.2 (5.6)
S: 1.8 (4.6)
T: 2.5 (6.3)
T: 2.5 (6.3)
T: 2.4 (6.1)
T: 2.4 (6.1)
MA: 0.9 (2.4)
BR: 0.6 (1.5)
BE: 1.1 (2.9)
RU: 0.9 (2.2)
WC: 0.8 (2.0)

Black-throated Gray Warbler *(Dendroica nigrescens)*

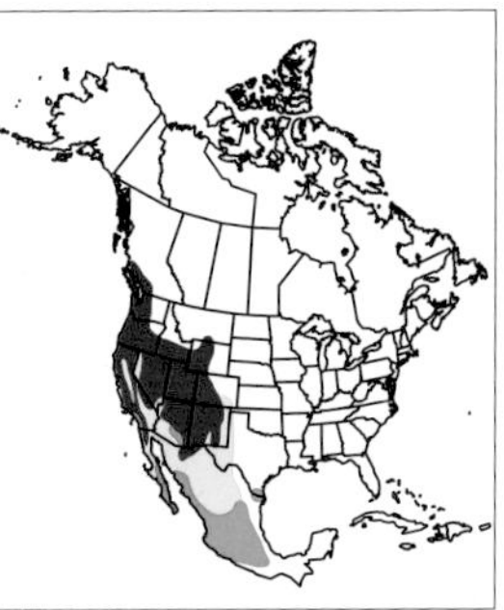

male
P: 2.2 (5.5)
P: 2.0 (5.0)
S: 1.6 (4.0)
TS: 1.4 (3.6)
T: 2.2 (5.5)
T: 2.2 (5.5)
NA: 0.6 (1.5)
BR: 0.5 (1.2)
FL: 1.1 (2.8)

Black-throated Green Warbler *(Dendroica virens)*

male
P: 2.1 (5.3)
P: 1.9 (4.7)
P: 1.8 (4.6)
P: 1.7 (4.4)
S: 1.6 (4.1)
TS: 1.5 (3.8)
T: 2.1 (5.3)
T: 2.0 (5.2)
MA: 1.0 (2.5)
FL: 0.8 (2.1)
BE: 0.8 (2.1)
TH: 0.5 (1.3)

Blackburnian Warbler *(Dendroica fusca)*

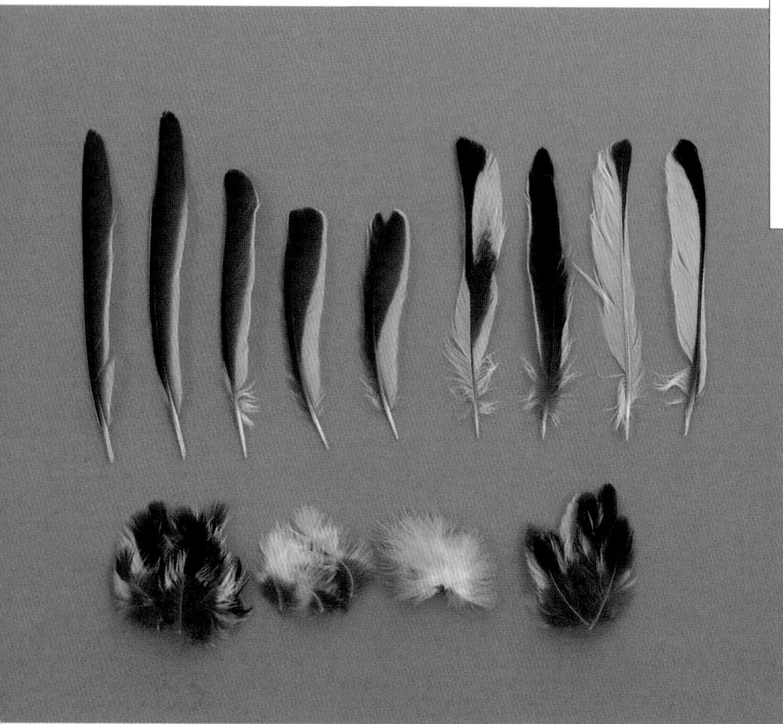

female
P: 2.4 (6.0)
P: 2.4 (6.2)
P: 2.0 (5.1)
S: 1.7 (4.3)
S: 1.6 (4.0)
T: 2.1 (5.3)
T: 2.1 (5.3)
T: 2.1 (5.3)
T: 2.1 (5.3)
MA: 0.6 (1.6)
BR: 0.7 (1.8)
TH: 0.6 (1.4)
RU: 0.8 (2.1)

Yellow-throated Warbler *(Dendroica dominica)*

unknown
P: 1.9 (4.7)
P: 1.7 (4.4)
P: 1.7 (4.2)
S: 1.5 (3.9)
S: 1.5 (3.7)
T: 1.9 (4.9)
T: 1.9 (4.9)
BR: 0.7 (1.7)
RU: 1.0 (2.6)

Pine Warbler *(Dendroica pinus)*

male
P: 2.5 (6.3)
P: 2.5 (6.4)
S: 1.8 (4.5)
TS: 1.6 (4.1)
T: 2.3 (5.9)
T: 2.3 (5.9)
NA: 0.7 (1.7)
MA: 0.7 (1.8)
TH: 0.5 (1.3)
BE: 0.9 (2.3)

Palm Warbler *(Dendroica palmarum)*

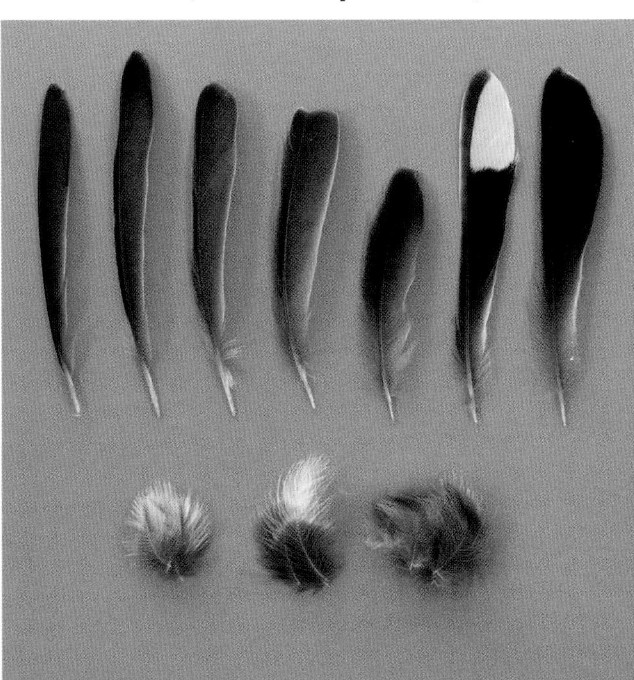

unknown
P: 2.0 (5.1)
P: 2.2 (5.6)
P: 2.0 (5.1)
S: 1.8 (4.6)
TS: 1.5 (3.9)
T: 2.2 (5.5)
T: 2.2 (5.5)
BR: 0.5 (1.3)
BE: 0.7 (1.9)
MA: 0.6 (1.5)

Bay-breasted Warbler *(Dendroica castanea)*

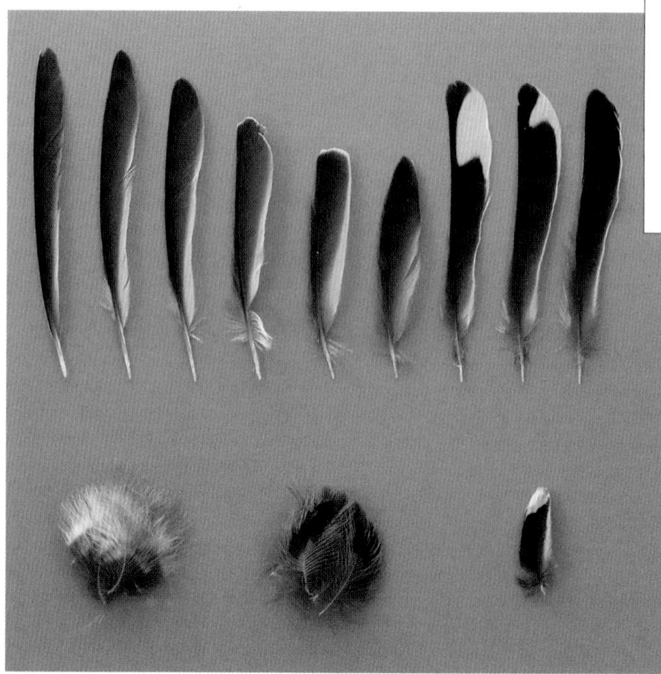

unknown
P: 2.6 (6.6)
P: 2.6 (6.6)
P: 2.4 (6.1)
P: 2.1 (5.3)
S: 1.9 (4.7)
TS: 1.8 (4.5)
T: 2.4 (6.1)
T: 2.4 (6.1)
T: 2.4 (6.0)
BR: 0.9 (2.4)
MA: 0.9 (2.4)
WC: 0.9 (2.2)

Blackpoll Warbler *(Dendroica striata)*

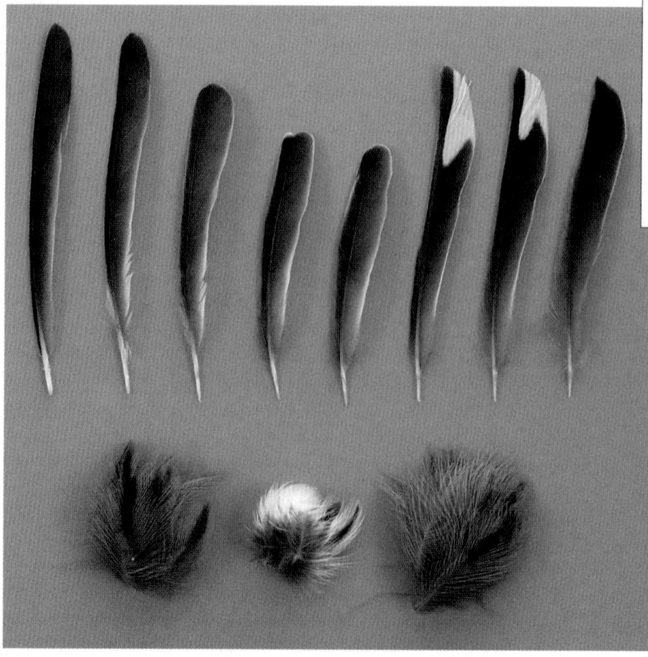

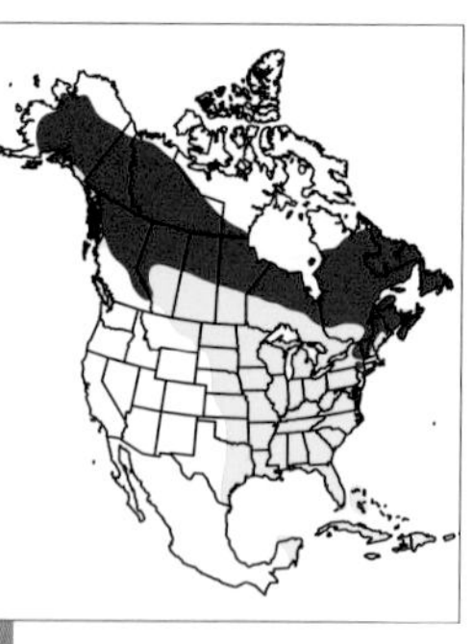

unknown
P: 2.5 (6.4)
P: 2.4 (6.2)
P: 2.1 (5.4)
S: 1.8 (4.5)
S: 1.8 (4.5)
T: 2.2 (5.7)
T: 2.2 (5.7)
T: 2.1 (5.4)
MA: 0.9 (2.3)
BR: 0.6 (1.4)
RU: 1.1 (2.9)

Black-and-white Warbler *(Mniotilta varia)*

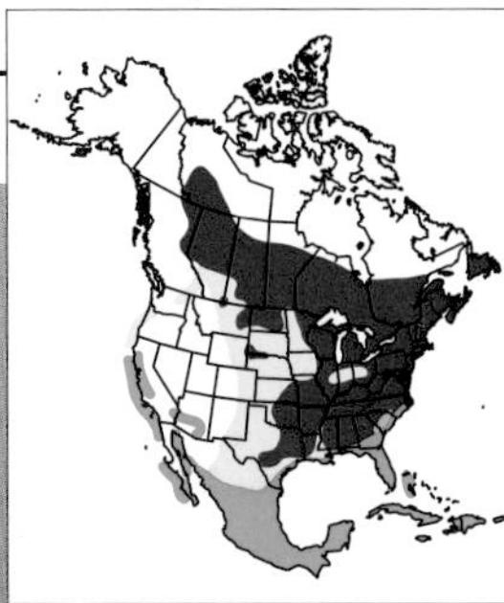

male, breeding
P: 2.3 (5.9)
P: 2.4 (6.2)
P: 2.4 (6.1)
P: 2.0 (5.2)
S: 1.8 (4.5)
S: 1.6 (4.1)
T: 2.2 (5.5)
T: 2.2 (5.5)
T: 2.1 (5.3)
BR: 1.2 (3.1)
MA: 1.1 (2.7)
RU: 0.9 (2.2)
WC: 0.7 (1.9)

American Redstart *(Setophaga ruticilla)*

male, male flight feathers vary from orange to yellow

P: 2.1 (5.3)
P: 2.2 (5.7)
P: 2.1 (5.3)
P: 2.0 (5.0)
S: 1.9 (4.8)
S: 1.8 (4.6)
T: 2.4 (6.1)
T: 2.6 (6.5)
T: 2.5 (6.4)
BR: 0.9 (2.2)
MA: 1.1 (2.7)
UNTC: 1.3 (3.2)

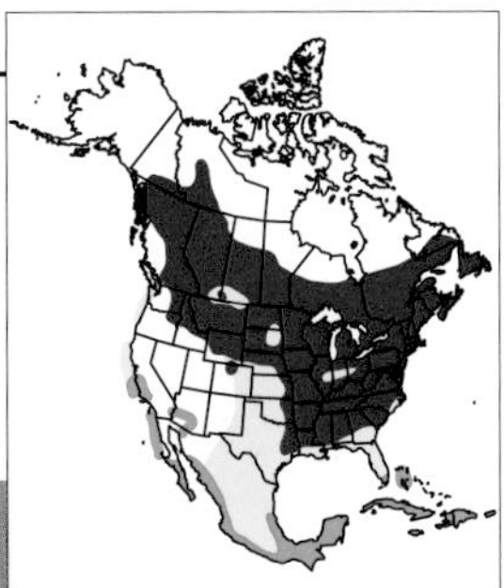

Worm-eating Warbler *(Helmitheros vermivorum)*

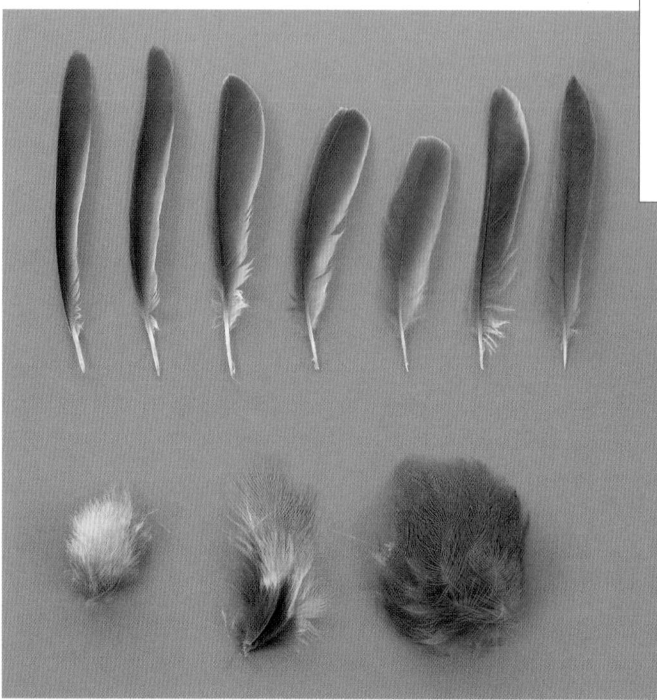

unknown
P: 2.3 (5.8)
P: 2.4 (6.0)
P: 2.1 (5.4)
S: 1.9 (4.8)
S: 1.7 (4.3)
T: 2.0 (5.2)
T: 2.1 (5.3)
NA: 0.6 (1.6)
BR: 1.5 (3.7)
MA: 1.3 (3.2)

Ovenbird *(Seiurus aurocapilla)*

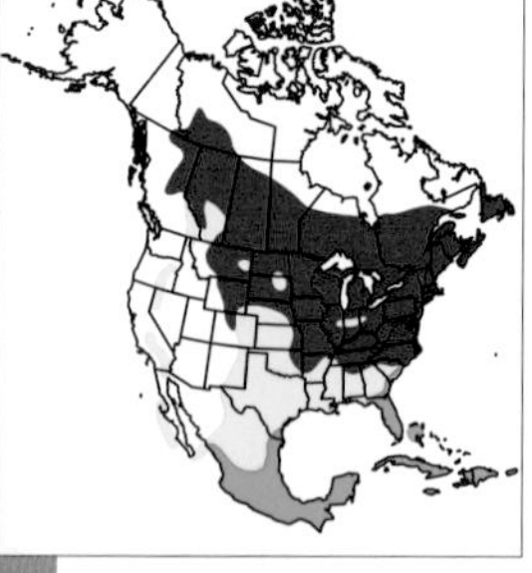

unknown
P: 2.3 (5.8)
P: 2.4 (6.2)
P: 2.1 (5.3)
P: 2.0 (5.2)
S: 2.0 (5.1)
S: 1.9 (4.8)
T: 2.2 (5.7)
T: 2.2 (5.6)
MA: 0.7 (1.8)
BR: 0.7 (1.8)
BE: 1.1 (2.7)

Northern Waterthrush *(Seiurus noveboracensis)*

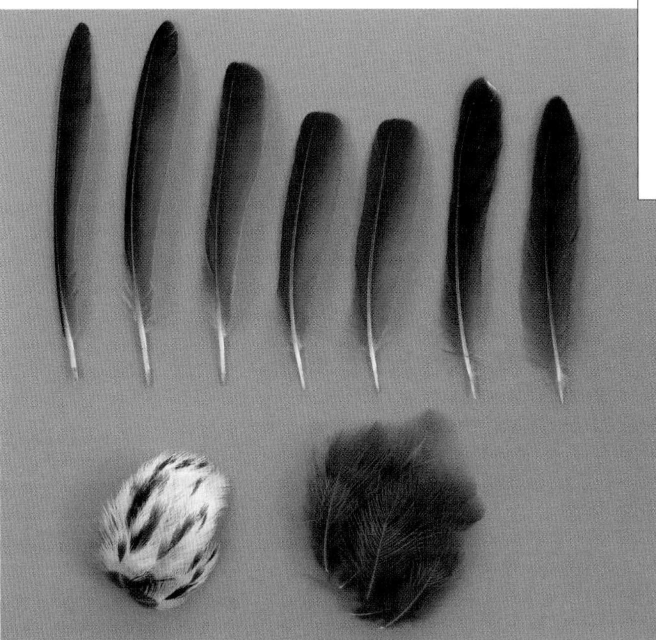

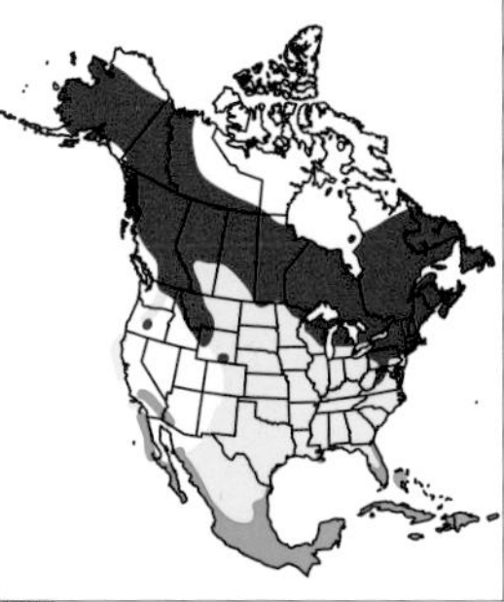

unknown
P: 2.6 (6.5)
P: 2.6 (6.6)
P: 2.3 (5.9)
S: 2.0 (5.1)
S: 2.0 (5.0)
T: 2.3 (5.8)
T: 2.2 (5.6)
BR: 0.7 (1.9)
MA: 1.1 (2.7)

Mourning Warbler *(Oporornis philadelphia)*

unknown
P: 2.0 (5.1)
P: 2.1 (5.4)
P: 2.1 (5.4)
P: 1.8 (4.6)
S: 1.5 (3.9)
TS: 1.6 (4.1)
T: 2.0 (5.0)
T: 2.0 (5.0)
BR: 1.0 (2.6)
BE: 1.1 (2.7)
MA: 0.9 (2.2)

MacGillivray's Warbler *(Oporornis tolmiei)*

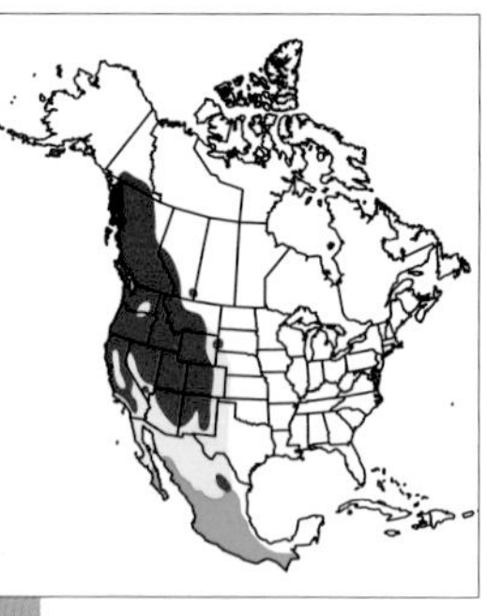

unknown
P: 2.1 (5.3)
P: 1.9 (4.7)
S: 1.7 (4.4)
S: 1.7 (4.2)
TS: 1.2 (3.1)
T: 2.2 (5.7)
T: 2.2 (5.7)
MA: 0.9 (2.2)
BE: 0.6 (1.5)
UPTC: 1.1 (2.7)

Common Yellowthroat *(Geothlypis trichas)*

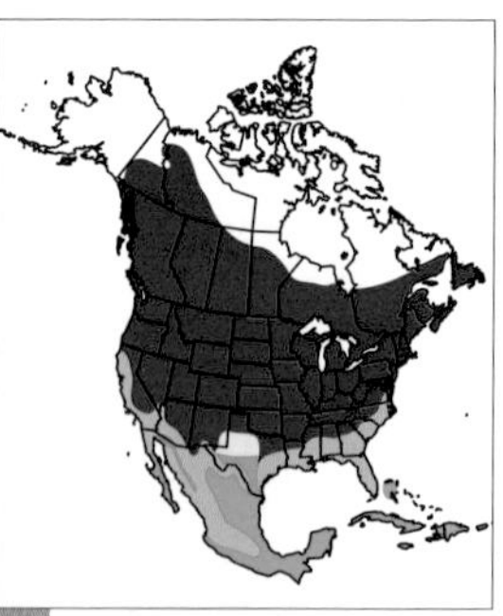

unknown
P: 1.7 (4.4)
P: 1.9 (4.8)
S: 1.7 (4.2)
S: 1.6 (4.0)
T: 2.1 (5.3)
T: 2.1 (5.3)
T: 2.1 (5.3)
BR: 0.6 (1.6)
BE: 0.9 (2.2)
MA: 0.7 (1.7)

Hooded Warbler *(Wilsonia citrina)*

male
P: 2.0 (5.2)
P: 2.2 (5.6)
P: 2.1 (5.3)
P: 2.0 (5.0)
S: 1.7 (4.2)
TS: 1.5 (3.9)
T: 2.3 (5.9)
T: 2.2 (5.6)
BR: 0.6 (1.5)
BE: 1.1 (2.7)
MA: 1.1 (2.8)

Wilson's Warbler *(Wilsonia pusilla)*

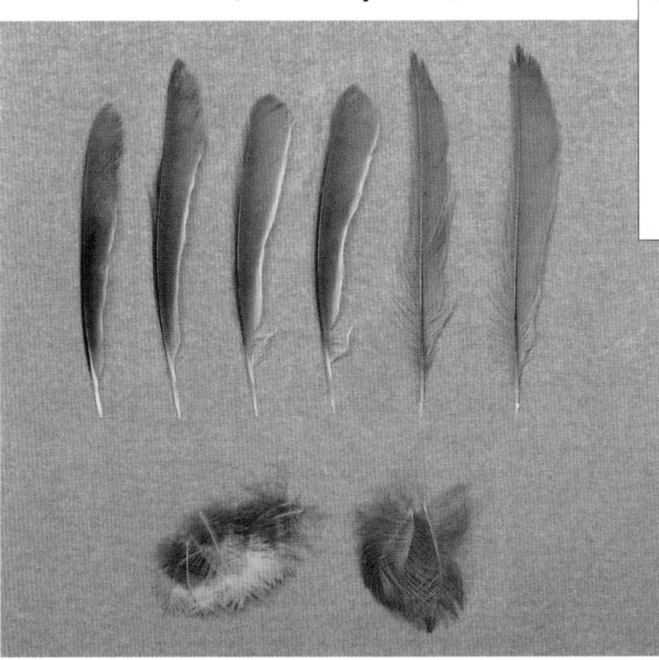

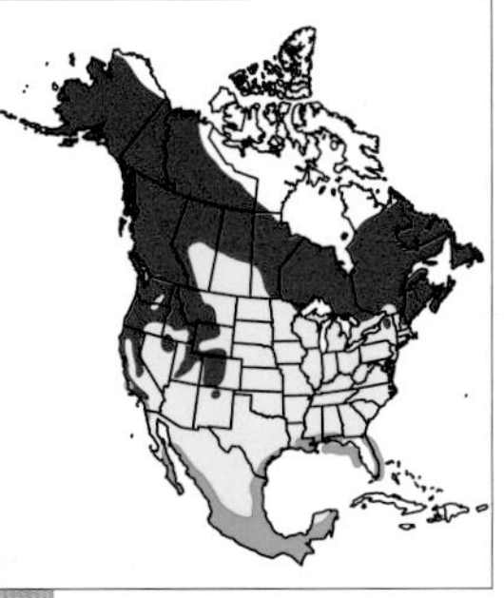

unknown
P: 1.7 (4.3)
P: 1.9 (4.9)
S: 1.7 (4.3)
S: 1.7 (4.3)
T: 2.0 (5.0)
T: 2.0 (5.0)
BR: 0.6 (1.4)
MA: 0.6 (1.6)

Canada Warbler *(Wilsonia canadensis)*

unknown
P: 2.0 (5.2)
P: 2.2 (5.6)
P: 2.0 (5.2)
P: 1.9 (4.8)
S: 1.8 (4.6)
S: 1.7 (4.4)
T: 2.2 (5.7)
T: 2.3 (5.8)
BR: 0.7 (1.9)
BE: 0.9 (2.4)
RU: 0.7 (1.8)

Yellow-breasted Chat *(Icteria virens)*

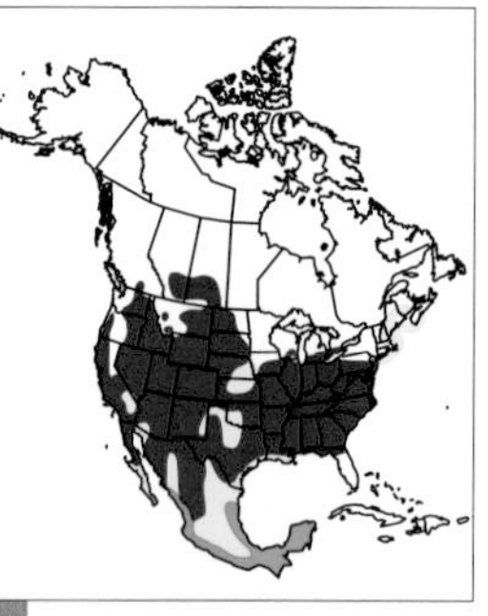

unknown
P: 2.0 (5.2)
P: 2.3 (5.9)
P: 2.2 (5.5)
S: 2.0 (5.1)
S: 1.9 (4.9)
T: 3.1 (7.8)
T: 3.0 (7.7)
BR: 0.4 (1.1)
BE: 0.8 (2.1)
TH: 0.5 (1.2)
MA: 0.9 (2.4)

Tanagers *(Thraupidae)*

Summer Tanager *(Piranga rubra)*

female
P: 3.0 (7.6)
P: 3.1 (8.0)
S: 2.2 (5.5)
TS: 2.2 (5.5)
T: 3.0 (7.7)
T: 2.4 (6.1) [sheathed]
T: 3.0 (7.7)
T: 3.0 (7.7)
NA: 0.8 (2.1)
MA: 1.2 (3.1)
UPTC: 1.8 (4.6)
TH: 0.6 (1.5)
BR: 0.7 (1.8)
BE: 1.3 (3.2)
CR: 0.6 (1.5)

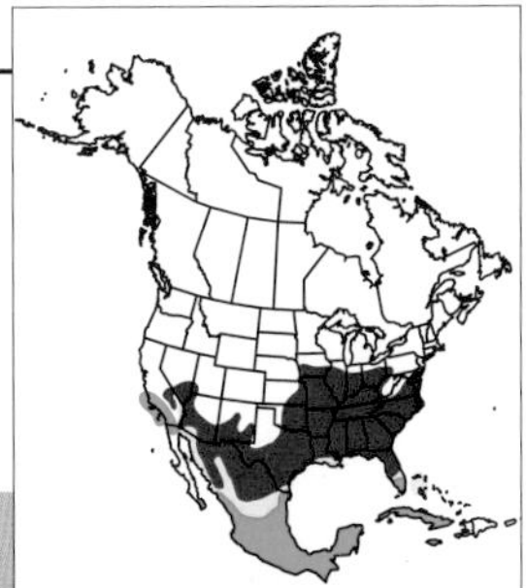

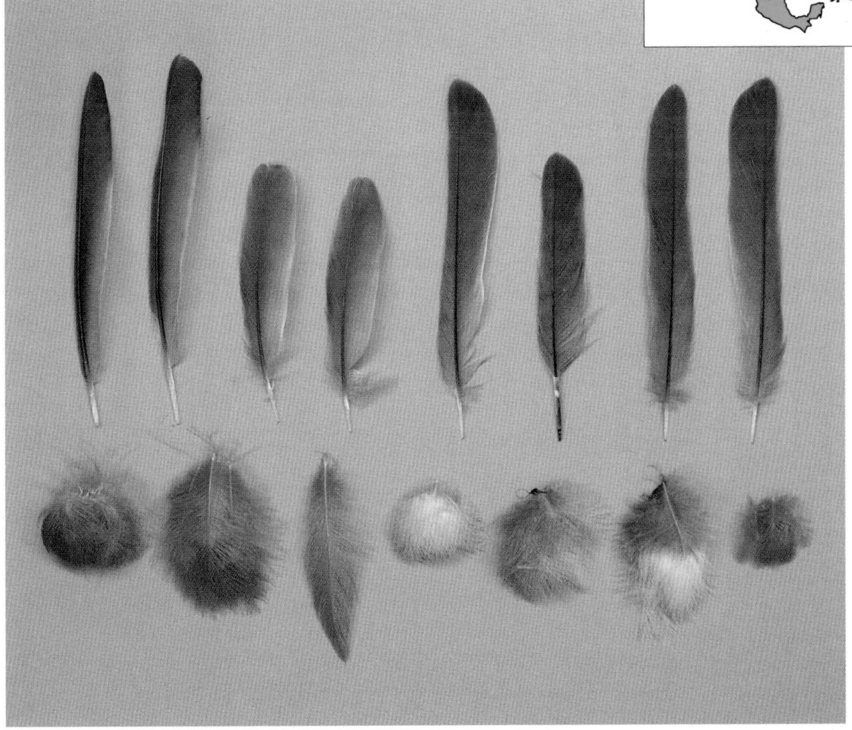

Scarlet Tanager *(Piranga olivacea)*

male
P: 3.3 (8.5)
P: 3.5 (8.8)
P: 2.8 (7.0)
S: 2.5 (6.3)
S: 2.4 (6.1)
T: 3.0 (7.5)
T: 2.9 (7.3)
T: 3.0 (7.7)
RU: 1.1 (2.9)
BE: 1.2 (3.1)
BR: 0.9 (2.3)
NA: 0.6 (1.6)

Western Tanager *(Piranga ludoviciana)*

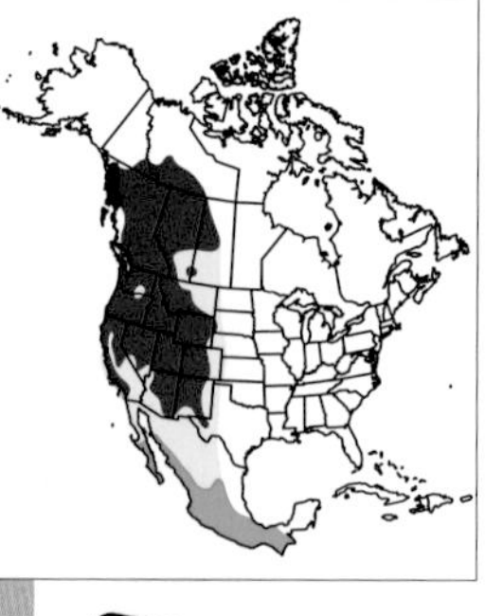

male
P: 3.1 (7.8)
P: 3.2 (8.1)
S: 2.4 (6.2)
S: 2.5 (6.4)
T: 2.7 (6.9)
T: 2.8 (7.1)
TH: 0.4 (1.1)
BE: 0.8 (2.1)
MA: 1.0 (2.6)
UNTC: 1.6 (4.1)
CR: 0.2 (0.5)

Emberizids: Sparrows, Old World Buntings, and Relatives *(Emberizidae)*

Some towhee flight feathers are similar to those of thrashers and cuckoos.

Green-tailed Towhee *(Pipilo chlorurus)*

male
P: 2.9 (7.3)
P: 2.9 (7.4)
S: 2.7 (6.9)
S: 2.7 (6.9)
TS: 2.1 (5.4)
T: 3.8 (9.6)
T: 3.8 (9.6)
WC: 1.1 (2.8)
MC: 0.3 (0.8)
NA: 0.8 (2.1)
CR: 0.5 (1.3)
UPTC: 0.9 (2.3)
TH: 0.5 (1.2)

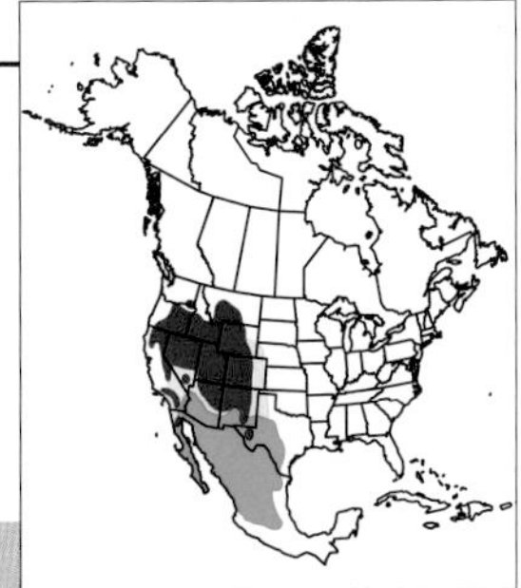

Spotted Towhee *(Pipilo maculatus)*

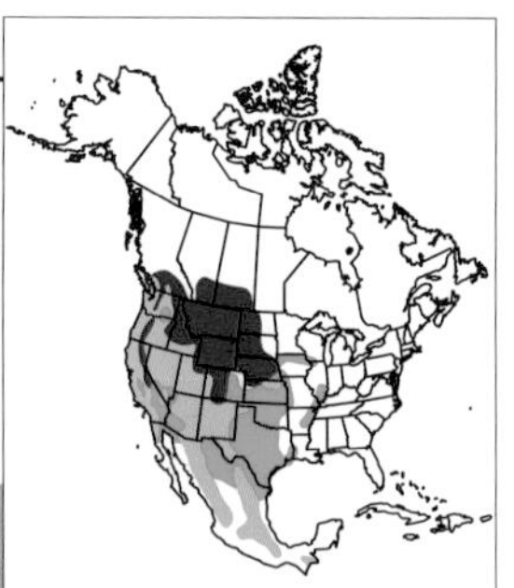

unknown
P: 2.4 (6.1)
P: 3.1 (7.8)
S: 2.8 (7.1)
S: 2.6 (6.6)
T: 3.9 (9.8)
T: 4.3 (10.8)
T: 3.9 (9.8)
WC: 1.2 (3.1)
FL: 0.8 (2.0)
BE: 1.0 (2.5)
RU: 1.4 (3.5)

Eastern Towhee *(Pipilo erythrophthalmus)*

unknown
P: 2.4 (6.0)
P: 3.0 (7.7)
P: 2.8 (7.0)
S: 2.5 (6.3)
T: 3.7 (9.4)
T: 4.0 (10.1)
T: 3.9 (10.0)
BR: 0.6 (1.6)
FL: 0.9 (2.4)
FL: 1.7 (4.2)
RU: 0.9 (2.3)

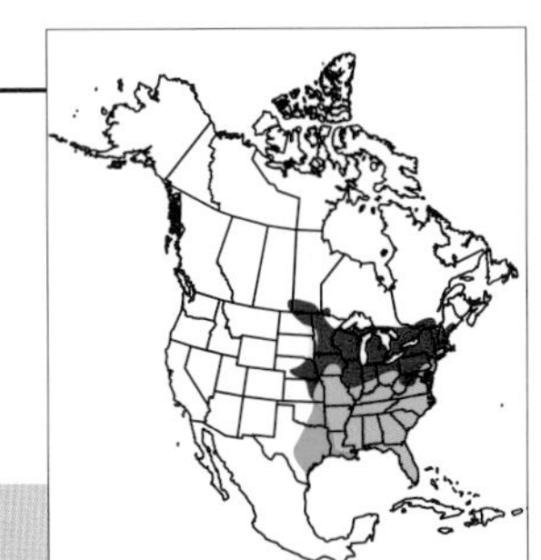

California Towhee *(Pipilo crissalis)*

unknown
P: 2.5 (6.3)
P: 3.4 (8.7)
P: 3.4 (8.7)
P: 3.0 (7.7)
S: 2.6 (6.7)
T: 4.5 (11.5)
T: 4.5 (11.5)
TH: 0.8 (2.1)
UNTC: 1.9 (4.7)
UPTC: 2.0 (5.2)
BE: 1.4 (3.5)
NA: 0.8 (2.1)
CR: 0.5 (1.3)

Abert's Towhee *(Pipilo aberti)*

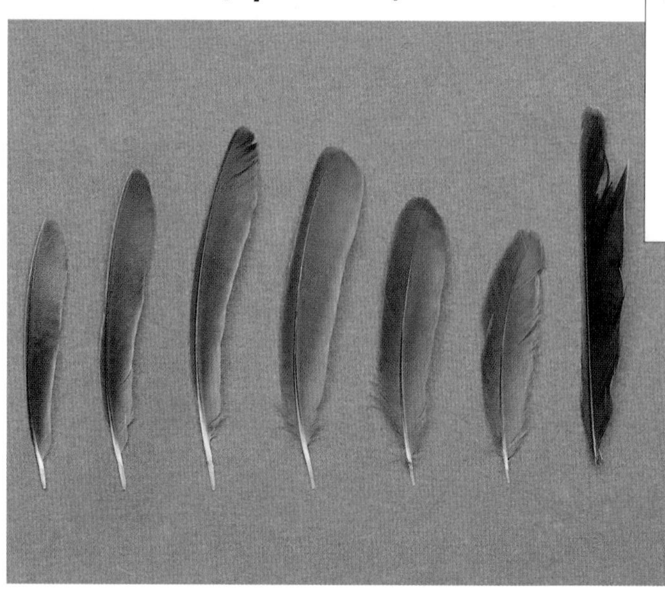

unknown
P: 2.3 (5.9)
P: 2.8 (7.1)
P: 3.1 (8.0)
P: 3.0 (7.6)
TS: 2.5 (6.4)
TS: 2.2 (5.7)
T: 3.1 (7.8)
[broken quill]

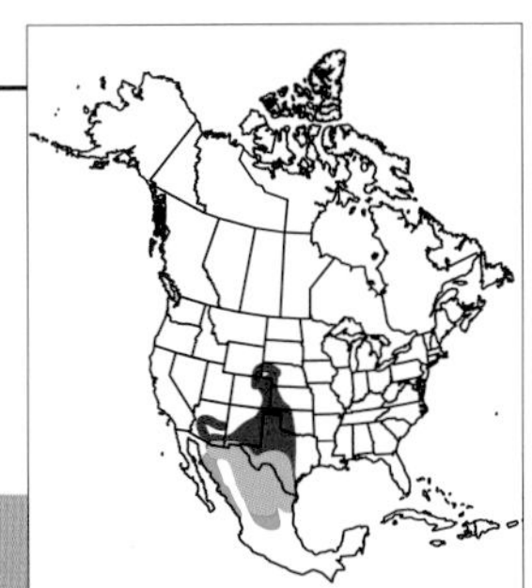

Cassin's Sparrow *(Aimophila cassinii)*

unknown
P: 2.0 (5.2)
P: 2.2 (5.7)
P: 2.0 (5.1)
S: 1.9 (4.9)
S: 1.9 (4.7)
T: 2.4 (6.1)
T: 2.7 (6.8)
BR: 0.6 (1.5)
BE: 0.9 (2.3)
MA: 0.8 (2.0)

Rufous-crowned Sparrow *(Aimophila ruficeps)*

unknown
P: 2.0 (5.1)
P: 2.0 (5.1)
S: 2.0 (5.1)
S: 1.9 (4.7)
T: 2.2 (5.6)
[sheathed]

T: 2.5 (6.4)
T: 2.5 (6.4)
CR: 0.5 (1.2)
MA: 1.2 (3.1)
TH: 0.5 (1.2)
BE: 0.9 (2.2)
MC: 0.4 (1.1)

American Tree Sparrow *(Spizella arborea)*

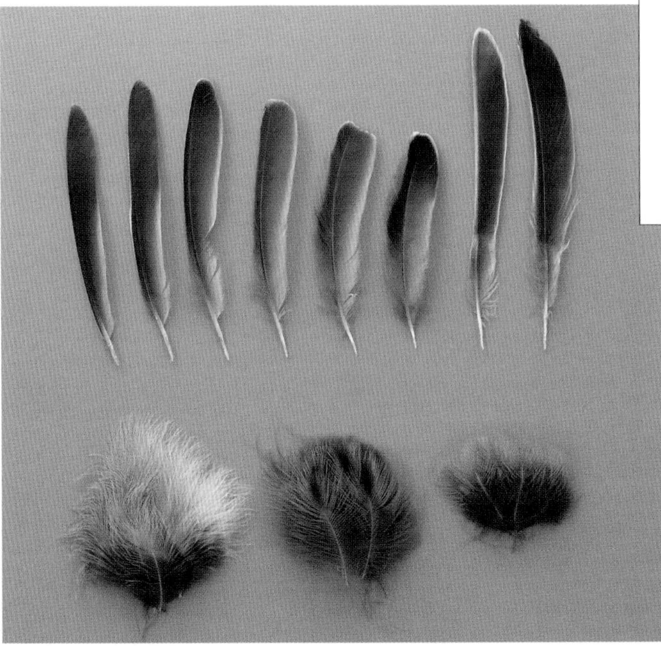

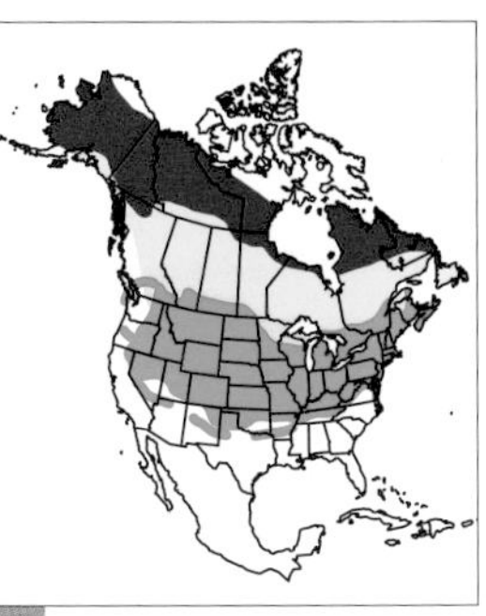

unknown
P: 2.2 (5.7)
P: 2.4 (6.1)
P: 2.4 (6.1)
P: 2.2 (5.5)
S: 1.9 (4.9)
TS: 1.8 (4.6)
T: 2.7 (6.9)
T: 2.8 (7.1)
BR: 1.4 (3.5)
MA: 1.1 (2.7)
RU: 0.9 (2.2)

Chipping Sparrow *(Spizella passerina)*

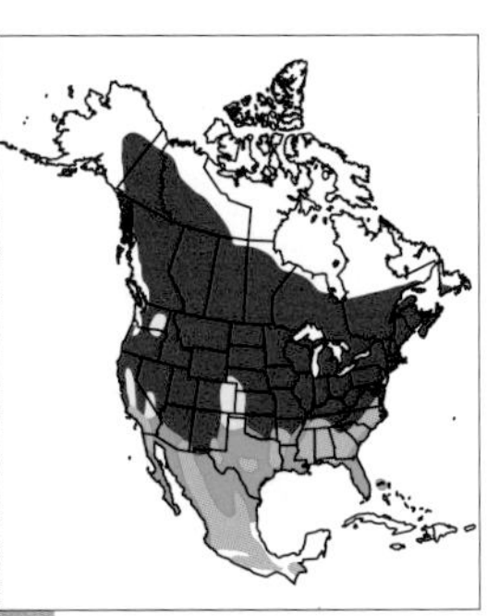

unknown
P: 2.5 (6.3)
P: 2.4 (6.1)
P: 2.1 (5.3)
S: 1.9 (4.7)
TS: 1.5 (3.9)
T: 2.6 (6.7)
T: 2.6 (6.6)
WC: 0.9 (2.3)
MA: 0.7 (1.9)
BR: 0.7 (1.7)
CR: 0.4 (1.1)

Brewer's Sparrow *(Spizella breweri)*

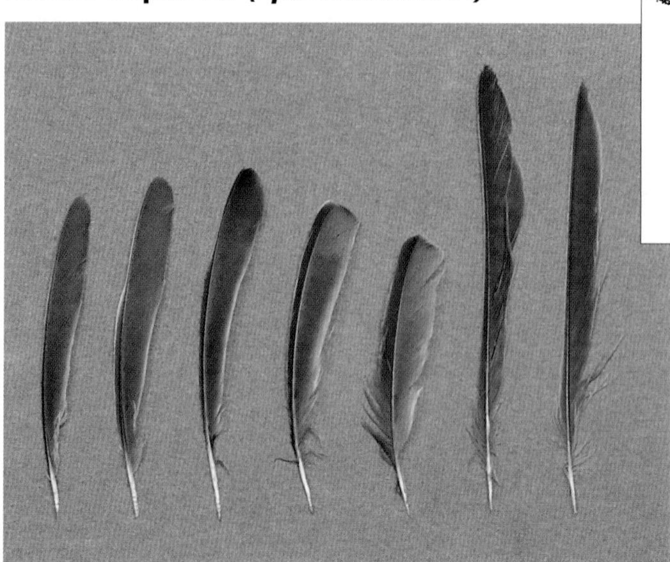

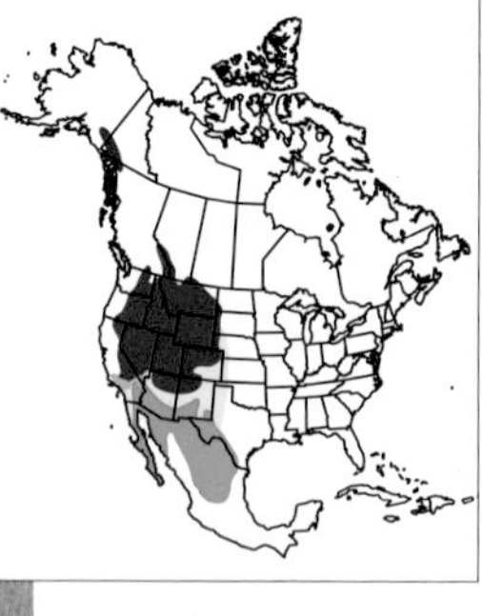

unknown
P: 2.0 (5.1)
P: 2.1 (5.3)
P: 2.2 (5.6)
S: 2.0 (5.0)
S: 1.7 (4.4)
T: 2.7 (6.9)
T: 2.7 (6.8)

Field Sparrow *(Spizella pusilla)*

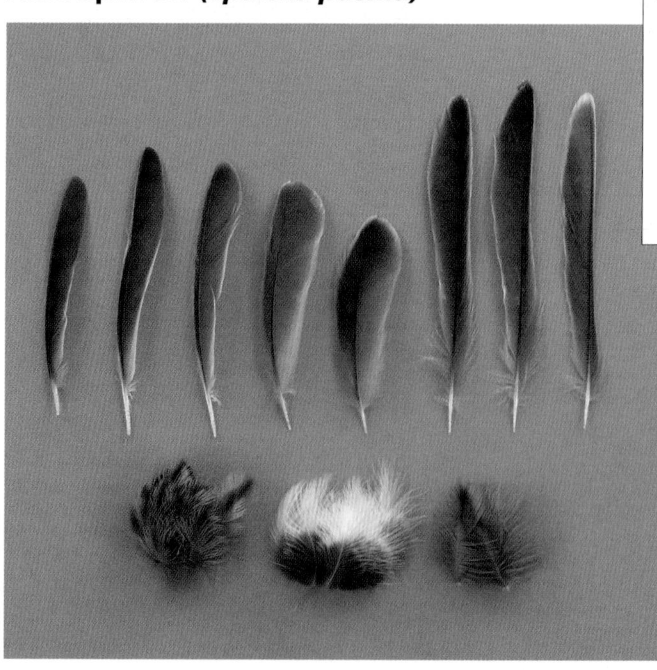

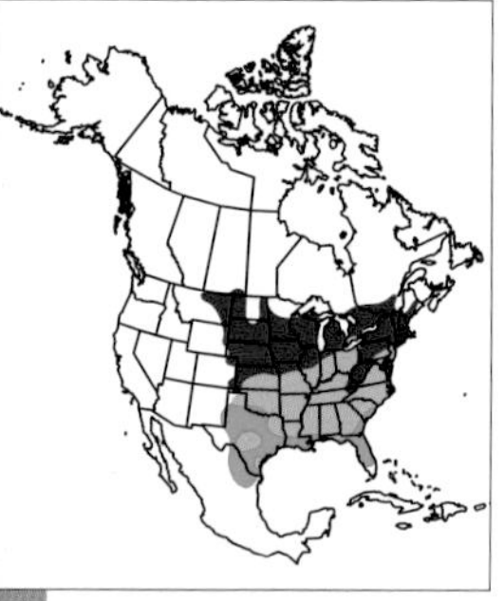

unknown
P: 1.9 (4.7)
P: 2.2 (5.6)
P: 2.1 (5.4)
S: 1.9 (4.8)
TS: 1.7 (4.2)
T: 2.6 (6.5)
T: 2.6 (6.7)
T: 2.6 (6.5)
MA: 0.5 (1.2)
BR: 0.7 (1.9)
RU: 0.9 (2.3)

Vesper Sparrow *(Pooecetes gramineus)*

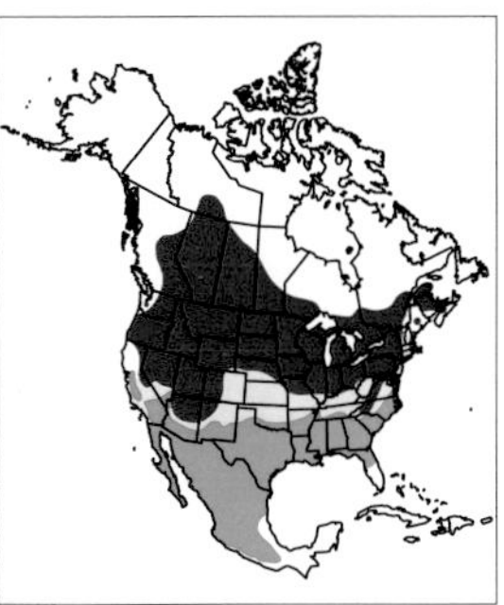

unknown
P: 2.7 (6.8)
P: 2.6 (6.5)
S: 2.4 (6.0)
T: 2.6 (6.6)
T: 2.5 (6.4)
T: 2.6 (6.6)
MA: 0.8 (2.1)
RU: 1.1 (2.7)
TH: 0.5 (1.3)
MC: 0.6 (1.4)
WC: 1.3 (3.4)

Lark Sparrow *(Chondestes grammacus)*

unknown
P: 2.8 (7.0)
P: 2.9 (7.3)
S: 2.4 (6.1)
S: 2.2 (5.6)
T: 2.8 (7.2)
T: 3.0 (7.5)
MA: 0.7 (1.9)
BE: 0.8 (2.0)
FL: 1.1 (2.7)

Black-throated Sparrow *(Amphispiza bilineata)*

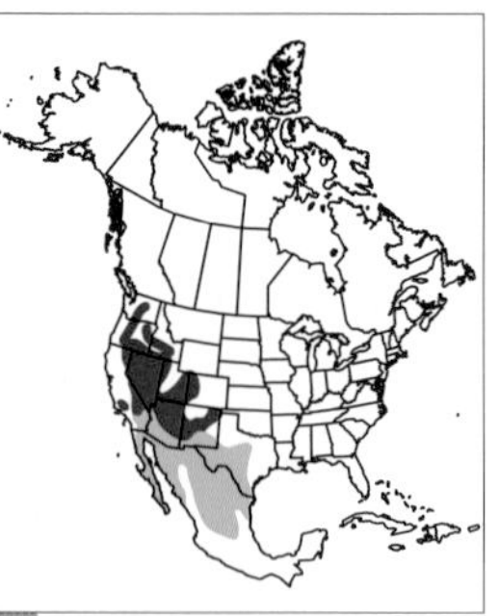

unknown
P: 2.0 (5.1)
P: 2.2 (5.5)
S: 1.9 (4.8)
S: 1.9 (4.9)
T: 2.4 (6.0)
T: 2.6 (6.5)
WC: 0.4 (1.1)
RU: 0.7 (1.7)
BR: 1.3 (3.4)

Savannah Sparrow *(Passerculus sandwichensis)*

unknown
P: 2.2 (5.7)
P: 2.3 (5.8)
P: 2.1 (5.4)
S: 2.0 (5.1)
TS: 2.0 (5.2)
T: 1.9 (4.7)
T: 2.3 (5.8)
BR: 1.1 (2.9)
MA: 1.1 (2.7)
RU: 1.3 (3.2)

Grasshopper Sparrow *(Ammodramus savannarum)*

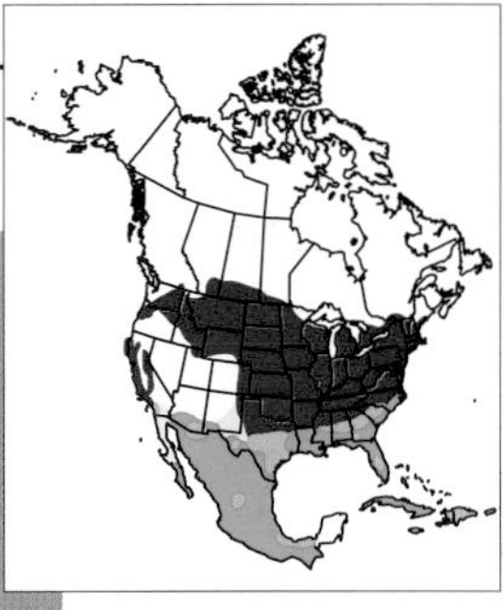

unknown
P: 2.0 (5.1)
P: 2.0 (5.2)
P: 1.9 (4.8)
S: 1.7 (4.4)
TS: 1.8 (4.5)
T: 1.9 (4.9)
T: 1.8 (4.5)
NA: 0.6 (1.4)
BE: 1.2 (3.0)
MA: 0.7 (1.7)

Seaside Sparrow *(Ammodramus maritimus)*

unknown
P: 1.9 (4.7)
P: 2.1 (5.3)
P: 2.1 (5.3)
S: 1.8 (4.6)
S: 1.9 (4.7)
T: 2.2 (5.6)
T: 2.3 (5.8)
BR: 0.8 (2.0)
BE: 1.0 (2.5)
MA: 1.1 (2.7)

Fox Sparrow *(Passerella iliaca)*

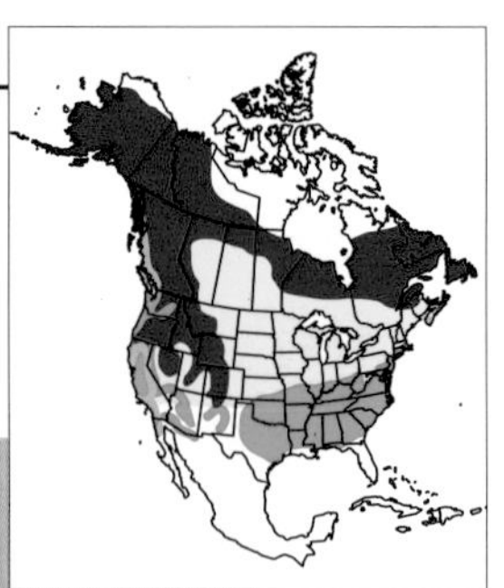

unknown
P: 2.8 (7.2)
P: 2.7 (6.8)
S: 2.4 (6.1)
T: 3.5 (8.8)

T: 3.5 (8.8)
WC: 0.9 (2.3)
UPTC: 1.1 (2.7)
MA: 0.8 (2.1)
BR: 0.6 (1.6)

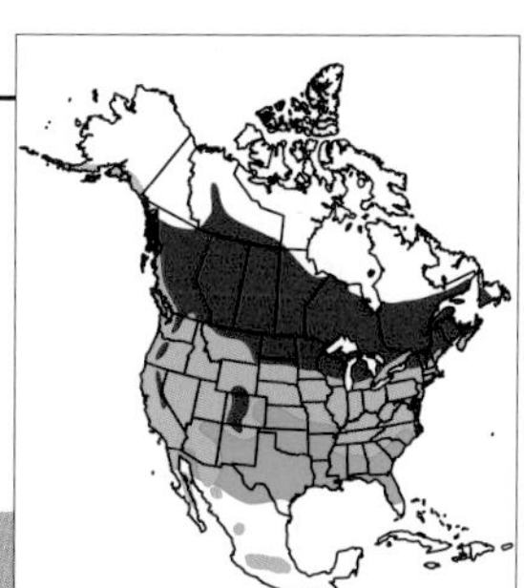

Song Sparrow *(Melospiza melodia)*

unknown
P: 2.3 (5.9)
P: 2.2 (5.5)
S: 2.1 (5.3)
S: 2.0 (5.2)

T: 2.5 (6.4)
T: 2.8 (7.1)
BR: 0.7 (1.7)
BE: 0.9 (2.4)
NA: 0.6 (1.6)
MA: 0.9 (2.4)

Lincoln's Sparrow *(Melospiza lincolnii)*

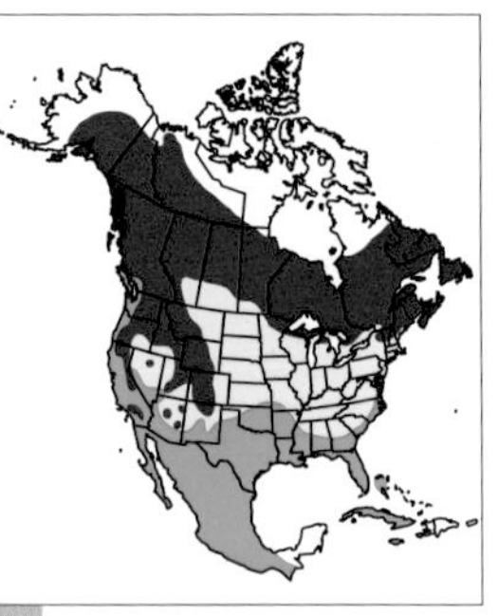

unknown
P: 2.2 (5.5)
P: 2.2 (5.5)
S: 1.8 (4.5)
S: 1.8 (4.5)
TS: 1.4 (3.6)
T: 2.4 (6.1)
T: 2.4 (6.1)
BE: 0.9 (2.4)
MA: 1.1 (2.8)
CR: 0.5 (1.2)

Swamp Sparrow *(Melospiza georgiana)*

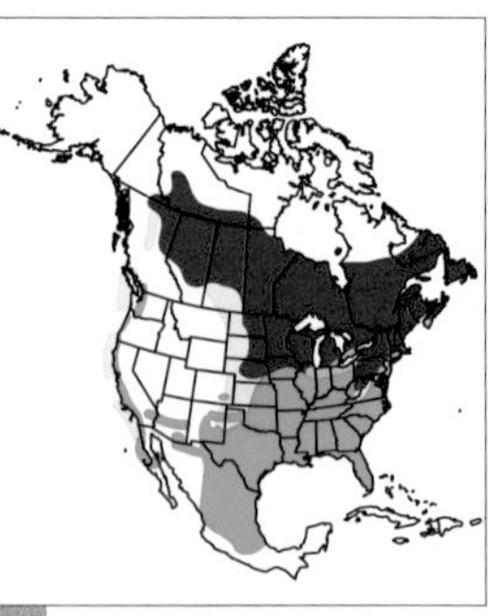

unknown
P: 1.8 (4.5)
P: 2.1 (5.3)
P: 2.1 (5.4)
S: 1.9 (4.8)
TS: 1.6 (4.1)
T: 2.2 (5.7)
T: 2.4 (6.1)
WC: 1.1 (2.7)
CR: 0.4 (1.1)
BE: 1.0 (2.5)
MA: 1.1 (2.9)

White-throated Sparrow *(Zonotrichia albicollis)*

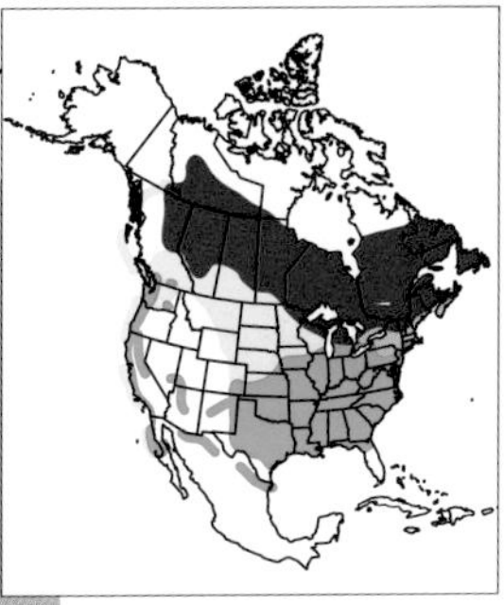

unknown
P: 2.2 (5.6)
P: 2.6 (6.6)
P: 2.5 (6.3)
S: 2.3 (5.9)
T: 3.0 (7.7)
T: 3.0 (7.5)
CR: 0.4 (0.9)
TH: 0.3 (0.7)
MA: 1.1 (2.7)
BE: 1.1 (2.8)
RU: 1.0 (2.5)

White-crowned Sparrow *(Zonotrichia leucophrys)*

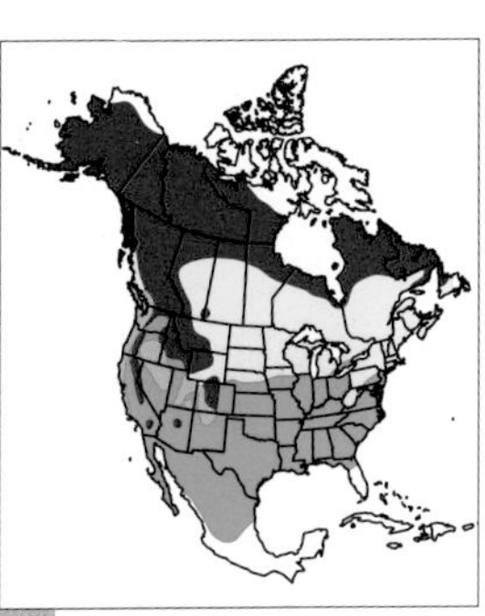

unknown
P: 2.5 (6.4)
P: 2.8 (7.1)
S: 2.2 (5.6)
TS: 2.1 (5.4)
T: 3.2 (8.1)
T: 3.1 (8.0)
WC: 1.0 (2.6)
MA: 0.9 (2.4)
RU: 1.1 (2.8)
TH: 0.7 (1.7)
CR: 0.5 (1.2)

Golden-crowned Sparrow *(Zonotrichia atricapilla)*

unknown
P: 2.2 (5.7)
P: 2.6 (6.6)
P: 2.4 (6.0)
S: 2.0 (5.1)
TS: 1.9 (4.9)
T: 3.0 (7.6)
T: 2.9 (7.4)
BR: 0.6 (1.4)
BE: 0.7 (1.9)
RU: 1.2 (3.0)

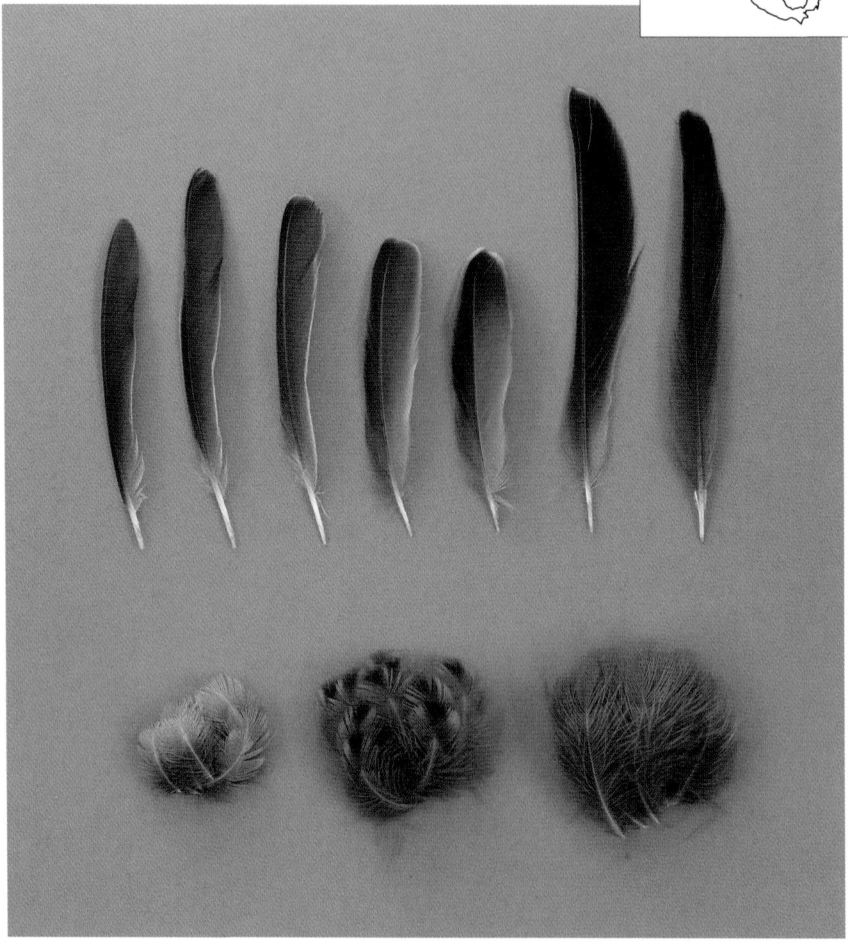

Dark-eyed Junco *(Junco hyemalis)*

(Slate-colored)

unknown	**S:** 1.9 (4.9)	**BR:** 1.0 (2.5)
P: 2.2 (5.6)	**TS:** 1.8 (4.6)	**BE:** 1.0 (2.6)
P: 2.6 (6.6)	**T:** 2.6 (6.6)	**MA:** 0.9 (2.4)
P: 2.4 (6.0)	**T:** 2.6 (6.7)	**NA:** 0.9 (2.2)
P: 2.1 (5.3)	**T:** 2.6 (6.6)	

Lapland Longspur *(Calcarius lapponicus)*

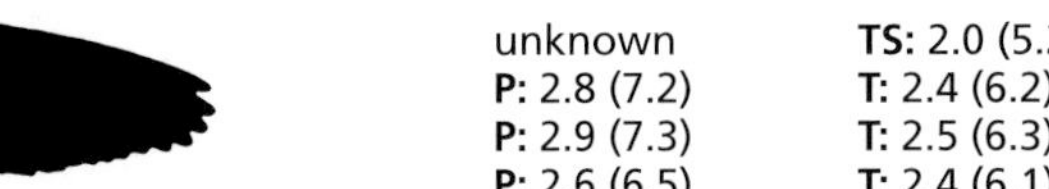

unknown	**TS:** 2.0 (5.2)
P: 2.8 (7.2)	**T:** 2.4 (6.2)
P: 2.9 (7.3)	**T:** 2.5 (6.3)
P: 2.6 (6.5)	**T:** 2.4 (6.1)
S: 2.0 (5.2)	

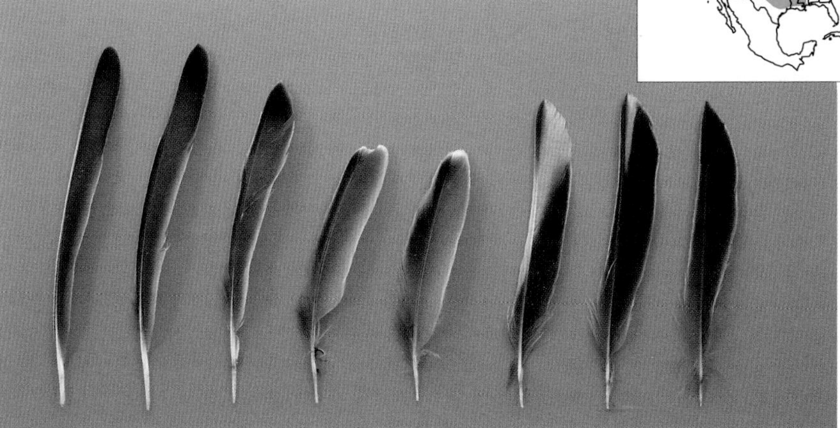

Cardinals and Allies *(Cardinalidae)*

Male Northern Cardinal feathers can be mistaken for those of the male Summer Tanager. Grosbeak primaries are similar to those of kingfishers, mockingbirds, woodpeckers, shrikes, and Lesser Goldfinch.

Northern Cardinal *(Cardinalis cardinalis)*

male
P: 3.0 (7.6)
P: 3.3 (8.5)
P: 3.3 (8.3)
P: 3.0 (7.7)
S: 2.9 (7.4)
TS: 2.6 (6.5)

T: 3.9 (9.9)
T: 4.3 (10.8)
BR: 0.9 (2.2)
BE: 1.5 (3.9)
MA: 1.5 (3.7)
CR: 1.2 (3.0)

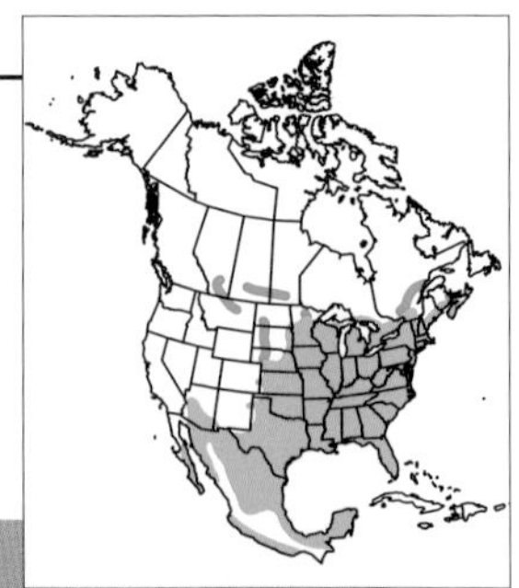

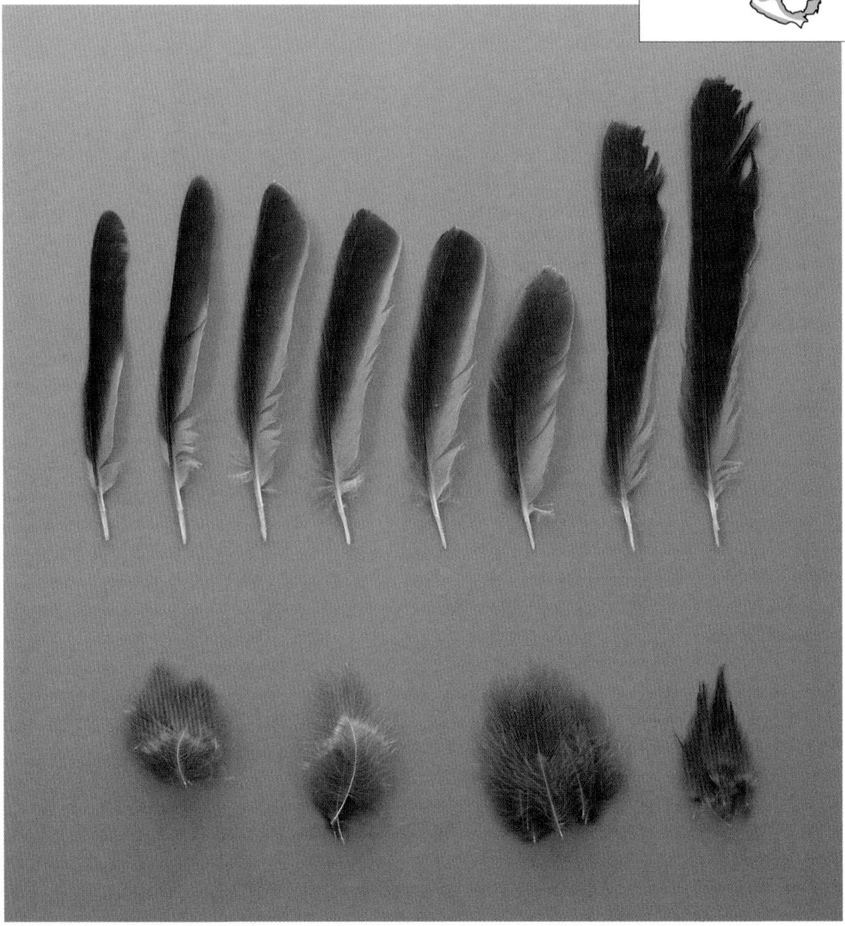

Rose-breasted Grosbeak *(Pheucticus ludovicianus)*

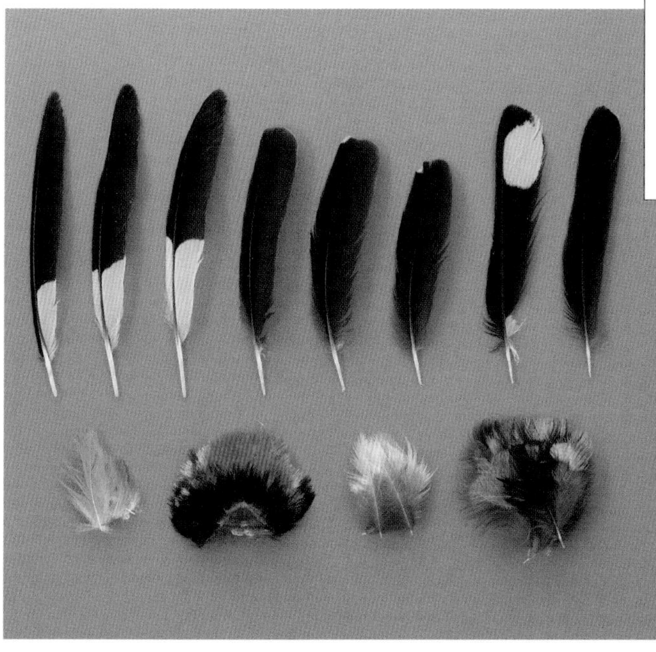

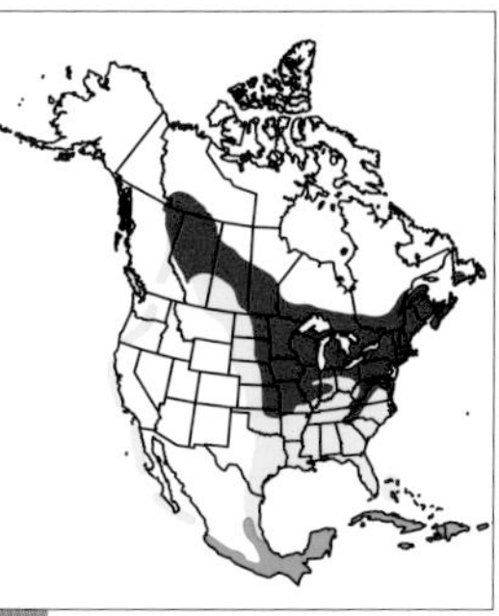

male
P: 3.1 (7.9)
P: 3.2 (8.1)
P: 3.2 (8.1)
P: 2.8 (7.0)
S: 2.6 (6.6)
TS: 2.3 (5.9)
T: 2.8 (7.2)
T: 2.8 (7.1)
UNWC: 1.1 (2.7)
BR: 1.0 (2.6)
BE: 1.0 (2.6)
MA: 1.4 (3.6)

Black-headed Grosbeak *(Pheucticus melanocephalus)*

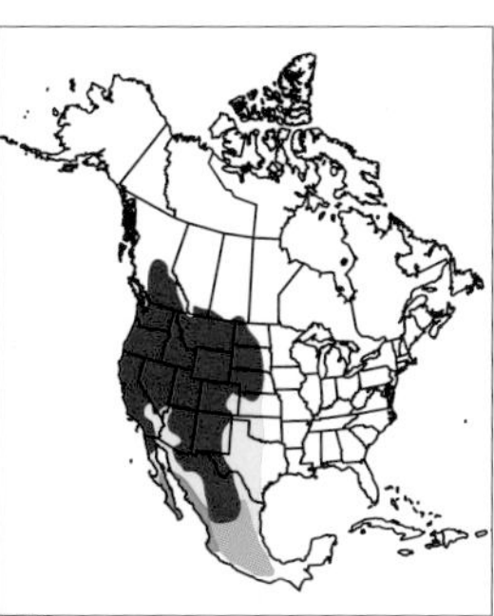

male
P: 3.1 (8.0)
P: 3.4 (8.6)
P: 3.0 (7.6)
S: 2.8 (7.1)
S: 2.5 (6.4)
S: 2.4 (6.0)
T: 3.3 (8.3)
T: 3.1 (8.0)
BR: 0.7 (1.7)
BE: 1.4 (3.5)
MA: 0.9 (2.2)
UNWC: 1.1 (2.8)

Blue Grosbeak *(Passerina caerulea)*

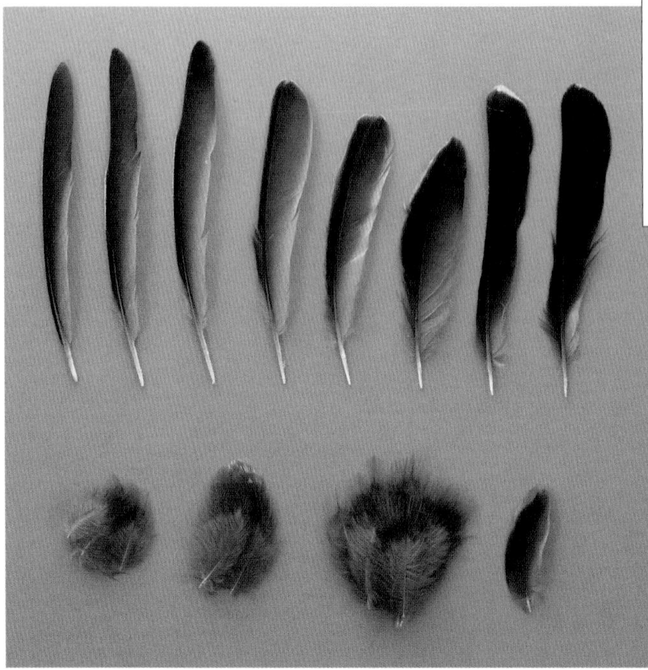

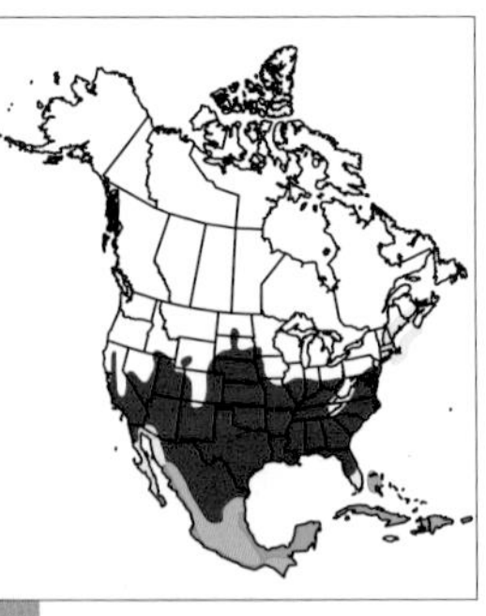

male
P: 3.0 (7.5)
P: 3.1 (7.9)
P: 3.1 (8.0)
P: 2.8 (7.1)
S: 2.5 (6.4)
TS: 2.3 (5.9)
T: 2.8 (7.1)
T: 2.9 (7.3)
BR: 0.6 (1.6)
BE: 1.2 (3.0)
RU: 1.4 (3.6)
WC: 1.2 (3.1)

Lazuli Bunting *(Passerina amoena)*

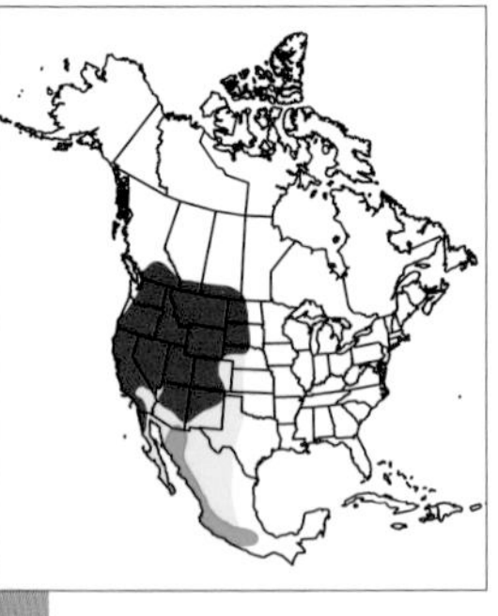

male
P: 2.3 (5.8)
P: 2.4 (6.1)
P: 2.1 (5.4)
S: 1.9 (4.9)
S: 1.9 (4.7)
T: 2.3 (5.9)
T: 2.1 (5.4)
NA: 0.5 (1.2)
BR: 0.6 (1.4)
BE: 0.9 (2.4)
MA: 0.8 (2.1)

Indigo Bunting *(Passerina cyanea)*

male
P: 2.3 (5.9)
P: 2.2 (5.6)
S: 2.0 (5.0)
S: 1.8 (4.6)
T: 2.2 (5.6)
T: 2.2 (5.6)
T: 2.1 (5.3)
WC: 0.8 (2.0)
WC: 0.7 (1.9)
TH: 0.4 (1.0)
BR: 0.6 (1.5)
RU: 0.9 (2.4)

Painted Bunting *(Passerina ciris)*

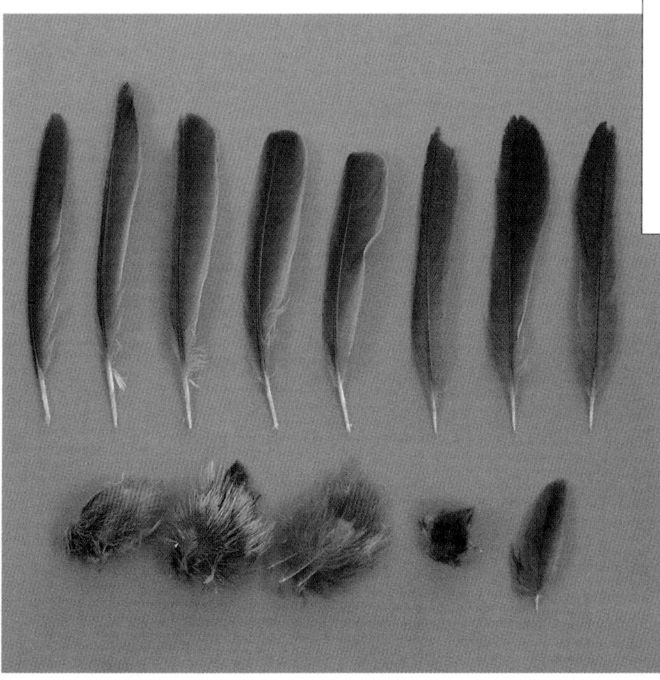

male
P: 2.2 (5.7)
P: 2.5 (6.4)
P: 2.3 (5.8)
S: 2.2 (5.5)
S: 2.0 (5.2)
T: 2.2 (5.7)
T: 2.3 (5.8)
T: 2.2 (5.7)
BR: 0.8 (2.1)
MA: 0.9 (2.2)
RU: 1.0 (2.6)
CR: 0.5 (1.2)
WC: 1.0 (2.5)

Icterids *(Icteridae)*

Bobolink *(Dolichonyx oryzivorus)*

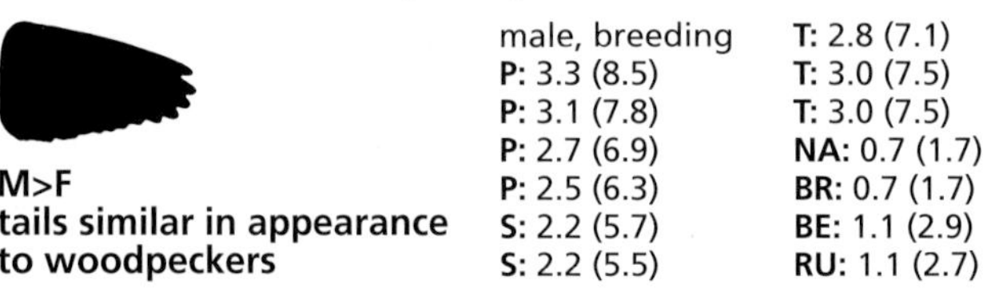

M>F
tails similar in appearance to woodpeckers

male, breeding
P: 3.3 (8.5)
P: 3.1 (7.8)
P: 2.7 (6.9)
P: 2.5 (6.3)
S: 2.2 (5.7)
S: 2.2 (5.5)
T: 2.8 (7.1)
T: 3.0 (7.5)
T: 3.0 (7.5)
NA: 0.7 (1.7)
BR: 0.7 (1.7)
BE: 1.1 (2.9)
RU: 1.1 (2.7)

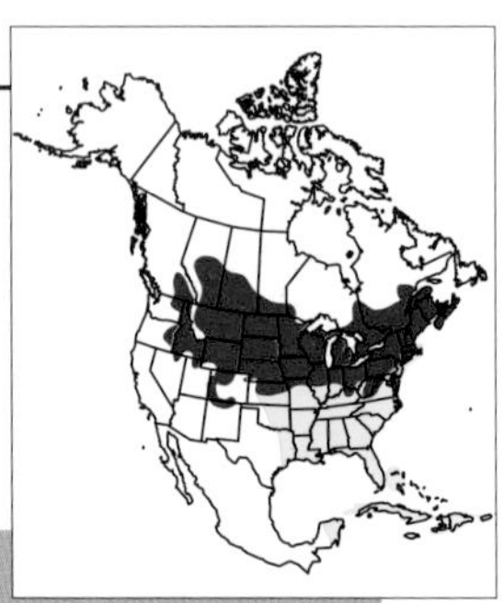

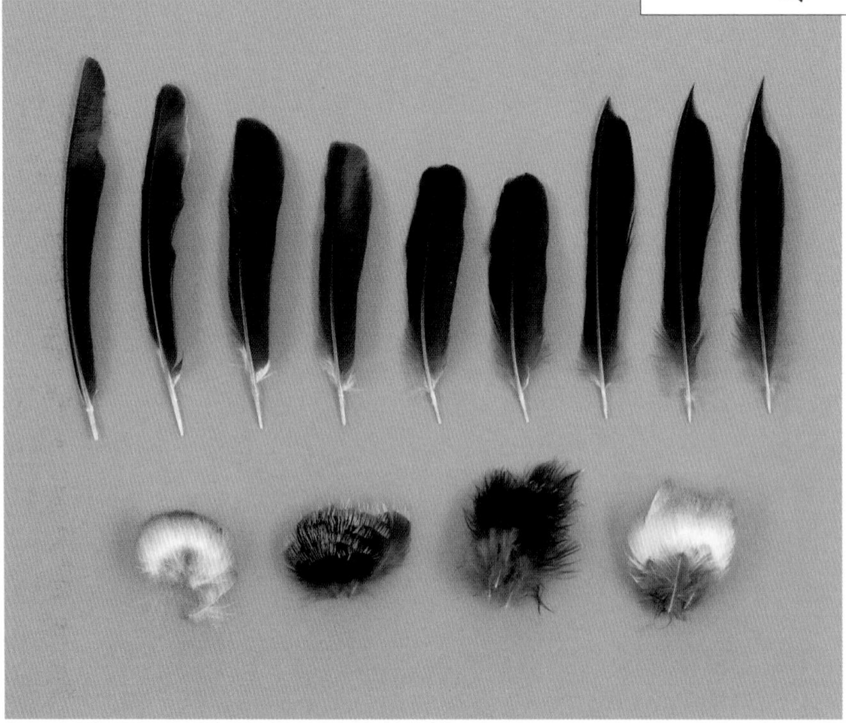

Red-winged Blackbird *(Agelaius phoeniceus)*

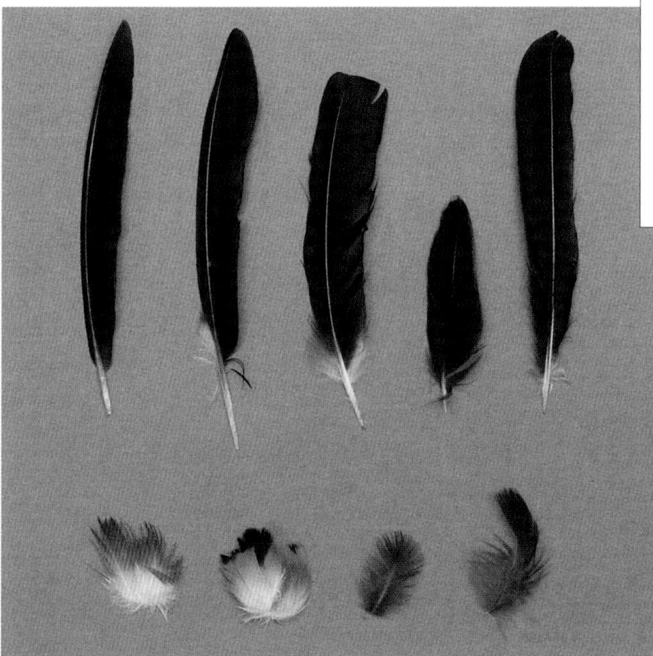

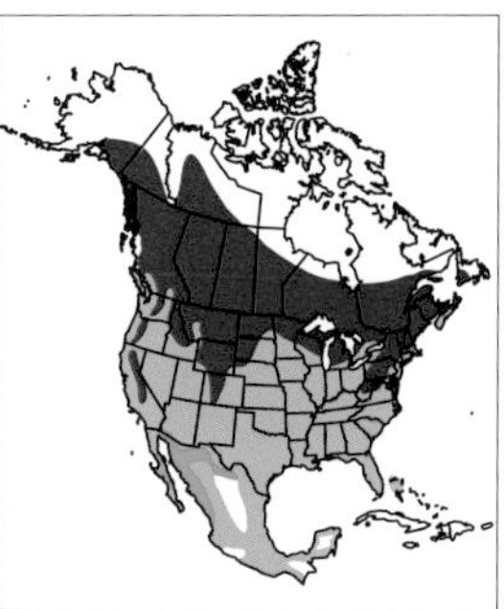

M>F

male
P: 4.1 (10.4)
P: 4.3 (11.0)
S: 3.6 (9.2)
TS: 2.3 (5.8)
T: 3.9 (10.0)
MC: 0.9 (2.3)
MC: 0.9 (2.2)
BE: 0.9 (2.4)
NA: 1.4 (3.6)

Eastern Meadowlark *(Sturnella magna)*

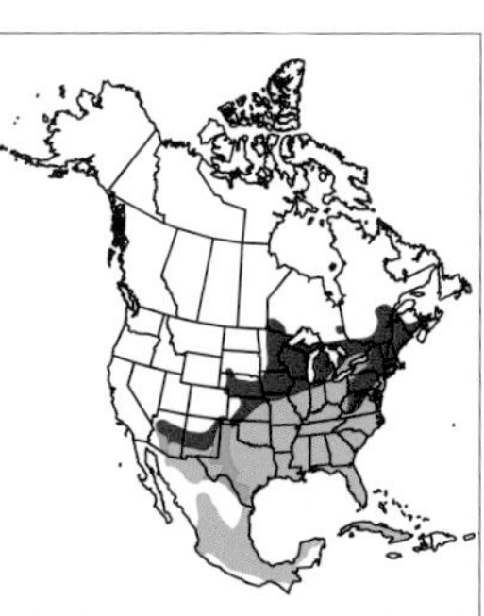

M>F

male
P: 4.0 (10.2)
P: 4.0 (10.2)
S: 3.5 (8.9)
T: 3.1 (8.0)
T: 3.2 (8.2)
T: 3.3 (8.3)
T: 3.3 (8.3)
WC: 1.6 (4.1)
FL: 1.4 (3.5)
BR: 1.3 (3.2)
MA: 1.5 (3.7)

Western Meadowlark *(Sturnella neglecta)*

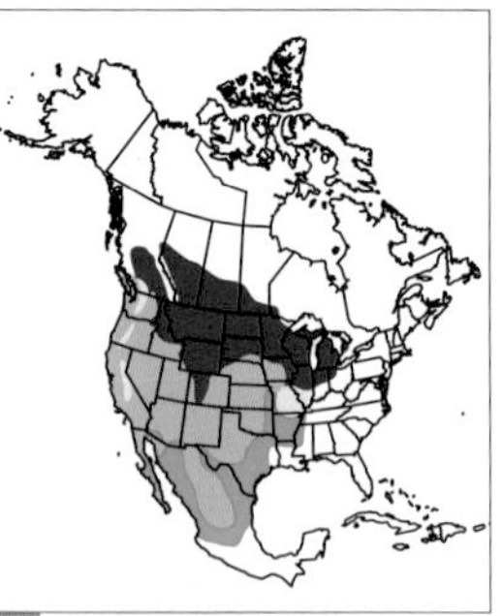

M>F

male
P: 4.0 (10.2)
P: 4.3 (10.8)
S: 3.6 (9.1)
S: 3.7 (9.3)
T: 3.5 (8.8)
T: 3.3 (8.4)
T: 3.3 (8.3)
BR: 1.0 (2.5)
BR: 0.8 (2.0)
BE: 1.3 (3.3)
MA: 1.8 (4.6)

Yellow-headed Blackbird *(Xanthocephalus xanthocephalus)*

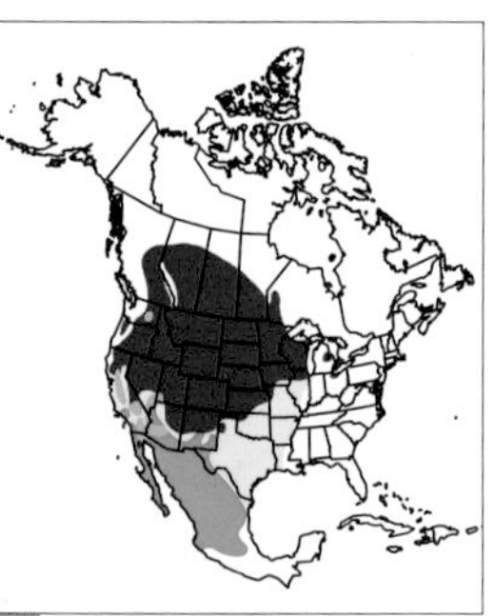

M>F

male
P: 5.0 (12.7)
P: 5.1 (13.0)
P: 4.5 (11.4)
S: 4.0 (10.2)
TS: 3.7 (9.3)
T: 4.6 (11.6)
T: 4.5 (11.5)
MA: 1.7 (4.3)
BR: 1.2 (3.0)
WC: 1.5 (3.7)

Rusty Blackbird *(Euphagus carolinus)*

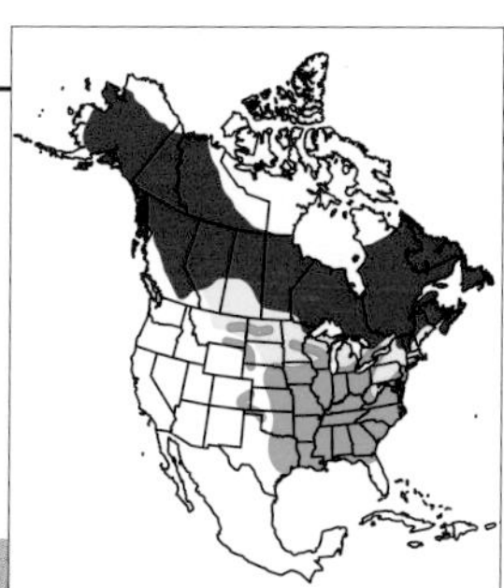

M>F

male,
non-breeding
P: 3.7 (9.3)
P: 4.0 (10.2)
P: 3.6 (9.1)
P: 3.3 (8.3)
S: 3.1 (7.9)
S: 3.0 (7.6)
T: 3.7 (9.5)
T: 3.8 (9.7)
BR: 1.4 (3.6)
MA: 1.7 (4.4)
CR: 0.9 (2.3)

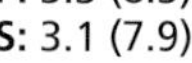

Common Grackle *(Quiscalus quiscula)*

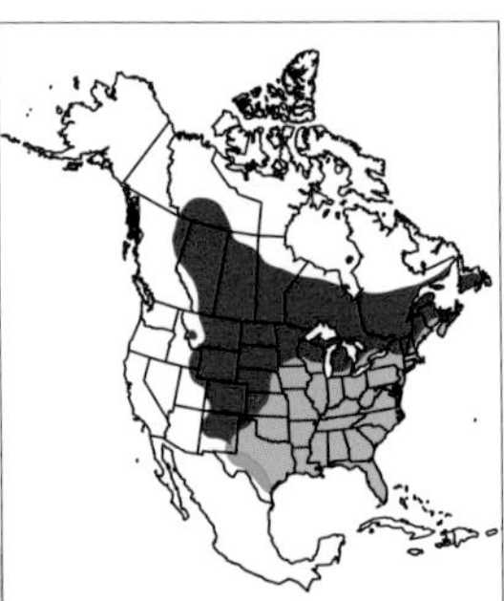

M>F

male, iridescent
P: 4.3 (11.0)
P: 4.6 (11.7)
P: 4.6 (11.7)
P: 4.2 (10.7)
S: 3.5 (9.0)

TS: 3.4 (8.6)
T: 4.3 (10.9)
T: 4.9 (12.4)
T: 5.4 (13.8)
BR: 1.4 (3.5)
BE: 1.5 (3.9)
RU: 1.9 (4.9)

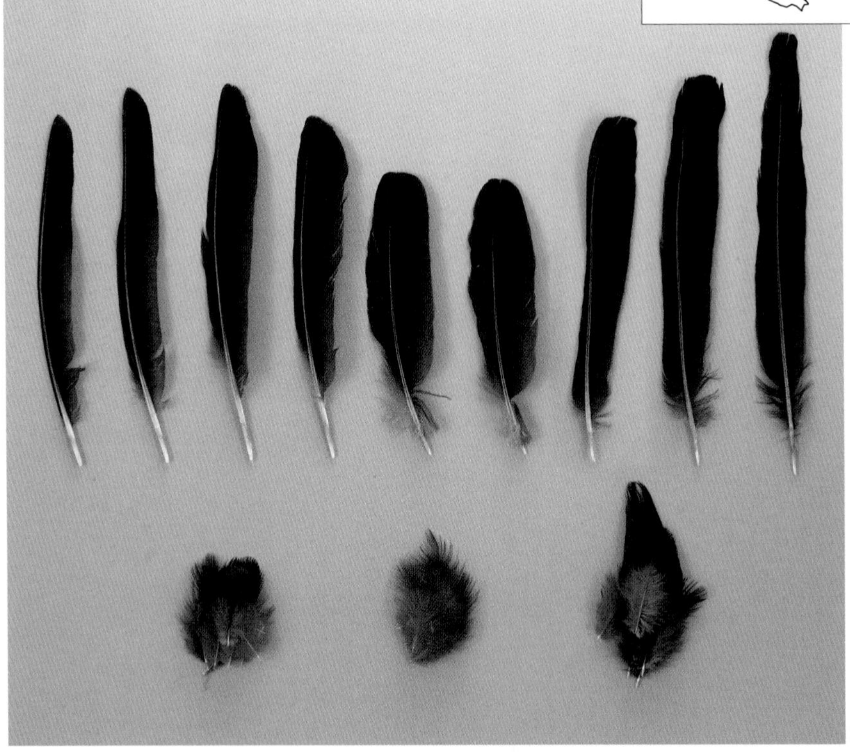

Great-tailed Grackle *(Quiscalus mexicanus)*

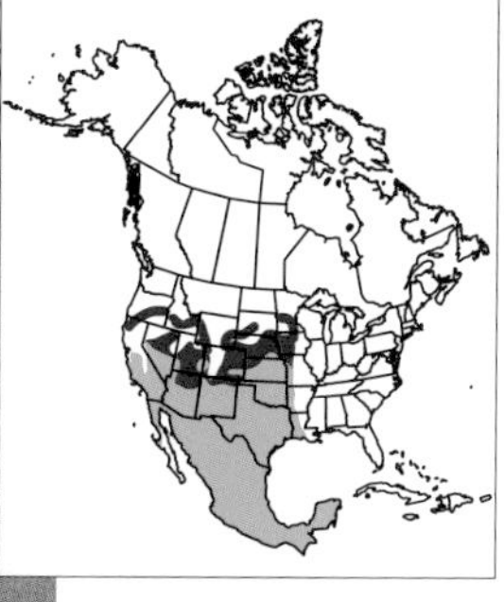

M>F

male
P: 6.4 (16.2)
P: 6.4 (16.2)
S: 5.0 (12.7)
TS: 3.1 (8.0)
T: 6.0 (15.3)
T: 7.3 (18.6)
RU: 1.9 (4.7)
BR: 1.1 (2.7)
BE: 2.7 (6.8)

female
P: 5.0 (12.6)
P: 4.9 (12.5)
P: 4.2 (10.6)
T: 5.6 (14.3)
T: 5.9 (15.0)
BE: 1.5 (3.8)
BR: 1.2 (3.0)
MA: 1.6 (4.1)

Brown-headed Cowbird *(Molothrus ater)*

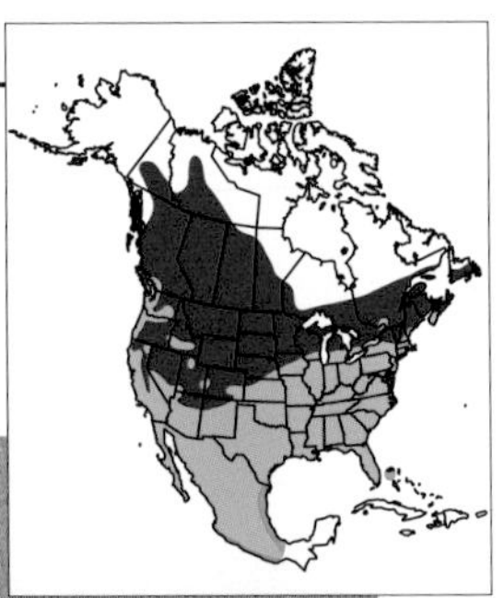

M>F

male
P: 3.7 (9.3)
P: 3.4 (8.7)
S: 2.5 (6.4)

T: 3.1 (7.8)
T: 3.1 (8.0)
NA: 1.1 (2.7)
BR: 1.0 (2.6)
WC: 1.2 (3.1)

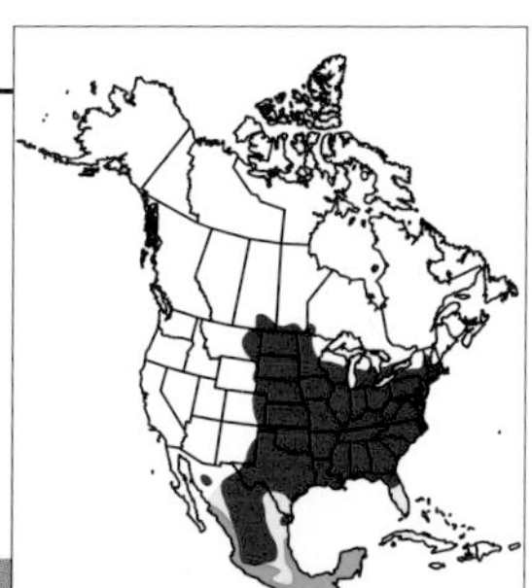

Orchard Oriole *(Icterus spurius)*

male
P: 2.8 (7.2)
P: 3.0 (7.6)
P: 2.4 (6.2)
S: 2.3 (5.9)
S: 2.2 (5.5)

T: 3.0 (7.5)
T: 3.0 (7.7)
T: 2.9 (7.4)
MA: 1.0 (2.6)
BR: 0.7 (1.8)
BE: 1.1 (2.7)
RU: 1.0 (2.6)

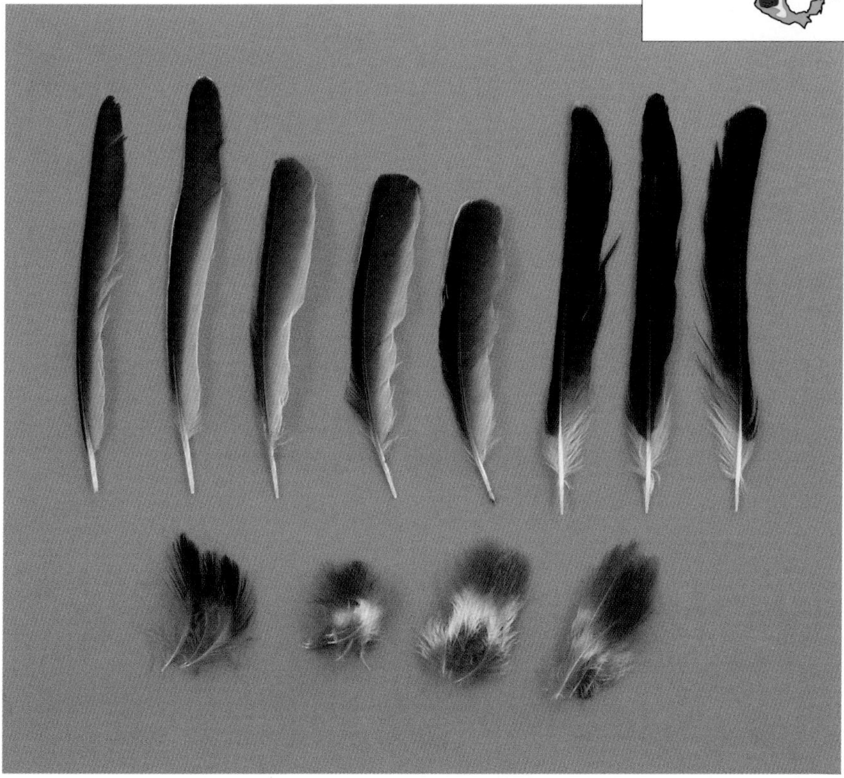

Bullock's Oriole *(Icterus bullockii)*

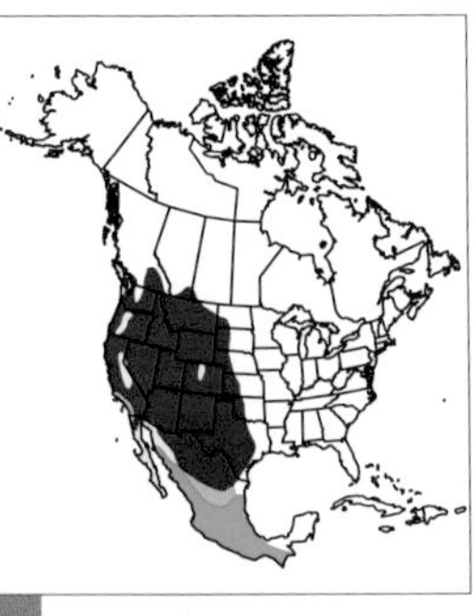

male
P: 3.3 (8.5)
P: 3.4 (8.6)
P: 2.9 (7.4)
S: 2.7 (6.8)
S: 2.6 (6.5)
S: 2.5 (6.3)
T: 3.4 (8.6)
T: 3.4 (8.6)
T: 3.4 (8.6)
NA: 0.7 (1.8)
BR: 0.8 (2.1)
BE: 1.4 (3.5)
MA: 1.1 (2.8)
RU: 1.0 (2.5)

Baltimore Oriole *(Icterus galbula)*

male
P: 3.0 (7.7)
P: 3.2 (8.2)
S: 2.6 (6.7)
S: 2.6 (6.5)
T: 3.1 (7.8)
T: 3.1 (7.8)
T: 3.1 (7.8)
NA: 0.7 (1.7)
BR: 0.7 (1.9)
BE: 0.9 (2.4)
RU: 1.1 (2.8)

Finches and Allies *(Fringillidae)*

Lesser Goldfinch and Evening Grosbeak primaries similar to those of kingfishers, mockingbirds, woodpeckers, and shrikes.

Pine Grosbeak *(Pinicola enucleator)*

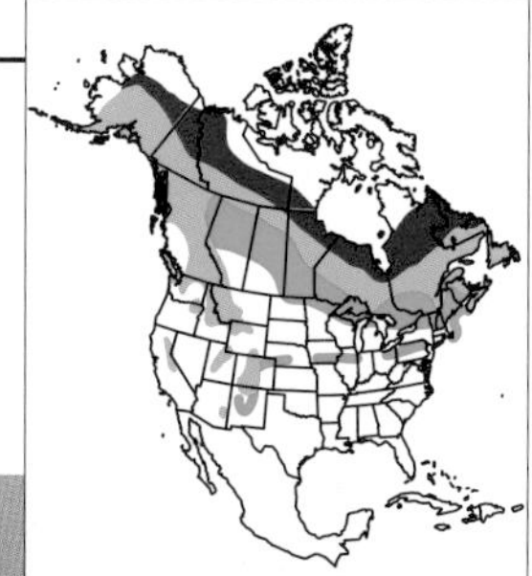

male

P: 3.9 (9.8)	**TS:** 2.7 (6.8)
P: 4.2 (10.6)	**T:** 4.1 (10.5)
P: 4.0 (10.1)	**T:** 4.1 (10.3)
P: 3.3 (8.4)	**BR:** 1.5 (3.9)
S: 3.0 (7.7)	**MA:** 1.4 (3.5)
	RU: 1.6 (4.1)

Purple Finch *(Carpodacus purpureus)*

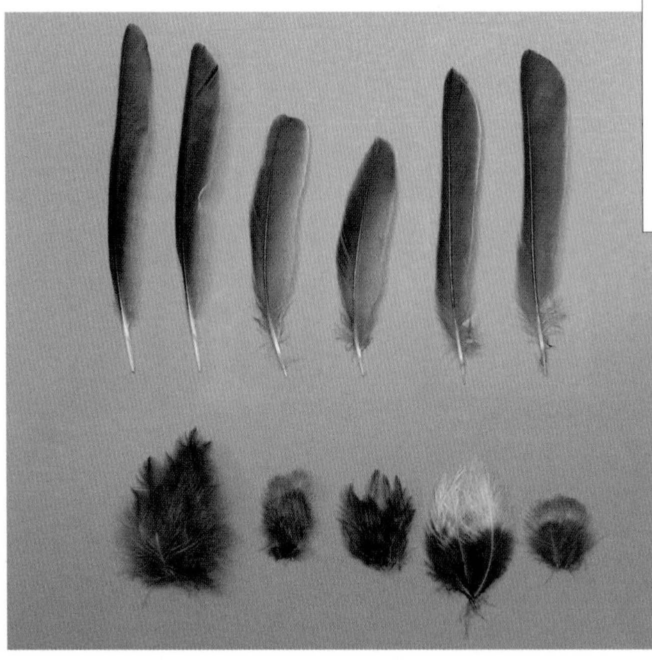

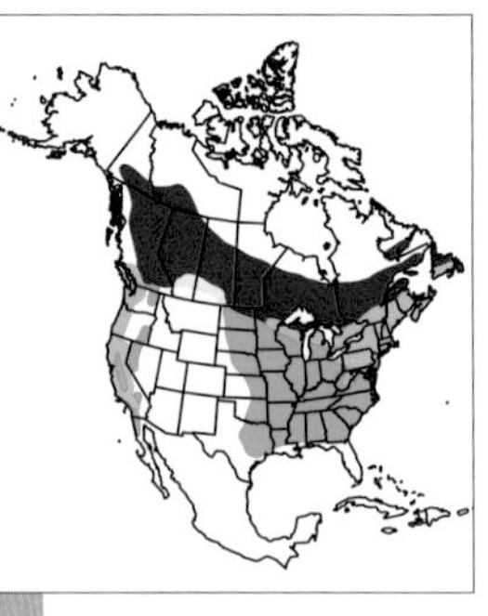

male
P: 2.8 (7.1)
P: 2.7 (6.8)
S: 2.1 (5.3)
TS: 1.9 (4.9)
T: 2.5 (6.4)
T: 2.6 (6.6)
MA: 1.0 (2.6)
TH: 0.3 (0.8)
CR: 0.6 (1.4)
BE: 2.8 (7.2)
BR: 0.6 (1.4)

Cassin's Finch *(Carpodacus cassinii)*

male
P: 3.0 (7.5)
P: 3.1 (7.8)
P: 2.8 (7.0)
P: 2.4 (6.0)
S: 2.2 (5.5)
S: 2.0 (5.2)
T: 2.6 (6.6)
T: 2.6 (6.7)
T: 2.3 (5.9)
RU: 1.1 (2.9)
MA: 1.3 (3.3)
BE: 1.3 (3.3)
BR: 0.7 (1.9)

House Finch *(Carpodacus mexicanus)*

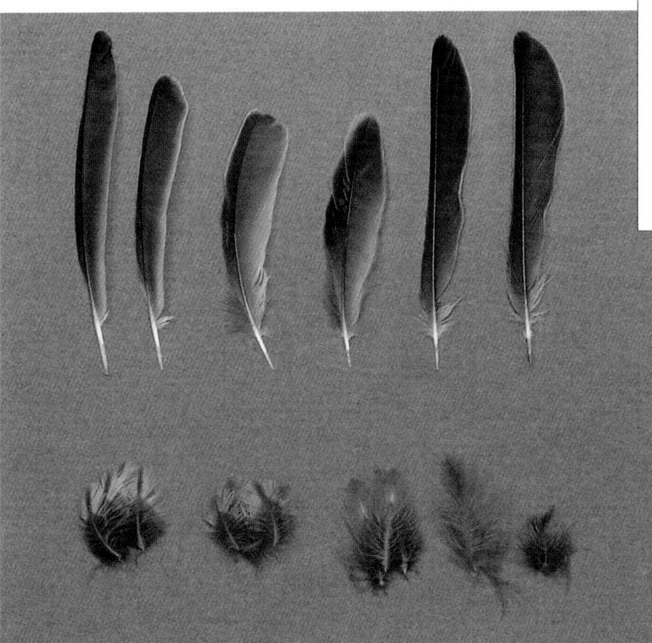

male
P: 2.7 (6.9)
P: 2.4 (6.0)
S: 2.0 (5.2)
TS: 2.0 (5.2)
T: 2.6 (6.6)
T: 2.6 (6.7)
BR: 0.8 (2.0)
TH: 0.6 (1.5)
UPTC: 0.8 (2.1)
MA: 1.0 (2.6)
NA: 0.4 (1.1)

Red Crossbill *(Loxia curvirostra)*

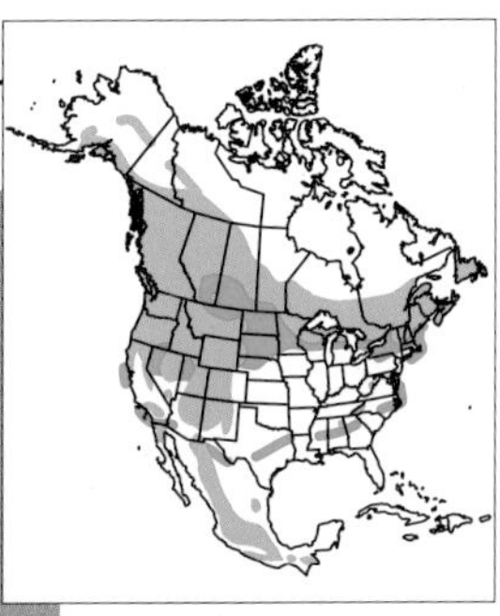

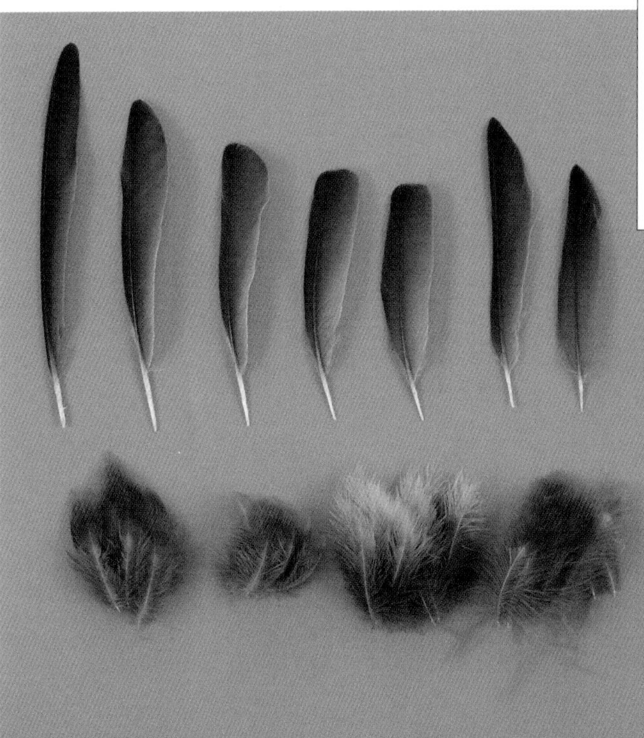

male
P: 3.0 (7.7)
P: 2.6 (6.7)
P: 2.2 (5.7)
S: 2.0 (5.1)
S: 1.9 (4.8)
T: 2.3 (5.9)
T: 2.0 (5.0)
UNTC: 1.4 (3.5)
BR: 0.8 (2.0)
BE: 1.2 (3.0)
RU: 1.1 (2.8)

White-winged Crossbill *(Loxia leucoptera)*

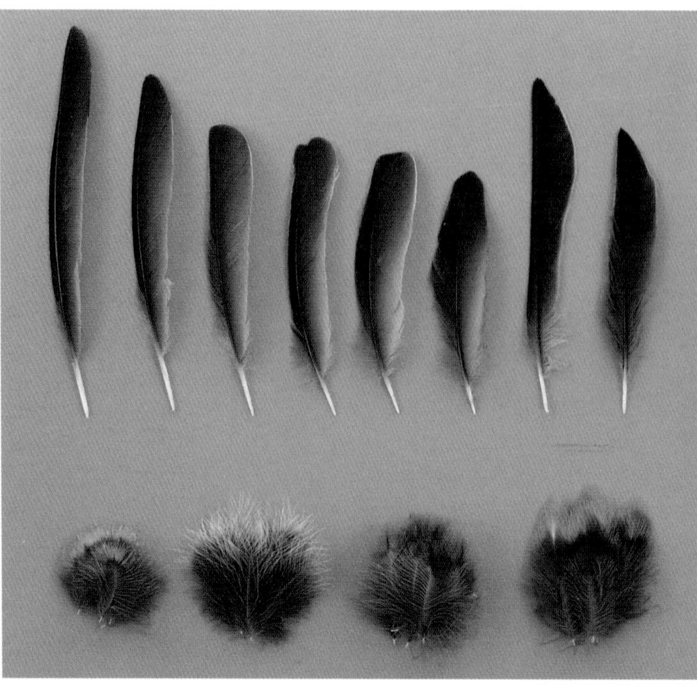

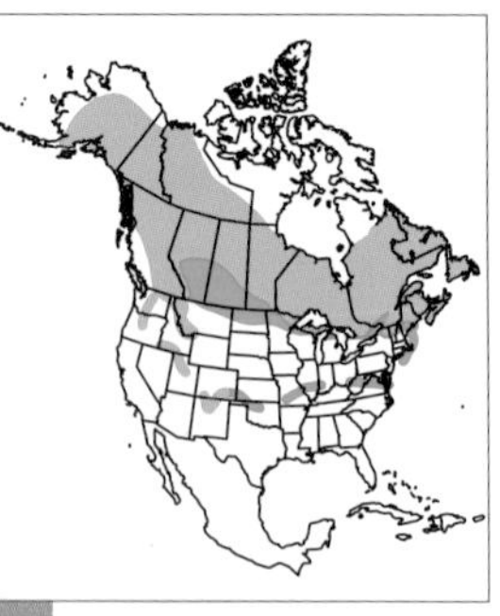

male
P: 3.0 (7.5)
P: 2.5 (6.4)
P: 2.2 (5.6)
P: 2.1 (5.4)
S: 2.0 (5.0)
TS: 1.9 (4.8)
T: 2.6 (6.5)
T: 2.2 (5.6)
BR: 0.6 (1.6)
BE: 0.9 (2.2)
MA: 0.7 (1.9)
RU: 1.0 (2.6)

Pine Siskin *(Carduelis pinus)*

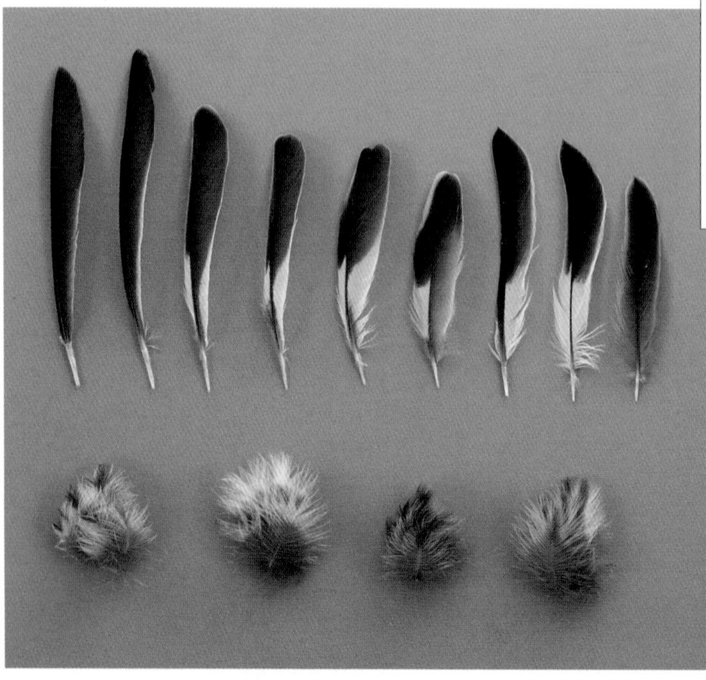

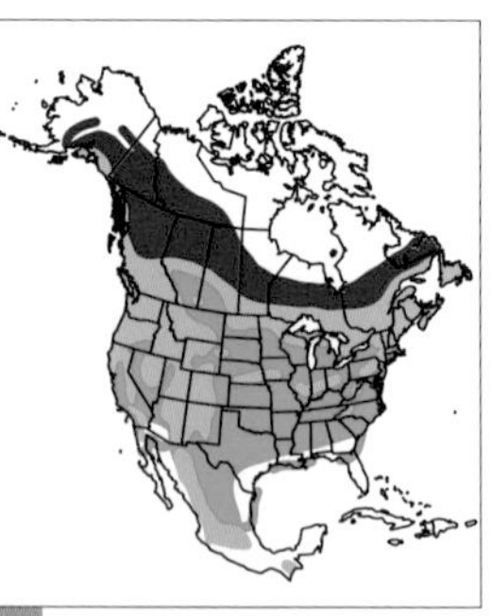

unknown
P: 2.4 (6.0)
P: 2.5 (6.3)
P: 2.1 (5.4)
P: 1.9 (4.8)
S: 1.7 (4.4)
TS: 1.6 (4.1)
T: 1.9 (4.9)
T: 1.9 (4.8)
T: 1.6 (4.1)
BR: 0.6 (1.6)
BE: 0.7 (1.8)
MA: 0.6 (1.4)
RU: 0.8 (2.1)

Lesser Goldfinch *(Carduelis psaltria)*

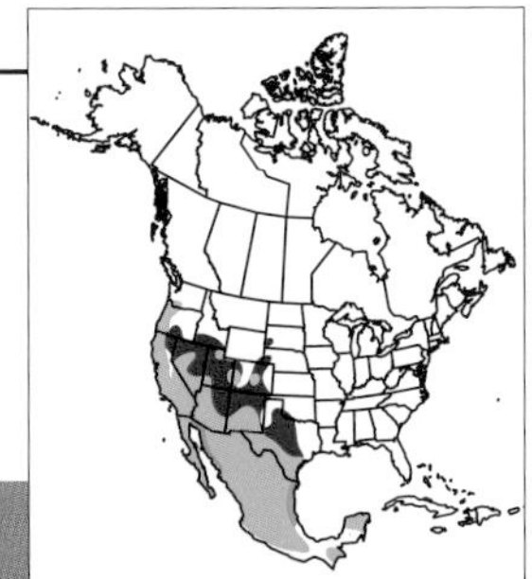

unknown
P: 2.1 (5.4)
P: 2.3 (5.8)
P: 1.9 (4.9)
S: 1.7 (4.2)
S: 1.6 (4.1)
TS: 1.4 (3.6)
T: 1.9 (4.8)
T: 1.7 (4.4)
NA: 0.4 (1.1)
BR: 0.6 (1.4)

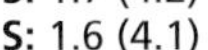

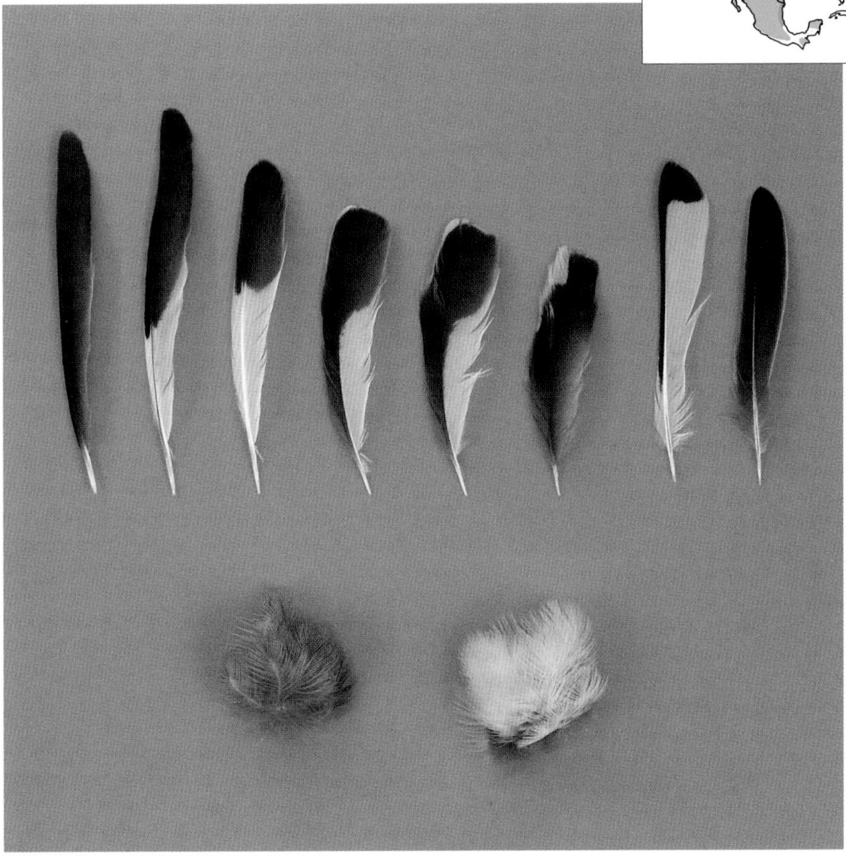

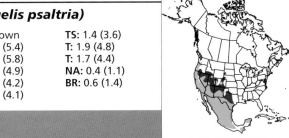

American Goldfinch *(Carduelis tristis)*

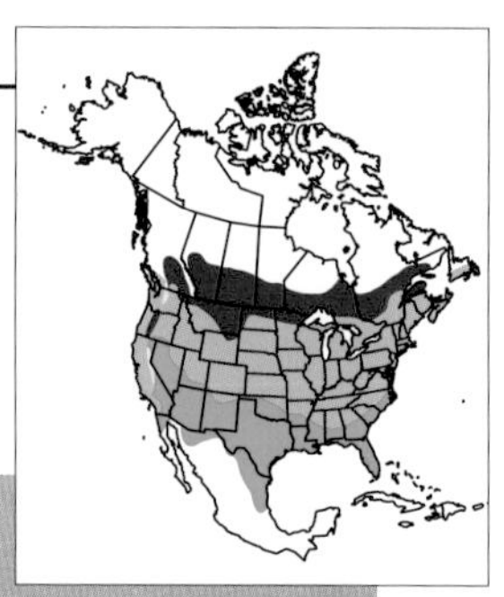

unknown
P: 2.0 (5.0)
P: 2.0 (5.2)
S: 1.8 (4.6)
S: 1.7 (4.3) [feather vane distortion on both secondaries]
T: 1.7 (4.3)
T: 1.5 (3.7)
T: 1.5 (3.9)
TS: 1.1 (2.9)
NA: 0.5 (1.2)
BE: 0.7 (1.9)
BR: 0.7 (1.8)

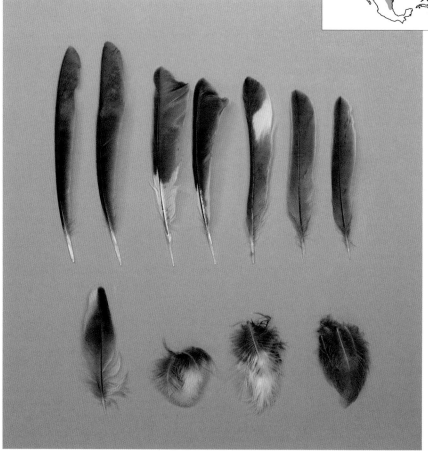

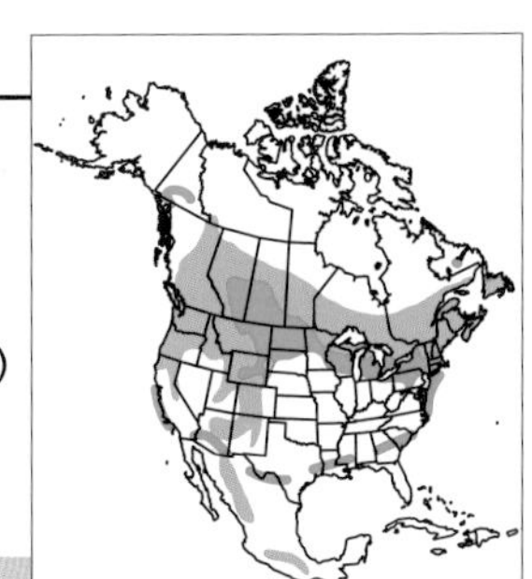

Evening Grosbeak *(Coccothraustes vespertinus)*

female, sometimes no white is present on the primaries of male birds

P: 3.3 (8.3)
P: 3.3 (8.5)
P: 2.8 (7.1)
S: 2.6 (6.6)
T: 2.8 (7.2)
T: 2.5 (6.3)

WC: 1.1 (2.9)
CR: 0.4 (0.9)
UPTC: 1.1 (2.7)
BR: 1.1 (2.8)
BE: 1.3 (3.2)

Old World Sparrows *(Passeridae)*

House Sparrow *(Passer domesticus)*

male
P: 2.5 (6.3)
P: 2.6 (6.5)
S: 2.1 (5.4)
TS: 2.0 (5.0)
T: 2.4 (6.2)
T: 2.4 (6.0)
[sheathed]
T: 2.4 (6.1)
TH: 0.6 (1.4)
WC: 1.0 (2.6)
MA: 0.8 (2.0)

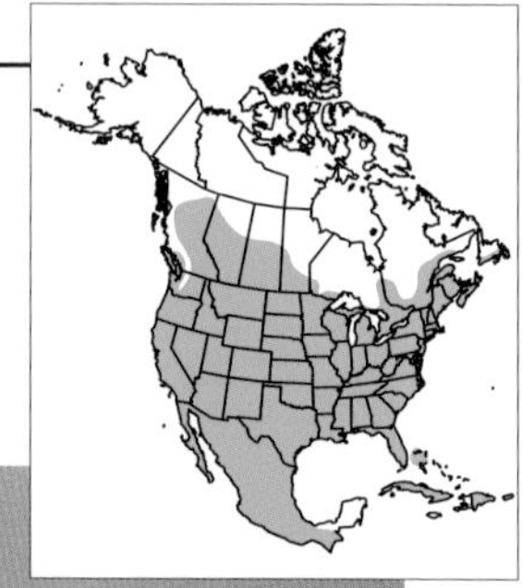

ADDITIONAL RESOURCES

We recommend the following resources for additional exploration or aid in feather identification.

Bird Tracks and Sign: A Guide to North American Species. Mark Elbroch with Eleanor Marks, Stackpole Books, 2001. A fantastic resource for general bird sign (tracks, scat, feeding sign, skulls, and so on) including a limited but aesthetic section on feather identification.

The Feather Atlas. A U.S. National Fish and Wildlife Forensics Laboratory online collection of wing and tail feather scans. An excellent resource for larger bird species. Especially useful for viewing clear markings and morphology across the wing or tail. http://www.lab.fws.gov/featheratlas/

The Sibley Guide to Birds. David Allen Sibley, Knopf, 2003. A good comprehensive guide to North American bird species. An authors' recommendation.

Slater Museum of Natural History, online wing collection. A brilliant idea, enabling the public to view thousands of wing images from hundreds of bird species. Especially useful for viewing upper and underside wings of a particular species to determine markings, morphology, and feather counts. http://digitalcollections.ups.edu/slater/

ACKNOWLEDGMENTS

The making of this guide was an immense undertaking that required the cooperation of over 120 organizations, many of whom are nonprofit wildlife rehabilitation centers. Below you will find a list of museums, wildlife rehabilitators, and universities that donated feathers or other assets to this project. Special thanks to the following organizations:

Birdfellow Corporation (birdfellow.com)
San Francisco State University, Department of Biology
Tri-State Bird Rescue & Research, Inc.
Field Museum of Natural History
Delaware Museum of Natural History
Centennial Museum, University of Texas at El Paso
Livingston Ripley Waterfowl Conservancy
Project Wildlife
Sierra Wildlife Rescue
Sarvey Wildlife Care Center
Audubon Center for Birds of Prey
Wolf Hollow Wildlife Rehabilitation Center
University of Missouri-Columbia, Department of Fisheries and Wildlife Sciences
Wildlife Center of Virginia
Pelican Harbor Seabird Station
West Sound Wildlife Shelter
Burke Museum, University of Washington
Alabama Wildlife Center
Broward Wildlife Center
Cascades Raptor Center
Conservancy of Southwest Florida Wildlife Rehabilitation Clinic
Delaware Valley Raptor Center
Hawks Aloft, Inc.

Hawk Creek Wildlife
Tufts Cummings School of Veterinary Medicine
WildCare Bay Area
Progressive Animal Welfare Society (PAWS)
Rocky Mountain Raptor Program
U.S. Fish and Wildlife Service
California State University Bakersfield
Native Bird Connections
South Bay Wildlife Rehab
Walden's Puddle Wildlife Rehabilitation and Education Center
Brukner Nature Center Vermont
Vermont Institute of Natural Science
Carolina Waterfowl Rescue
Chicago Bird Collision Monitors
Willowbrook Wildlife Center
Flint Creek Wildlife Rescue
Minnesota Department of Natural Resources
Bay Beach Wildlife Sanctuary
Slater Museum of Natural History, University of Puget Sound
The Blue Ridge Wildlife Center

Particular thanks goes out to Paul Lehman for creating all of the incredibly detailed range maps that grace the pages of this book; to Birdfellow.com, a remarkable new social networking website for birders, for digitizing the maps; and the architect of Birdfellow, Bjorn Hinrichs, for coordinating it all.

Great thanks is due also to the following individuals who contributed their time, energy, and resources to the making of this guide. In their extremely busy schedules, these people somehow found time to help us with the compilation of the information presented here. There were no rewards or payments for their efforts, other than an understanding that they were contributing to the cause of helping *Bird Feathers* become a reality. For the help of these selfless individuals we are very grateful, for this project would not have been possible without them:

Dr. Erica Miller, Dr. David Willard, Dr. Jean L. Woods , Scott Cutler, the Clinic Supervisors at Tri-State Bird Rescue & Research Inc., Greg S. Spicer, Tara Roth, Mikki Scott, Linda Scott, Steve Smith, Jessica Haapala, Travis Haapala, Bjorn Hinrichs, Emily Gibson, Angie Jordan, Zora O'Neill, Ian Gereg, Linda King, Nancy Barbachano, Leslie R. Henry, Dianna Flynt, Samantha Little, Shona Aitken, Walter Wehtje, Dave McRuer, Robert C. Faucett, Wendy Fox, Kristen Washer, Mike Pratt, Lynne Weber, Sandra Allinson, Joseph DiDonato, Richard Seever, Jennifer Keller, Mariangelique Diaz, Dr. Stefan Harsch, Louise Shimmel, Joanna Fitzgerald, Jayne Amico, Bill Streeter, Stephanie Streeter, Jan Lucciola, Gail Garber, Loretta Jones, Matt Zymanek, Olivia Lattanzi, Glennon Beresin, Dr. Flo Tseng, Melanie

Piazza, Kevin Mack, Gail Kratz, Marc Weitzel, Marlene Hensley-Benton, Lois Hoy, Jenny Papka, LouAnn Partington, Ann C. Lynch, Bettina Bowers Schwan, Karen Reddemann, Becky Crow, Allison Stark, Jennifer Gordon, David Moskowitz, Aaron Jessup, Kyle Koloini, David Christopher Doung, Jill Cooper, Laura Gunion, Dan Martinelli, Gary Shugart, Dr. Belinda Burwell, Alexia Allen, Gavin Cotterill, Michel Scott, Hayley Riach, Ryan McCliment, Wes Childers, and Sharon Rose.

—the authors

I would like to start by thanking my wonderful wife, Mikki, for her encouragement and constant support and for putting up with all of the feathers that floated around our house for two years. You are amazing, Mikki, and I could not have done it without you. I, of course, would also like to thank my incredible family for always believing in me and for providing the strong roots in my life that have allowed me to venture forth in pursuit of my dreams with confidence. Special thanks must also go out to my daughter, Haven, for always reminding me of the things that are really important in life. To my good friend and coauthor, Casey McFarland; for your partnership, perspective, input, and company on those endless days of sampling and photographing, I am so very grateful. To all of my teachers, I am forever indebted for the knowledge and stories you have shared with me, and for your inspiration—specifically, Jon Young, Tom Brown Jr., Ingwe, Grandfather Stalking Wolf, Billy McConnell, Kristy McFetridge, Joe Lau, Robert Humphrey, Chris Laliberte, Angie Jordan, Casey McFarland, David Moskowitz, Eddie Starnater, and Nate Kempton. An additional thanks goes out to the Wilderness Awareness School, P.A.S.T. Skills Wilderness School, and the Tracker Wilderness Survival School, for the incredible work you do reconnecting people to nature. And finally, I would like to reserve my biggest thanks for the natural world, which provides us with endless beauty and natural mystery.

—D. S.

My first thanks of course go to those that truly made this book possible—the birds. They have always given life an extra dose of magic, have always been a source of awe and inspiration. Many thanks to the particular birds whose feathers fill these pages; we hope these photographs serve many a curious mind for years to come.

I'm grateful to my family for their love and support, in times past and during the making of this guide. To my sister Zora, who I still look up to though I'm taller than she is, for her encouragement, advice, and editing of our work through each stage of this endeavor. To Patrick O'Neill, who nourished my curiosity of all things wild when I was little and introduced me to the wonderful world of tomato worms, centipedes, mantises, and jumping spiders.

And to my parents, Beverly and Gary McFarland, who walked me through canyons, up mountain trails, and along creeks and rivers for as long as I can remember. They never once batted an eye at my "less-than-traditional" interests, never worried about me on my long, solo wanderings into the desert woods as long as I was back in time to set the table. There is no greater gift than that.

Thanks to friends who've offered suggestions, helped me laugh when the going got rough, and whose genuine excitement helped carry me along: Angie Jordan (who also dreamed up and created the first version of the quick reference charts), Ryan Salmon, Todd Thompson, Jonah Evans, Neal Wight, Marcus Reynerson, Natalie Dawson, Brent Coyle, Beatriz Mendoza, and all of the "East Mountain Boys" crew. There are many, many more—thank you all!

Dave Moskowitz shared of his own recent book-writing experience and gave patient tutelage and support with photography throughout this process.

Thanks to Mark Elbroch for all of his encouragement and book-writing advice. He also gave Photoshop instruction, leads for acquiring birds, book suggestions, and put in a good word for us to Mark Allison of Stackpole Books. That all has meant so much.

It was Dave that drummed up the idea to write this book; without his inspiration, this work would not exist. He invited me to join him in the making of this guide, and it has been exceptional time, one I'll always be fond of. Thanks Dave, for your dedication, friendship, your wonderful work ethic; thanks for sharing the exhaustion, rough times, and elated moments. And special thanks to Mikki, Dave's wife, who helped us in so many ways—organizing zillions of details and reworking messy data sheets—and for her excellent advice on anything we brought to her. Thanks to Mikki also for sharing her home for the last months of the writing of this guide. She put up with much: countless dead things, cluttered work spaces, and tired men who were always up too late.

My sincere thanks go also to the folks at Stackpole Books that have made this guide a reality: thanks to Mark Allison for giving us this opportunity and Kathryn Fulton for her careful, patient editing of the heap of material we gave to her.

—C. M.

BIBLIOGRAPHY

Attenborough, D. *The Life of Birds*. Princeton, NJ: Princeton University Press, 1998.

Bortolotti, G. R., R. D. Dawson, and G. L. Murza. "Stress during Feather Development Predicts Fitness Potential." *Journal of Animal Ecology*. 2002; 71(2): 333–342.

Brown, R., J. Ferguson, M. Lawrence, and D. Lees. *Tracks and Signs of the Birds of Britain and Europe: An Identification Guide*. Kent: Christopher Helm, 1987.

Brush, A. H. "Evolving a Protofeather and Feather Diversity." *American Zoologist*. 2000; 40(4): 631–639.

Bundle, M. W., and K. P. Dial. "Mechanics of wing-assisted incline running." *Journal of Experimental Biology*. 2003; 210: 1742–1751.

Burton, R. *Bird Flight*. New York, NY: Facts On File, 1990.

Burton, R. *Bird Migration*. New York, NY: Facts On File, 1992.

Coues, E. *A Key to North American Birds*. Boston: Dana Estes and Company, 1903.

Elbroch, M., with E. Marks. *Bird Tracks and Sign: A Guide to North American Species*. Mechanicsburg, PA: Stackpole Books, 2001.

Feduccia, A. "Evidence from Claw Geometry Indicating Arboreal Habits of *Archaeopteryx*." *Science*. 1993; 259(5096): 790–793.

Feduccia, A. *The Origin and Evolution of Birds*. New Haven, CT: Yale University Press, 1996.

Freedman, R. *How Birds Fly*. New York: Holiday House Books, 1977.

Freethy, R. *How Birds Work*. London: Blandford Press, 1982.

Goslow, G. E., Jr., K. P. Dial, and F. A. Jenkins, Jr. "Bird Flight: Insights and Complications." *BioScience*. 1990; 40(2): 108–115.

Hedenstrom, A. "Migration by Soaring or Flapping Flight in Birds: The Relative Importance of Energy Cost and Speed." *Philosophical Transactions: Biological Sciences*. 1993; 342(1302): 353–361.

Henderson, C. L. *Birds in Flight: The Art and Science of How Birds Fly*. Minneapolis, MN: Voyageur Press, 2008.

Hilty, A. J., W. Z Lidicker, Jr., A. M. Merenlender. *Corridor Ecology: The Science and Practice of Linking Landscapes for Biodiversity Conservation*. Washington, DC: Island Press, 2006.

Humphrey, P. S., and K. S. Parkes. “An approach to the study of molts and plumages.” *The Auk*. 1959; 76: 1–31.

Jenkins, F. A., Jr., K. P. Dial, and G. E. Goslow, Jr. “A Cineradiographic Analysis of Bird Flight: The Wishbone in Starlings Is a Spring.” *Science*. 1988; 24(4872): 1495–1498.

Jones, T. D, J. A. Ruben, L.D. Martin, E. N. Kurochkin, A. Feduccia, P. F. A. Maderson, W. J. Hillenius, N. R. Geist, and V. Alifanov. “Nonavian Feathers in a Late Triassic Archosaur.” *Science*. 2000; 288(5474): 2202–2205.

Jovani, R., and J. Blas. “Adaptive allocation of stress-induced deformities on bird feathers.” *Journal of Evolutionary Biology*. 2004; 17(2): 294–301.

Keast, A. “Wing Shape in Insectivorous Passerines Inhabiting New Guinea and Australian Rain Forests and Eucalypt Forest/Eucalypt Woodlands.” *The Auk*. 1996; 113(1): 94–104.

Kerlinger, P. *How Birds Migrate*. Mechanicsburg, PA: Stackpole Books, 1996.

Maurer, B. “The evolution of body size in birds. I. Evidence for non-random diversification.” *Evolutionary Ecology*.1998; 12(8): 925–934.

Mayr, G., B. Pohl, and S. D. Peters. “A Well-Preserved Archaeopteryx Specimen with Theropod Features.” *Science*. 2005; 310(5753): 1483–1486.

Moller, A. P., and F. De Lope. “Senescence in a Short-Lived Migratory Bird: Age-Dependent Morphology, Migration, Reproduction and Parasitism.” *Journal of Animal Ecology*. 1999; 68(1): 163–171.

Mönkkönen, M. “Do Migrants have more pointed wings? A comparative study.” *Evolutionary Ecology*. 1995; 9l: 520–528.

Norberg, U. M. “How a Long Tail and Changes in Mass and Wing Shape Affect the Cost for Flight in Animals.” *Functional Ecology*. 1995; 9(1): 48–54.

Norell, M. A., and X. Xing. “Feathered Dinosaurs.” *Annual Review of Earth and Planetary Sciences*. 2005; 33: 277–299.

Page, J., and E. S. Morton. *Lords of the Air: The Smithsonian Book of Birds*. Avenel, NJ: Wings Books, 1995.

Pérez-Tris, J., and J. L. Tellería. “Age-Related Variation in Wing Shape of Migratory and Sedentary Blackcaps *Sylvia atricapilla*.” *Journal of Avian Biology*. 2001; 32(3): 207–213.

Peterson, R. T. *Field Guide to Birds of North America*. New York: Houghton Mifflin, 2008.

Podulka, S., R. W. Rohrbaugh, Jr., and R. Bonney. *Cornell Lab of Ornithology Handbook of Bird Biology*. Princeton, NJ: Princeton Press, 2004.

Poore, S. O, A. Sánchez-Haiman, and G. E. Goslow, Jr. “Wing Upstroke and the Evolution of Flapping Flight. *Nature*. 1997; 387:799–802.

Proctor, N., and P. Lynch. *Manual of Ornithology: Avian Structure and Form*. New Haven, CT: Yale University Press, 1993.

Prum, R. O., R. H. Torres, S. Williamson, and J. Dyck. “Coherent light scattering by blue feather barbs.” *Nature*. 1998; 396: 28–29.

Prum, R. O. "Development and Evolutionary Origin of Feathers." *Journal of Experimental Zoology*. 1999; 285: 291–306.

Prum, R. O. "Why ornithologists should care about the theropod origin of birds." *The Auk*. 2002; 119(1): 1–17.

Roots, C. *Flightless Birds*. Westport, CT: Greenwood Press, 2006.

Senar, J. C., J. Lleonart, and N. B. Metcalfe. "Wing-Shape Variation between Resident and Transient Wintering Siskins *Carduelis spinus*." *Journal of Avian Biology*. 1994; 25(1): 50–54.

Sibley, D. A. *NAS The Sibley Guide to Birds*. New York: Alfred A. Knopf, 2000.

Sibley, D. A. *NAS The Sibley Guide to Bird Life and Behavior*. New York: Afred A. Knopf, 2001.

Sibley, D. A. *Sibley's Birding Basics*. New York: Alfred A. Knopf, 2001.

Stettenheim, P. R. "The Integumentary Morphology of Modern Birds—An Overview." *American Zoologist*. 2000; 40(4): 461–477.

Sumida, S. S., and C. A. Brochu. "Phylogenetic Context for the Origin of Feathers." *American Zoologist.* 2000 40(4): 486–503.

Susanna, K., and S. Hall. "Do nine-primaried passerines have nine or ten primary feathers? The evolution of a concept." *Journal of Ornithology*. 2005; 146(2): 121–126.

Swaddle, J. P., and R. Lockwood. "Wingtip shape and flight performance in the European Starling *Sturnus vulgaris*." *Ibis*. 2003; 145(2003): 457–464.

Terres, J. K. *The Audubon Society Encyclopedia of North American Birds*. New York: Alfred A. Knopf, 1980.

Tucker, V. A., and G. C. Parrot. "Aerodynamics of Gliding Flight in a Falcon and Other Birds." *Journal of Experimental Biology*. 1970; 52: 345–367.

Van Tyne, J., and A. J. Berger. *Fundamentals of Ornithology*. New York: John Wiley and Sons, 1959.

Vance, A. T. "The Effect of Molting on the Gliding Performance of a Harris' Hawk *Parabuteo unicinctus*." *The Auk*. 1991; 108(1):108–113.

Videler, J. J. *Avian Flight*. Oxford, NY: Oxford University Press, 2005.

Voitkevich, A. A. *The Feathers and Plumages of Birds*. London: Sidgwick and Jackson, 1966.

Wells, J. V. *Birder's Conservation Handbook: 100 North American Birds at Risk*. Princeton, NJ: Princeton University Press, 2007.

Wilkinson, M. T. "Sailing the skies: the improbable aeronautical success of the pterosaurs." *Journal of Experimental Biology*. 2007; 210: 1663–1671.

Williams, N. *How Birds Fly*. Tarrytown, NY: Benchmark Books, 1997.

Wilson, J. A. "Sweeping flight and soaring by albatrosses." *Nature*. 1975; 257: 307–308.

Witter, M. S., and S. J. Lee. "Habitat Structure, Stress and Plumage Development." *Proceedings: Biological Sciences*. 1995; 261(1362): 303–308.

Wood, H. B. "Growth Bars in Feathers." *The Auk*. 1950; 67(4): 486–491.

Yong, W., and F. R. Moore. "Flight Morphology, Energetic Condition, and the Stopover Biology of Migrating Thrushes." *The Auk*. 1994; 111(3): 683–692.

INDEX

ABOUT THE AUTHORS

S. David Scott is a skilled wildlife tracker and naturalist who has been involved in environmental education since 2003. His focus as an educator is to help students reconnect with the natural world through wildlife tracking, increased sensory awareness, wilderness survival skills, and the development of a strong sense of place.

Scott is an instructor at the Wilderness Awareness School in Duvall, Washington, where he teaches courses on bird feather identification, wildlife tracking, wilderness survival, and ecology. He also designs custom workshops to meet the specific goals of course participants. To find out more about classes or to schedule a custom course, visit www.SDavidScott.com.

Casey McFarland has been involved in environmental education for more than a decade, instructing in venues ranging from wilderness schools and state agencies to the Sierra Club. He currently trains and certifies biologists, research teams, and the general public across the country in wildlife tracking field skills through the CyberTracker Conservation evaluation system, an international standard for gauging and enhancing in-field knowledge of sign and behavior of mammals, birds, reptiles, amphibians, and invertebrates alike.

McFarland also contributes to wildlife connectivity research, providing consultation, training, and field methodology design to help researchers and monitoring teams assess and improve fragmented landscapes for wildlife permeability. He can be contacted at wildlifetrackingsouthwest.com.